D1297109

Wetlands and Urbanization

Implications for the Future

Wetlands and Urbanization

Implications for the Future

Edited by
Amanda L. Azous
Environmental Scientist
Azous Environmental Sciences

Richard R. Horner
Water Resources Consultant
Research Associate Professor
Landscape Architecture and
Civil and Environmental Engineering
University of Washington, Seattle

Library of Congress Cataloging-in-Publication Data

Wetlands and urbanization : implications for the future / edited by Amanda L. Azous and Richard R. Horner.

p. cm.

Includes bibliographical references and index.

ISBN 1-56670-386-7

1. Wetland ecology—Washington (State)—Puget Sound Region. 2. Wetland management—Washington (State)—Puget Sound Region. 3. Wetlands—Washington (State)—Puget Sound Region. 4. Urbanization—Washington (State)—Puget Sound Region. I. Azous, Amanda L. II. Horner, Richard R. (Richard Ray)

QH105.W2 W38 2000
577.68′09797′7—dc21 00-032730
CIP

Direct all inquiries to CRC Press LLC, 2000 N.W. Corporate Blvd., Boca Raton, Florida 33431.

The research described in this document has been funded by the U.S. Environmental Protection Agency and the Washington State Department of Ecology with funds obtained from the National Oceanic Atmospheric Administration, and appropriated for Section 306b of the Coastal Zone Management Act of 1972. King County Department of Natural Resources and Surface Water Management Division also funded this research. Mention of trade names or commercial products does not constitute endorsement or recommendation for use. The views expressed herein are those of the authors and do not necessarily reflect the views of the publisher or any of the funding agencies.

Lewis Publishers is an imprint of CRC Press LLC

International Standard Book Number 1-56670-386-7
Library of Congress Card Number 00-032730
Printed in the United States of America 1 2 3 4 5 6 7 8 9 0
Printed on acid-free paper

Preface

In the early 1980s stormwater managers and developers were proposing to store urban runoff in wetlands to reduce flooding impacts and to protect stream channels from erosion. There was also interest in exploiting the known ability of wetlands to capture and retain pollutants in stormwater. In response to these proposals, natural resource managers argued that flood storage and pollutant trapping were only two of the numerous functions attributable to wetlands. Among other values were groundwater recharge and discharge, shoreline stabilization, food chain support, and habitat for wildlife. It was further claimed that using wetlands for stormwater management would likely damage these other vital functions.

Both stormwater and natural resources managers came together in early 1986 in the Puget Sound area of Washington State to consider how best to resolve these national issues concerning wetlands and stormwater management. Together, representatives from federal, state, and local agencies; academic institutions; and other local interests determined information and management needs required to guide policy and management of wetlands. Out of these requests was born the design of a research program that would produce such information to guide policy and management of wetlands. The research was to identify the short- and long-term impacts of urban stormwater on palustrine wetlands; develop management criteria by wetland type; recommend stormwater management strategies that avoided or minimized negative effects on wetlands; and to identify features critical to improving urban runoff water quality prior to entering wetlands.

Early in our studies, it became apparent that wetlands in urbanizing watersheds would inevitably be impacted by clearing, development, and other anthropogenic activities even if there was no intention to use them for stormwater management. We also learned it was essential to identify the characteristics of wetland watersheds and the surrounding landscapes in order to understand the relationships between urban stormwater discharge and wetland ecology. In this book you will learn what we found monitoring and analyzing the five major structural and ecological components of wetlands: hydrology, water quality, soils, plants, and animals in wetlands over an eight-year period. These pages provide a thorough descriptive ecology of studied wetlands, discussions of urbanization influences affecting these wetlands, and substantive recommendations for minimizing potential adverse stormwater impacts from urbanization. This information is developed from comparisons of wetlands located in watersheds undergoing urbanization to wetlands in watersheds remaining mostly undeveloped during our studies.

Continued urbanization of the natural landscape is an ongoing ever-increasing activity, driving efforts to protect remaining wetlands. The goal of this book is to support protection efforts by increasing professional and public knowledge of wetland functions in urbanizing environments and to improve the management of both wetland resources and urban stormwater management.

Puget Sound Wetlands and Stormwater Management Research Program

Richard R. Horner

Amanda L. Azous

Klaus O. Richter

Lorin E. Reinelt

Sarah S. Cooke

About the Puget Sound Wetlands and Stormwater Management Program Team

Amanda L. Azous is an environmental scientist, consultant, and sole proprietor of Azous Environmental Sciences, a private consulting firm established in Seattle, Washington in 1990 and located on Orcas Island, Washington since 1993. Ms. Azous received a Bachelor of landscape architecture and an M.S. in environmental engineering and science, from the University of Washington, Seattle and she is also a registered professional wetland scientist.

Amanda Azous has worked on a broad range of projects including development of environmental policy, writing environmental protection regulations, and developing performance standards for wetland restoration and mitigation projects. Her recent experience includes wetland design, enhancement and restoration, watershed analysis, environmental impact assessment, water quality studies, GIS analysis of landscapes, land use surveys, and environmental impact evaluations. Her firm specializes in land management plans for wetlands, forestry, conservation, and stewardship as well as evaluations of environmental factors as a basis for community planning. She has authored and co-authored several journal articles and numerous technical reports addressing community planning, urban stormwater impacts, and management of biological communities in urbanizing watersheds.

Richard R. Horner received his B.S. and M.S. degrees from the University of Pennsylvania and a Ph.D. in environmental engineering from the University of Washington in 1978. Following 13 years of college teaching and professional practice, he joined the University of Washington faculty in 1981. Since 1993 he has split his time between private consulting and teaching at the University of Washington, where he is a research associate professor with appointments in civil and environmental engineering, landscape architecture, and urban horticulture. Dr. Horner's principal interests involve analyzing the effects of human activities, especially diffuse landscape sources of water pollution, on natural freshwater resources, and solutions that protect those resources. He founded the Center for Urban Water Resources Management at the University of Washington in 1990 to advance applied research and education in these areas, directed the center for three years, and continues as an affiliated faculty member. Dr. Horner coordinated the Puget Sound Wetlands and Stormwater Management Research Program, the basis for this book, from its beginning in 1986 to its conclusion in 1997. He has also been a principal investigator on more than 30 other research projects in his area of interest, funded by agencies such as National Science Foundation, National Research Council, U.S. Environmental Protection Agency, the Washington State Departments of Ecology and Transportation, and a number of Puget Sound regional and local governments. His consulting clients have included many of these same agencies, the Natural Resources Defense Council, the Santa Monica and San Diego Baykeeper organizations, the Greater Vancouver (Canada) Regional District, and many of the national environmental consulting firms.

Dr. Horner is the author or co-author of one previously published book and more than 50 refereed journal publications, book chapters, and papers in conference proceedings. His current University of Washington teaching includes a graduate watershed analysis and design studio in landscape architecture and a number of continuing education courses through engineering professional programs.

Klaus O. Richter is the senior wetland ecologist in King County's Department of Natural Resources. He received his B.S. in forestry/wildlife biology at the State University of New York (SUNY) College of Environmental Sciences and Forestry in Syracuse, an M.S. in science education at SUNY in Oswego, and a Ph.D. in forest zoology at the University of Washington, Seattle. For the past 15 years Dr. Richter has specialized in freshwater wetland science, management, protection, and regulation. As co-recipient of the 1996 National Wetlands Award in Science Research sponsored by the Environmental Law Institute and the EPA, he was honored for his research of amphibian ecology and reproduction in urban areas. His models to account for amphibian declines in urbanizing landscapes are being applied to reduce amphibian losses through improved stormwater management and in implementing site-specific wetland enhancement, restoration and creation practices throughout the Puget Sound and the U.S.

Dr. Richter has authored numerous papers on the monitoring, distribution, and decline of amphibians as well as habitat mitigation criteria. Additionally, he helped develop Washington State's Hydrogeomorphic Wetlands Functional Assessment and is currently developing methods and metrics (i.e., biocriteria) applicable to employing amphibians as bioindicators of wetland/watershed condition for EPA's Wetland Division in Washington, D.C. Dr. Richter is also a popular instructor at local universities and colleges and leads numerous training courses on wetland monitoring, management, and restoration.

Lorin E. Reinelt is a Senior Water Resources Engineer for the Public Works Engineering Department at the City of Issaquah. He is responsible for stormwater capital projects, including stream rehabilitation projects for flood mitigation and habitat enhancement, water quality treatment, and local drainage control. He also works in a regional role as the Sammamish watershed coordinator, supporting the Sammamish Watershed Forum, and implementing fish habitat, water quality, and flood protection projects.

Dr. Reinelt has been involved in public service, research, education, and consulting on water resource issues for the past 15 years. This includes nonpoint source pollution management, wetlands and stormwater management, basin planning, aquatic resource monitoring, flood hazard mitigation, water quality assessment, and groundwater management. Previously, he was employed by the King County Surface Water Management Division, the Center for Urban Water Resources Management at the University of Washington, and in private consulting.

Dr. Reinelt has a B.S. in civil engineering from the University of the Pacific, a M.S. in environmental engineering and science from the University of Washington, and a Ph.D. in water and environmental studies from Linköping University in Sweden. He is a registered professional wetland scientist and engineer-in-training.

Sarah S. Cooke, Ph.D. (edaphic ecology), M.S. (plant taxonomy), M.S. (geology, biology), is a registered professional wetland scientist, soils scientist (SS), and has a national reputation in wetlands and plant-soil interactions (geobotany) research. Dr. Cooke has 20 years experience in ecological and geological research, with 14 years in wetlands ecological research and environmental consulting in the Pacific Northwest.

She specializes in wetland restoration design and implementation and has conducted wetland inventories, delineations, baseline studies, monitoring programs, rare plant surveys, soil assessments, vegetation mapping, and watershed analysis in the Pacific Northwest. Dr. Cooke developed the Semi-Quantitative Assessment Methodology (SAM) functional assessment used by many wetlands professionals; she was on the development team for the Washington State Wetlands Functional Assessment Methodology. Her most recent publication was *A Field Guide to the Common Wetland Plants of Western Washington and Northwestern Oregon* (Seattle Audubon Society, 1997), the definitive guide to the region's wetland plants.

Dr. Cooke has taught assessment methodologies, wetlands science and ecology, delineation protocols, hydric soil science, and wetland plant identification at the University of Washington and Portland State University. She has provided expert training for agency personnel and private consultants in soils and botany throughout the Pacific Northwest.

Contributors

Amanda L. Azous
Azous Environmental Sciences
P.O. Box 530
Olga, WA 98279

Richard R. Horner, Ph.D.
Civil and Environmental Engineering and Landscape Architecture Department
University of Washington
Seattle, WA 98195

Klaus O. Richter, Ph.D.
King County Department of Natural Resources
Water and Land Resources Division
201 S. Jackson Street Suite 600
Seattle, WA 98104

Lorin E. Reinelt, Ph.D.
King County Department of Natural Resources
201 S. Jackson Street, Suite 600
Seattle, WA 98104

Sarah Spear Cooke, Ph.D.
Cooke Scientific Services, Inc.
4231 N.E. 110th St
Seattle, WA 98125

Marion Valentine
U.S. Army Corps of Engineers
4673 East Marginal Way South
Seattle, WA 98134

Ken A. Ludwa
Parametrix, Inc.
5808 Lake Washington Boulevard N.E.
Kirkland, WA 98033

Brian L. Taylor
Entranco Engineers
10900 NE 8th Street, Suite 300
Bellevue, WA 98004

Nancy T. Chinn
Montgomery Water Group
P.O. Box 2517
Kirkland, WA 98083

Jeff Burkey
King County Department of Natural Resources
201 S. Jackson Street, Suite 600
Seattle, WA 98104

Acknowledgments

The editors thank co author and fellow scientist Dr. Klaus Richter for his commitment to wetland science and unending dedication to producing this book. The resulting publication has been significantly enhanced thanks to his efforts.

The authors would like to thank the many individuals who contributed to this research effort. This project would not have been possible without the help of Dr. Derek Poon who understood the importance of developing environmental policy and regulations based on science and proved it by consistently finding funding for this research. We gratefully acknowledge his contribution to the on-going success of our efforts. We would especially like to thank Erik Stockdale who was one of the visionaries behind the Puget Sound Wetlands and Stormwater Research Program and a major contributor to our early studies. Our thanks to Dr. Kern Ewing for his continuing dedication and thoughtful contributions to this work. We also want to acknowledge Dr. Fred Weinmann for his many helpful comments and unending support. Our technical advisory committee members are too numerous to mention but we owe each of them great thanks for the direction and institutional support they provided. Our heartfelt thanks also to Tina Rose, graphic designer, who capably produced many of the technical illustrations used for this publication. Finally, we appreciate the hard work of many graduate students who gathered data and contributed important research, some of whom we may not have adequately acknowledged in individual chapters, but to whom we are indebted.

The research that serves as the basis for the chapters in this book was made possible by the organizations that funded and supported this long-term study. Our thanks to:

Washington State Department of Ecology
U.S. Environmental Protection Agency, Region 10
King County Department of Development and Environmental Services
King County Department of Natural Resources
King County Surface Water Management Division
University of Washington

Table of Contents

Section I: Overview of the Puget Sound Wetlands and Stormwater Management Research Program

Introduction 3
Richard R. Horner

Section II: Descriptive Ecology of Freshwater Wetlands in the Central Puget Sound Basin

Chapter 1
Morphology and Hydrology 31
Lorin E. Reinelt, Brian L. Taylor, and Richard R. Horner

Chapter 2
Water Quality and Soils 47
Richard R. Horner, Sarah S. Cooke, Lorin E. Reinelt, Kenneth A. Ludwa, and Nancy T. Chin

Chapter 3
Characterization of Central Puget Sound Basin Palustrine Wetland Vegetation 69
Sarah S. Cooke and Amanda L. Azous

Chapter 4
Macroinvertebrate Distribution, Abundance, and Habitat Use 97
Klaus O. Richter

Chapter 5
Amphibian Distribution, Abundance, and Habitat Use 143
Klaus O. Richter and Amanda L. Azous

Chapter 6
Bird Distribution, Abundance, and Habitat Use 167
Klaus O. Richter and Amanda L. Azous

Chapter 7
Terrestrial Small Mammal Distribution, Abundance, and Habitat Use 201
Klaus O. Richter and Amanda L. Azous

Section III: Functional Aspects of Freshwater Wetlands in the Central Puget Sound Basin

Chapter 8
Effects of Watershed Development on Hydrology 221
Lorin E. Reinelt and Brian L. Taylor

Chapter 9
Effects of Watershed Development on Water Quality and Soils 237
Richard R. Horner, Sarah S. Cooke, Lorin E. Reinelt, Kenneth A. Ludwa, Nancy T. Chin, and Marian Valentine

Chapter 10
Wetland Plant Communities in Relation to Watershed Development 255
Amanda L. Azous and Sarah S. Cooke

Chapter 11
Emergent Macroinvertebrate Communities in Relation to Watershed Development 265
Kenneth A. Ludwa and Klaus O. Richter

Chapter 12
Bird Communities in Relation to Watershed Development 275
Klaus O. Richter and Amanda L. Azous

Section IV: Management of Freshwater Wetlands in the Central Puget Sound Basin

Chapter 13
Managing Wetland Hydroperiod: Issues and Concerns 287
Amanda L. Azous, Lorin E. Reinelt, and Jeff Burkey

Chapter 14
Wetlands and Stormwater Management Guidelines 299
Richard R. Horner, Amanda L. Azous, Klaus O. Richter, Sarah S. Cooke, Lorin E. Reinelt, and Kern Ewing

Index 325

Section I

Overview of the Puget Sound Wetlands and Stormwater Management Research Program

Introduction

Richard R. Horner

CONTENTS

The Issues 4
Potential Impacts of Urbanization on Wetlands 5
Sources of Impacts to Wetlands 6
Influence of Wetland and Watershed Characteristics on Impacts to Wetlands 6
Hydrologic Impacts 7
Water Quality Impacts 7
  Direct Water Quality Impacts 7
  Hydrologic Impacts on Water Quality 8
Impacts to Wetland Soils 8
  Hydrologic Impacts to Wetland Soils 8
  Water Quality Impacts to Wetland Soils 8
Impacts to Vegetation 9
  Hydrologic Impacts on Vegetation 9
  Water Quality Impacts on Vegetation 10
Impacts to Wetland Fauna 11
  Hydrologic Impacts on Wetland Fauna 11
  Water Quality Impacts on Wetland Fauna 12
Use of Wetlands for Stormwater Treatment 13
Puget Sound Wetlands and Stormwater Management Research Program13
Design 13
Literature Review and Management Needs Survey 13
Research Program Design 14
Wetlands Impacted by Urbanization in the Puget Sound Basin 14
Stormwater Impact Studies 15
Definition of Watershed and Surrounding Landscape Characteristics 20
Organization of the Monograph 21
References 21

The Puget Sound Wetlands and Stormwater Management Research Program (PSWSMRP) was a regional research effort intended to define the impacts of urbanization on wetlands. The wetlands chosen for the study were representative of those found in the Puget Sound lowlands and most likely to be impacted by urban devel-

1-56670-386-7/00/$0.00+$.50

opment. The program's goal was to employ the research results to improve the management of both urban wetland resources and stormwater.

This overview section begins by defining the issues facing the program at its inception. It then summarizes the state of knowledge on these issues existing at the beginning and in the early stages of the program. It concludes by outlining the general experimental design of the study. Subsequent sections present the specific methods used in the various monitoring activities.

THE ISSUES

The program was inspired by proposals of stormwater managers and developers in the 1980s to store urban runoff in wetlands to prevent flooding and to protect stream channels from the erosive effects of high peak flow rates.[1,2] Stormwater managers were also interested in exploiting the known ability of wetlands to capture and to retain pollutants in stormwater, interrupting their transport to downstream water bodies (see listed citations for discussion of the use of wetlands for runoff quality control).[1-6]

In response to proposals to use wetlands for urban runoff storage, natural resources managers argued that flood storage and pollutant trapping are only two of the numerous ecological and social functions filled by wetlands. Among the other values of wetlands are groundwater recharge and discharge; shoreline stabilization; and food chain, habitat, and other ecological support for fish, waterfowl, and other species.[7,8] Resource managers further contended that using wetlands for stormwater management could damage other important wetland functions.[6,9-12] They noted the general lack of information on the types and extent of impacts to wetlands used for stormwater treatment.[3,10-13]

Several researchers have suggested that findings about the impacts of municipal wastewater treatment in wetlands are relevant to stormwater treatment in wetlands.[3,14] In some cases, wastewater treatment in wetlands has caused severe ecological disruptions, particularly when wastewater delivery is uncontrolled.[15,16] A number of studies have raised concerns about possible long-term toxic metal accumulations, biomagnification of toxics in food chains, nutrient toxicity, adverse ecological changes, public health problems, and other impacts resulting from wastewater treatment in wetlands.[17-20]

Other researchers have reported negative impacts on wetland ecosystems from wastewater treatment. Wastewater additions can lead to reduced species diversity and stability, and a shift to simpler food chains.[21,22] Wastewater treatment in natural northern wetlands tended to promote the dominance of cattails (*Typha* spp.).[23] In addition, animal species diversity usually declined. Discharge of wastewater to a bog and marsh wetland eliminated spruce and promoted cattails in both the bog and marsh portions.[24] Thirty years of effluent discharge to a peat bog caused parts of the bog to become a monoculture cattail marsh.[25] Application of chlorinated wastewater to a freshwater tidal marsh reduced the diversity of annual plant species.[26] These findings on the effects of wastewater applications to wetlands have probable implications for the use of wetlands for stormwater treatment.

Despite the controversy over use of natural wetlands for stormwater treatment, it became apparent in early discussions on the subject that wetlands in urbanizing watersheds will inevitably be impacted by urbanization, even if there is no intention to use them for stormwater management. For example, the authors of a U.S. Environmental Protection Agency (USEPA) handbook on use of freshwater wetlands for stormwater management stated that the handbook was not intended to be a statement of general policy favoring the use of wetlands for runoff management, but acknowledged that some 400 communities in the Southeast were already using wetlands for this purpose.[15] Moreover, directing urban runoff away from wetlands in an effort to protect them can actually harm them. Such efforts could deprive wetlands of necessary water supplies, changing their hydrology and threatening their continued existence as wetlands.[2] In addition, where a wetland's soil substrate is subsiding, continuous sediment inputs are necessary to preserve the wetland in its current condition.[27] Directing runoff to wetlands can help to furnish nutrients that support wetland productivity.[2]

In its early years, the program focused on evaluating the feasibility of incorporating wetlands into urban runoff management schemes. Given this objective, the researchers initially viewed the issues more from an engineering perspective rather than natural science. However, in later years, an appreciation of the fact that urban runoff reaches wetlands, whether intended or not, led the researchers to shift their inquiry to more fundamental questions about the impact of urbanization on wetlands. Thereafter, the program's point of view ultimately merged natural science and engineering considerations. The information yielded by the program will, therefore, be useful to wetland and other scientists, as well as to stormwater managers.

IMPACTS OF URBANIZATION ON WETLANDS

Urbanization impacts wetlands in numerous direct and indirect ways. For example, construction reportedly impacts wetlands by causing direct habitat loss, suspended solids additions, hydrologic changes, and altered water quality.[28] Indirect impacts, including changes in hydrology, eutrophication, and sedimentation, can substantially alter wetlands in addition to direct impacts, such as drainage and filling.[29] Urbanization may affect wetlands on the landscape level, through loss of extensive areas, at the wetland complex level, through drainage or modification of some of the units in a group of closely spaced wetlands, and at the level of the individual wetland, through modification or fragmentation.[30]

Over the past several decades, it has become increasingly apparent that untreated runoff is a significant threat to the country's water quality. There has, consequently, been substantial research about the relationship between urbanization and runoff quality and quantity. However, this program focused on the impacts of runoff to wetlands themselves, and not on the effects of urbanization on runoff flowing to wetlands.

Runoff can alter four major wetland components: hydrology, water quality, soils, and biological resources.[31,32] Because impacts to individual wetland components affect the condition of others, it is difficult to distinguish between the effects of each impact or to predict the ultimate condition of a wetland component by simply aggregating the effects of individual impacts.[31,33] Moreover, processes within wetlands interact in complex ways. For example, wetland chemical, physical, and biological processes interact to influence the retention, transformation, and release of a large variety of substances in wetlands. Increased peak flows transport more sediment to wetlands that, in turn, may alter the wetlands' vegetation communities and impact animal species dependent on the vegetation.

SOURCES OF IMPACTS TO WETLANDS

Brief consideration of how urbanization affects runoff illustrates the potential for dramatic alteration of wetlands. Hydrologic change is the most visible impact of urbanization. Hydrology concerns the quantity, duration, rates, frequency, and other properties of water flow. It has been called the linchpin of wetland conditions because of its central role in maintaining specific wetland types and processes.[34,35] Moreover, impacts on water quality and other wetland components, to a considerable degree, are a function of hydrologic changes.[36]

Of all land uses, urbanization has the greatest ability to alter hydrology. Urbanization typically increases runoff peak flows and total flow volumes and damages water quality and aesthetic values. For example, one study comparing a rural and an urban stream found that the urban stream had a more rapidly rising and falling hydrograph, and exhibited greater bed scouring and suspended solids concentrations.[37]

Pollutants reach wetlands mainly through runoff.[38,39] Urbanized watersheds generate large amounts of pollutants, including eroded soil from construction sites, toxic metals and petroleum wastes from roadways and industrial and commercial areas, and nutrients and bacteria from residential areas. By volume, sediment is the most important nonpoint pollutant.[39] At the same time that urbanization produces larger quantities of pollutants, it reduces water infiltration capacity, yielding more surface runoff. Pollutants from urban land uses, therefore, are more vulnerable to transport by surface runoff than pollutants from other land uses. Increased surface runoff combined with disturbed soils can accelerate the scouring of sediments and the transport and deposition of sediments in wetlands.[11,40] Thus, there is an intimate connection between runoff pollution and hydrology.

INFLUENCE OF WETLAND AND WATERSHED CHARACTERISTICS ON IMPACTS TO WETLANDS

Watershed and wetland characteristics both influence how urbanization affects wetlands. For example, impacts of highways on wetlands are affected by such factors as highway location and design, watershed vulnerability to erosion, wetland flushing capacity, basin morphology, sensitivity of wetland biota, and wetland recovery capacity.[41] Regional storm patterns also have a significant influence on impacts to wetlands.[31] Hydrologic impacts are affected by such factors as watershed land uses; wetland to watershed area ratios; and wetland soils, bathymetry, vegetation, and inlet and outlet conditions.[31,42]

Clearly, an assessment of the impacts of urbanization on a wetland should take into account the landscape in which the wetland is located. Some have suggested that a landscape approach might be useful for evaluating the effect of cumulative impacts on a wetland's water quality function.[43] The rationale for such an approach is that most watersheds contain more than one wetland, and the influence of a particular wetland on water quality depends both on the types of the other wetlands present and their positions in the landscape.

HYDROLOGIC IMPACTS

The direct impacts of hydrologic changes on wetlands are likely to be far more dramatic, especially over the short term, than other impacts. Hydrologic changes can have large and immediate effects on a wetland's physical condition, including the depth, duration, and frequency of inundation of the wetland. It is fair to say that changes in hydrology caused by urbanization can exert complete control over a wetland's existence and characteristics. One study, using the Surface Water Management Model (SWMM), predicted that urbanization bordering a swamp forest would increase runoff volumes by 4.2 times.[44] Greater surface runoff is also likely to increase velocities of inflow to wetlands, which can disturb wetland biota and scour wetland substrates.[39] Increased amounts of stormwater runoff in wetlands can alter water level response times, depths, and duration of water detention.[31] Reduction of watershed infiltration capacity is likely to cause wetland water depths to rise more rapidly following storm events. Diminished infiltration in wetland watersheds can also reduce stream baseflows and groundwater supplies to wetlands, lengthening dry periods and impacting species dependent on the water column.[15,45]

WATER QUALITY IMPACTS

Direct Water Quality Impacts

Prior to the PSWSMRP study, there was very little information specifically covering the impacts of urban runoff on water quality within wetlands.[39] On the other hand, there have been extensive inquiries into the effects of urbanization on runoff and receiving water quality generally.[122] Much of this information undoubtedly is suggestive of the probable effects of urban runoff on wetland water quality. There have also been numerous "before and after" studies evaluating the effectiveness of wetlands for treatment of municipal wastewater and urban runoff.[3,4,10,12,20,46-51] Many of these studies have focused on the effectiveness of wetlands for water treatment rather than on the potential for such schemes to harm wetland water quality.

Nevertheless, data on the quality of inflow to and pollutant retention by wetlands are likely to give some indication of the effects of urban runoff on wetland water quality. Studies on the effects of wastewater and runoff on other wetland components, such as vegetation, also may provide indirect evidence of impacts on wetland water quality.[22,24-26,52-57] A number of researchers have warned of the risks of degradation of wetland water quality and other values from intentional routing of runoff through wetlands.[3,10-12,14,58] Subsequent sections in this monograph describe the results of water quality impact studies performed by the program.

Hydrologic Impacts on Water Quality

Hydrology influences how water quality changes will impact wetlands. Hydrologic changes can make a wetland more vulnerable to pollution.[59] Increased water depths or frequencies of flooding can distribute pollutants more widely through a wetland.[39] How wetlands retain sediment is directly related to flow characteristics, including degree and pattern of channelization, flow velocities, and storm surges.[10] Toxic materials can accumulate more readily in quiescent wetlands.[60] A study on use of wetlands for stormwater treatment found that wetlands with a sheet flow pattern retained more phosphorus, nitrogen, suspended solids, and organic carbon than channelized systems, which were found to be ineffective.[50]

Changes in hydroperiod can also affect nutrient transformations and availability and the deposition and flux of organic materials.[36,61] One study observed higher phosphorus concentrations in stagnant than in flowing water.[62] In wetland soils, the advent of anaerobic conditions can transform phosphorus to dissolved forms.[31] Another study reported that anaerobic conditions in flooded emergent wetlands increased nutrient availability to wetland plants, compared to infrequently flooded sites.[63]

IMPACTS TO WETLAND SOILS

Hydrologic Impacts to Wetland Soils

Flow characteristics within wetlands directly influence the rate and degree of sedimentation of solids imported by runoff.[22] If unchecked, excessive sedimentation can alter wetland topography and soils, and, ultimately result in the filling of wetlands. Alternatively, elevated flows can scour a wetland's substrate, changing soil composition, and leading to a more channelized flow.[40] Materials accumulated over several hundred years could, therefore, be lost in a matter of decades.[64]

Water Quality Impacts to Wetland Soils

The physical, chemical, and biological characteristics of wetland soils change as they are subjected to urban runoff.[31] The physical effects of runoff on wetland soils, including changes in texture, particle size distributions, and degree of saturation are not well documented.[31] However, a wetland's soil can be expected to acquire the physical characteristics of the sediments retained by the wetland.

Suspended matter has a strong tendency to absorb and adsorb other pollutants.[39] Sedimentation, therefore, is a major mechanism of pollutant removal in wetlands.[3,14] Chemical property changes in wetland soils typically reflect sedimentation patterns.[65,66] Materials are often absorbed by wetland soils after entering a wetland, as well.[67]

When nutrient inputs to wetlands rise, temporary or long-term storage of nutrients in ecosystem components, including soils, can increase.[23] Rates of nutrient transfer among ecosystem components and flow through the system may also accelerate. When chlorinated wastewater was sprayed onto a freshwater tidal marsh, surface litter accumulated nitrogen and phosphorus.[26] However, although wetland soils can retain nutrients, a change of conditions, such as the advent of anaerobiosis and changed redox potential, can transform stored pollutants from solid to dissolved forms, facilitating export from the soil.[31] The capacity of wetland soils to retain phosphorus becomes saturated over time.[68-70] If the soil becomes saturated with phosphorus, release is likely.

Wetland soils can also trap toxic materials, such as metals.[31] High toxic metals accumulations have been found in inlet zones of wetlands affected by urban runoff.[71] One study observed increased sediment metals concentrations in several locations in a wetland receiving wastewater.[56] The quantity of metals that a wetland can absorb without damage depends on the rate of metals accretion and degree of burial.[15] If stormwater runoff alters soil pH and redox potential, many stored toxic materials can become immediately available to biota.[72]

Water quality impacts on wetland soils can eventually threaten a wetland's existence. Where sediment inputs exceed rates of sediment export and soil consolidation, a wetland will gradually become filled. Filling by sediment is a particular concern for wetlands in urbanizing areas.[39] Many wetlands have an ability to retain large amounts of sediment. For example, it was reported that a wetland captured 94% of suspended solids from stormwater.[4] Other scientists observed that a stormwater treatment wetland lost 18% of permanent storage volume and 5% of total storage volume because of high rates of solids retention.[51]

IMPACTS TO VEGETATION

Impacts on wetland hydrology and water quality can, in turn, affect wetland vegetation. Emergent zones in Pacific Northwest wetlands receiving urban runoff are dominated by an opportunistic grass species, *Phalaris arundinacea*, while non-impacted wetlands contain more diverse groupings of species.[73] Marked changes in community structure, vegetation dynamics, and plant tissue element concentrations were observed in New Jersey Pine Barrens swamps receiving direct storm sewer inputs, compared to swamps receiving less direct runoff.[53] However, human impacts on wetland ecosystems can be quite subtle. For example, upon reconsidering data from two prior studies of ecological changes in wetlands, one inquiry concluded that human influences, and not natural succession, as originally believed, were the principal causes of change in the vegetation of two New England wetlands.[29]

HYDROLOGIC IMPACTS ON VEGETATION

Hydrologic changes can have significant impacts on the livelihood of the whole range of wetland flora, from bacteria to the higher plants. It was observed that microbial activity in wetland soils correlated directly to soil moisture.[47] However, surface microbial activity decreased when soils were submerged and became anaerobic.[4] To a greater or lesser degree, wetland plants are also adapted to specific hydrologic regimes. For example, the frequency and duration of flooding was documented to have determined the distribution of bottomland tree species.[74] Flood plain terraces with different flooding characteristics had distinct species compositions. Increased watershed imperviousness can cause faster runoff velocities during storms that can impact wetland biota.[39] However, as watersheds become more impervious, stream base flows and groundwater supplies can decline. As a result, dry periods in wetlands may become prolonged, impacting species dependent on the inundation.[15,45] Changes in average depths, duration, and frequency of inundation ultimately can alter the species composition of plant and animal communities.[39]

There have been numerous reports on the tolerance to flooding of wetland and non-wetland trees and plants.[75-94] While flooding can harm some wetland plant species, it promotes others.[31] There is little information available on the impacts of hydrologic changes on emergent wetland plants, although some species that can tolerate extended dry periods have been identified.[95] Hay yields in native wet meadows were reported to have increased with the length of flood irrigation if depths remained at 13 cm or less, and declined if depths stayed at 19 cm for 50 days or longer.[78]

Plant species often have specific germination requirements, and many are sensitive to flooding once established.[96] The life stage of plant species is an important determinant of their flood tolerances. While mature trees of certain species may survive flooding, the establishment of saplings could be retarded.[39] Where water levels are constantly high, wetland species may have a limited ability to migrate, and may be able to spread only through clonal processes because of seed bank dynamics.[97] The result may be reduced plant diversity in a wetland. However, anaerobic conditions can increase the availability of nutrients to wetland plants.[63]

Hydrologic impacts on individual plant species eventually translate into long-term alterations of plant communities.[15] Changes in hydroperiod can cause shifts in species composition, primary productivity, and richness.[15,72] It has been theorized that changes in hydroperiod were among the causes of a decline of indigenous plant species and an increase in exotic species in New Jersey Pine Barrens cedar swamps.[53] Early results of the PSWSMRP study indicated that wetlands with hydroperiods that fluctuated significantly between monthly high and low water levels have lower species richness than systems with lower monthly changes in water level.[45,100] (See Chapter 10, Wetland Plant Communities in Relation to Watershed Development, for the results of the PSWSMRP study on the effects of water level changes on wetland vegetation.)

In general, periodic inundation yields more plant diversity than either constantly wet or dry conditions.[98,99] Monitoring in a Cannon Beach, Oregon wastewater treatment wetland revealed little change in herbaceous and shrub plant cover after two years of operation, except in channelized and deeply flooded portions, where herbaceous cover decreased.[46] Slough sedge cover increased slightly in a shallowly flooded area. In 1986, flooding stress was observed in red alder trees in deeper parts of the wetland. In another wetland, part of which was drained and part of which was impounded to a greater depth, vegetation in the drained portion became more dense and diverse, but there was a marked decline in the number of species in the flooded portion after three years.[93]

Water Quality Impacts on Vegetation

High suspended solids inputs can reduce light penetration, dissolved oxygen, and overall wetland productivity.[39] Inflow containing high concentrations of nutrients can also promote plant growth. One study reported, for example, that in a wastewater treatment wetland, plants closer to the discharge point had greater biomass and higher concentrations of phosphorus in their tissues, and the cattails were taller.[57] When nutrient inputs to wetlands increase, they may be stored either temporarily or over the long-term in ecosystem components, including vegetation.[101] Rates of nutrient movement, by transfer among ecosystems components and through the system, may accelerate as a result.

Toxic materials in runoff can interfere with the biological processes of wetland plants, resulting in impaired growth, mortality, and changes in plant communities. The amount of metals absorbed by plants, for some species, is a function of supply. In cedar swamps in the New Jersey Pine Barrens, plants took up more lead when direct storm sewer inputs were present than when runoff was less direct.[53] The degree to which plants bioaccumulate metals is highly variable. Pickleweed (*Salicornia* sp.) was found to concentrate metals, especially zinc and cadmium, more than mixed marsh and upland grass vegetation.[52] However, plants in a brackish marsh that had received stormwater runoff for more than 20 years did not appear to concentrate copper, cadmium, lead, and zinc any more than plants in control wetlands not receiving storm water.[3]

While toxic metals accumulate in certain species, such as cattails, without causing harm, they interfere with the metabolism of other species.[39] Toxic metals can harm certain species by interfering with nitrogen fixation.[102] Metals can also impinge on photosynthesis in aquatic plants, such as waterweed (*Elodea* spp.).[103] Another study (1981) reported that roadway runoff containing toxic metals had an inhibitory effect on algae.[104] A bioassay study of the effects of stormwater on algae showed that nutrients did not stimulate growth as much as predicted because of the presence of metals in the stormwater.[105] The germination rates of wetland plants exposed to roadside snowmelt in several concentrations were found to vary inversely with the concentration of snowmelt.[54]

Pollution in wetlands may impact plant community composition the most. The major effect observed of residential and agricultural runoff with high pH and nitrate concentrations was to cause indigenous aquatic macrophytes of the New Jersey Pine Barrens to be replaced by non-native species.[55] Marked changes in plant community structure and vegetation dynamics in Pine Barrens cedar swamps were also reported where direct storm sewer inputs were present.[53] Wetland plants that were exposed to roadside snowmelt in several concentrations, showed differences in community biomass, species diversity, evenness, and richness after one month of growth that varied inversely with snowmelt concentration.[54] Impacts were not as severe where runoff was less direct.

IMPACTS TO WETLAND FAUNA

Hydrologic Impacts on Wetland Fauna

Hydrologic changes also greatly affect wetland animal communities. In two coastal marshes, animal species richness and abundance declined as hydrologic disturbance increased.[106] Shifts in plant communities as a result of hydrologic changes can have impacts on the preferred food supply and cover of such animals as waterfowl.

Increased imperviousness in wetland watersheds can reduce stream base flows and groundwater supplies, prolonging dry periods in wetlands and impacting species dependent on the water column. Many amphibians require standing water for breeding, development, and larval growth. Amphibians and reptile communities may experience changes in breeding patterns and species composition with changed water levels.[107] Because amphibians place their eggs in the water column, the eggs may be directly damaged by changes in water depth. Alterations in hydroperiod can be especially harmful to amphibian egg and larval development if water levels decline and eggs attached to emergent vegetation are exposed and desiccated.[108] Water temperature changes that accompany shifting hydrology may also impact egg development.[109]

Hydrologic changes have implications for other wetland animals, as well. Alterations to water quality and wetland soils caused by hydrologic changes may negatively affect animal species. For example, increased peak flows that accelerate sedimentation in wetlands or cause scouring can damage fish habitat.[11] Mortality of the eggs and young of waterfowl during nesting periods may rise if water depths become excessive.[31] Water level fluctuations resulting from an artificial impoundment in eastern Washington State caused a redistribution of bird populations. When potholes were flooded by the impoundment, waterfowl production was reduced, and breeding waterfowl were forced into the remaining smaller potholes.[110] Hydrologic changes may impact mammal populations in wetlands by diminishing vegetative habitat and by increasing the potential for proliferation of disease organisms and parasites as base flows become shallower and warmer.[108] Also, research has indicated a need to maintain habitat around wetlands that are receiving stormwater in order to permit free movement of animals during storm events.[31]

Water Quality Impacts to Wetland Fauna

Pollutants can have both direct and indirect effects on wetland fauna. Road runoff containing toxic metals had an inhibitory effect on zooplankton, in addition to algae.[104] A significant negative correlation between water conductivity (a general indicator of dissolved substance concentrations) and amphibian species richness was reported.[45] Aquatic organisms, particularly amphibians, readily absorb chemical contaminants.[111] Thus, the status of such organisms can be an effective indicator of a wetland's health. The degree of bioaccumulation of metals in wetland animals varies by species. In a brackish marsh that had received storm runoff for 20 years, there was no observed bioaccumulation of metals in benthic invertebrates.[112] However, a filter-feeding amphipod (*Corophium* sp.), known for its ability to store lead in an inert crystal form, accumulated significant amounts of lead. Water quality changes can indirectly harm fish and wildlife by reducing the coverage of plant species preferred for food and shelter.[35,108,113] (Please see Section III for discussions of amphibian, emergent aquatic insect, bird, and small mammal communities in relation to watershed development and habitat conditions, and for the results of the program's study on the effects of hydrologic and water quality changes on wetland animals.)

USE OF WETLANDS FOR STORMWATER TREATMENT

Impacts from intentional use of wetlands for stormwater management could be more harmful than those that would occur with incidental drainage from an urbanized watershed. For example, raising the outlet and controlling the outflow rate would, in general, change water depths and the pattern of rise and fall of water. Structural revisions to improve pollutant trapping ability would increase toxicant accumulations, in addition to the direct effects of construction. On the other hand, stormwater management actions could be linked with efforts to upgrade wetlands that are already highly damaged.

PUGET SOUND WETLANDS AND STORMWATER MANAGEMENT RESEARCH PROGRAM DESIGN

Representatives of the stormwater and resource management communities in the Puget Sound area of Washington State formed a committee in early 1986 to consider how to best resolve questions concerning wetlands and stormwater runoff. Committee members came from federal, state, and local agencies, academic institutions, and other local interests. The Resource Planning Section of the government of King County, Washington coordinated the committee's work. The committee's initial effort was to enumerate the wetland resources that are implicated in urban stormwater management decisions and to identify the general types of effects that runoff could have on these resources. The committee members also oversaw the preparation of a literature review, designed to determine the extent to which previous work could address the issues before them, and a survey of management needs.

LITERATURE REVIEW AND MANAGEMENT NEEDS SURVEY

The principal activity of the program's first year was a comprehensive literature review, which concluded with a report and an annotated bibliography covering the reported research and observations relevant to the issue of stormwater and wetlands.[114,115] The review was updated in 1991.[39] These reviews concentrated on what was known and what was not known about these issues at the time. Best known was the performance of wetlands in capturing pollutants, mostly derived from studies on their ability to provide advanced treatment to municipal wastewater effluents. Only a small body of information pertained to stormwater. The greatest shortcoming of the literature concerned the ecological impacts to wetlands created by any kind of waste stream. The literature reviews also made clear the dearth of research on any aspect of Pacific Northwest wetlands, in contrast to some other areas of the country. Many detailed aspects of the subject of stormwater and wetlands were very poorly covered, including the relative roles of hydrologic and water quality modifications in stressing wetlands and the transport and fate of numerous toxicants in wetlands.

On the basis of their discussions and the literature review, the committee members participated in a formal survey designed to identify the most important needs for reaching the goal of protecting wetlands in urban and urbanizing areas, while improving the management of urban stormwater. The survey involved rating a long list of candidate management needs with respect to certain criteria. Computer processing of the ratings led to the following list of consensus high-priority management needs:

- Definition of short- and long-term impacts of urban stormwater on palustrine wetlands;
- Management criteria by wetland type;
- Allowable runoff storage schedules that avoid or minimize negative effects on wetlands and their various functions; and
- Features critical to urban runoff water quality improvement in wetlands.

RESEARCH PROGRAM DESIGN

After completion of the literature review and management needs survey, the committee and staff assembled by King County turned to defining a research program to serve the identified needs. The program they developed included the following major components:

- Wetland survey;
- Water quality improvement study;
- Stormwater impact studies; and
- Laboratory and special field studies.

The purpose of the wetland survey was to provide a broad picture of freshwater wetlands representative of those in the Puget Sound lowlands. The survey covered 73 wetlands throughout lowland areas of King County. One important goal of the survey was to identify how urban wetlands differ from those that are lightly affected by human activity. The survey's design, results, and conclusions were published in previous reports.[73,74] The survey results assisted in designing the remainder of the research program.

The water quality improvement study was an intensive, two-year (1988–1990) effort to answer remaining questions about the water quality functioning of wetlands and is also discussed elsewhere.[116] The results from the various portions of the program were used to develop extensive guidelines for coordinated management of urban wetlands and stormwater. These guidelines were continuously updated and refined as more information became available.

WETLANDS IMPACTED BY URBANIZATION IN THE PUGET SOUND BASIN

The research program focused primarily on palustrine wetlands because urbanization in the Puget Sound region is impacting this wetland type more than other types. Palustrine wetlands are freshwater systems in headwater areas or isolated from other water bodies.[117] They typically contain a combination of water and vegetation zones. Some palustrine wetlands consist of open water with only submerged or floating plants, or with no vegetation. Others include shallow or deep marsh zones containing herbaceous emergent plants, shrub-scrub vegetation, and sometimes forested plant communities.

Two "poor fens" being impacted by urban development were also monitored during the study. Poor fens, commonly confused with true bogs, are a special wetland type that is of considerable interest in northern regions. Under natural conditions, water supply to poor fens consists only of precipitation and groundwater. The lack of surface water inflow restricts nutrient availability, resulting in a relatively unusual plant community adapted to low nutrition and the attendant acidic conditions. Such a community is vulnerable to increased nutrient supply and buffering by surface water additions.

STORMWATER IMPACT STUDIES

The stormwater impact studies formed the core of the program. This field research was supplemented by the laboratory and special field studies, which allowed investigation of certain specific questions under more control than offered by the broader field studies. A special effort was made to ensure that research was conducted according to sound scientific design, so that the results and their application in management would be defensible. In order to approximate the classic "before and after, control and treatment" experimental design approach, the impact study included "control" and "treatment" wetlands. Nineteen wetlands were included in the stormwater impact study, with approximately half the treatment sites and the remainder of the control sites (general locations are shown in Figure 1).

The treatment wetlands, located in areas undergoing urban development during the course of the study, were monitored before, during, and after urbanization. The goals of studying these wetlands were to characterize preexisting conditions and to assess the consequences of any changes accompanying urbanization and modification of stormwater inflow. The use of control sites was intended to make it possible to judge whether observed changes in treatment wetlands were the result of urbanization or of broader environmental conditions affecting all wetlands in the region. Control wetlands were paired with treatment sites on the basis of size, water and plant zone configuration, and vegetation habitat classifications.

Not all of the treatment watersheds developed as much as anticipated at the outset of the study. Only six watersheds developed 10% or more than the developed area at the start of the study. Of these six, only three wetlands had significant increases in watershed development of 100, 73, and 42% with the remaining three having increases of only 10.5, 10.3, and 10.2%. Fortunately, watersheds of most of the control wetlands were characterized by relative stability in land use during the study.

The unexpected slowness of development in the study watersheds affected our ability to identify differences between control and treatment pairs attributable to stormwater and urbanization. Also, the watersheds of control wetlands ranged from no urbanization to relatively high levels so no comparisons could be made unless the matched treatment wetland underwent significant urbanization in the watershed.

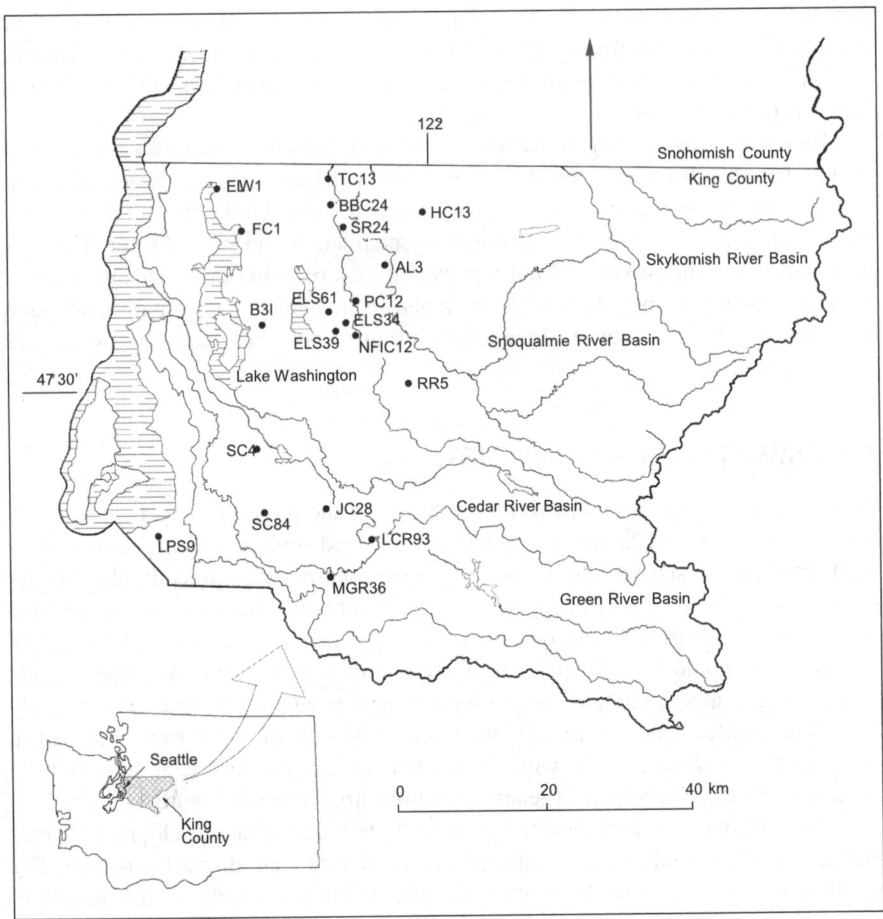

FIGURE 1 Puget Sound Wetlands and Stormwater Management Research Program study locations.

Under the circumstances, the plan to compare control and treatment pairs of wetlands was abandoned and revisions to the categories for data analyses were made.

Several categories of wetlands related to land use and watershed changes were used in the program's analyses and are shown in Table 1. Wetlands are identified as a control or a treatment wetland and land uses present in the watersheds of the wetlands at the start and completion of the study are listed. The table also lists watershed area, wetland area, and a categorization of wetland morphology.

Because the program was interested in long-term as well as short-term effects, the monitoring of impacts was continued for eight years. Research in 1988 and 1989 generally provided the baseline data for the treatment wetlands. Data from 1990 reflected the early phase of urbanization in these wetlands. Monitoring resumed in 1993, shortly after a phase of building in the watersheds ended. Monitoring in 1995 was intended to document effects that took longer to appear.

TABLE 1
Landscape Data for Study Wetlands

Site	Watershed Area (ha)	Wetland Area (ha)	Wetland Type OW = open water FT = flow through	Treatment (T) or Control (C)	Land Use Treatment and Controls[a]	Urbanization Category[b]	% Urban Cover 1989	% Urban Cover 1995	% Urban Cover Change	% Forest Cover 1989	% Forest Cover 1995	% Forest Cover Change	% Impervious Cover 1989	% Impervious Cover 1995	% Impervious Cover Change
AL3	47.35	0.81	OW	C	RC	N	13.3	13.3	0.0	73.9	73.9	0.0	4.1	4.1	0.0
B3I	183.73	1.98	FT	C	UC	H	74.7	75.2	0.5	0.0	0.0	0.0	54.9	55.4	0.5
BBC24	38.45	2.10	OW	T	T	L	10.5	52.7	42.2	89.5	47.4	–42.1	3.4	10.6	7.2
ELS39	69.20	1.74	OW	T	T	M	88.8	87.9	–0.9[c]	18.5	10.8	–7.7	24.6	24.2	–0.4
ELS61	27.11	2.02	OW	T	T	M	23.9	34.4	10.5	2.5	3.7	1.2	5.1	10.6	5.5
ELW1	54.63	3.84	FT	C	UC	M	56.6	56.6	0.0	0.0	0.0	0.0	19.9	19.9	0.0
FC1	357.34	7.28	FT	C	UC	M	81.2	81.2	0.0	14.7	14.7	0.0	30.8	30.8	0.0
HC13	359.36	1.62	OW	C	RC	N	1.5	1.5	0.0	76.6	75.1	–1.5	3.6	3.6	0.0
JC28	296.64	12.55	FT	T	T	M	54.7	64.9	10.2	34.4	19.8	–14.6	20.0	20.6	0.6
LCR93	198.22	6.09	FT	C	RC	N	12.8	11.0	–1.8	44.1	13.0	–31.1	5.8	6.1	0.3
LPS9	183.32	7.69	FT	C	UC	H	69.8	73.8	4.0	0.0	0.0	0.0	21.8	21.6	–0.2
MGR36	45.73	2.23	FT	C	RC	N	4.1	4.1	0.0	88.8	88.8	0.0	2.9	2.9	0.0
NFIC12	3.24	0.61	OW	T	T	H	0.0	100.0	100.0	100.0	0.0	–100.0	2.0	40.0	38.0
PC12	84.58	1.50	OW	T	T	N	23.5	34.0	10.5	75.2	64.7	–10.5	5.1	6.8	1.7
RR5	64.35	10.52	OW	C	RC	N	2.4	2.4	0.0	62.4	62.4	0.0	3.4	3.4	0.0
SC4	3.64	1.62	FT	C	RC	M	12.5	12.5	0.0	46.1	46.1	0.0	11.8	11.8	0.0
SC84	193.04	2.83	OW	C	UC	M	77.8	78.2	0.4	20.1	19.7	–0.4	18.5	17.0	–1.5
SR24	88.22	10.12	OW	C	RC	N	0.0	0.0	0.0	100.0	100.0	0.0	2.0	2.0	0.0
TC13	11.74	2.06	OW	C	RC	N	0.0	0.0	0.0	100.0	89.7	–10.3	2.0	2.3	0.3

[a] RC = rural control (less than 12% impervious area and greater than 40% forest); UC = urban control (greater than 12% impervious area and less than 40% forest); T = treatment (wetlands that changed during the study period).

[b] Column represents watershed urbanization at start of study: N = less than 4% impervious area and greater than 40% forest; H = greater than 20% impervious area and less than 7% forest; M = watershed conditions intermediate between N and H.

[c] The watershed of ELS39 was estimated to be 15% developed in 1988 when the study began. The watershed was developed prior to the GIS analysis of 1989 and no accurate cover data is available until 1995.

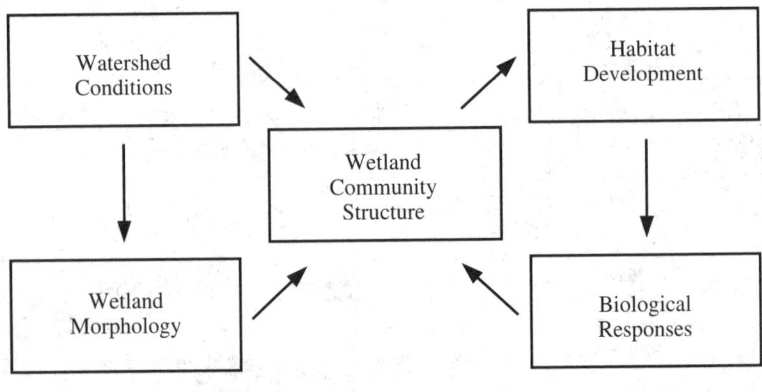

FIGURE 2 Puget Sound Wetlands and Stormwater Management Research Program experimental strategy.

Figure 2 illustrates the conceptual framework of the designs of the specific sampling programs pursued in the stormwater impact study to analyze and interpret the resulting data. The two blocks on the left of the diagram represent the driving forces determining a wetland's character (Watershed Conditions and Wetland Morphology). The term "surrounding landscape" signifies that not only a wetland's watershed (the area that is hydrologically contributory to the wetland) but also adjacent land outside of its watershed can influence the wetland. The surroundings include the wetland buffer, corridors for wildlife passage, and upland areas that provide for the needs of some wetland animals. Wetland morphology refers to form and structure and embraces shape, dimensions, topography, inlet and outlet configurations, and water pooling and flow patterns.

The central block (Wetland Community Structure) represents the physical and chemical conditions that develop within a wetland and constitute a basis for its structure. Included are both quantity and quality aspects of its water supply and its soil system. Together, these structural elements develop various habitats that can provide for living organisms, represented by the block at the upper right of the diagram. Biota will respond depending on habitat attributes, as illustrated by the block at the lower right. It is a fundamental goal of the Puget Sound Wetlands and Stormwater Management Research Program to describe these system components for the representative wetlands individually and collectively.

Connecting lines and arrows on Figure 2 depict the interactions among the components. It is a second fundamental goal of the program to understand and be able to express these interactions, to advance wetlands science and the management of urban wetlands and stormwater. Expression could come in the form of qualitative descriptions, relatively simple conceptual models, or more comprehensive mathematical algorithms. The extent to which definition of these interactions can be developed will determine the thoroughness with which management guidelines and new scientific knowledge can be generated by this research program.

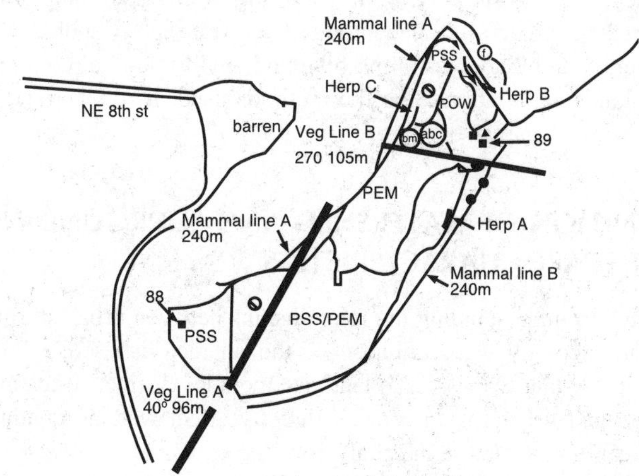

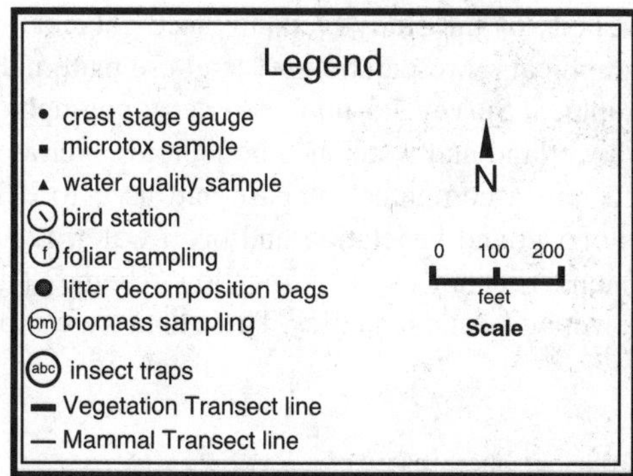

FIGURE 3 Typical monitoring plan (Patterson Creek 12 wetland).

The stormwater impact study examined the five major structural components of wetlands: (1) hydrology, (2) water quality, (3) soils, (4) plants, and (5) animals. Figure 3 presents a typical plan for monitoring of these components. A crest stage gauge was used to register the maximum water level since the preceding monitoring occasion, and a staff gauge gave the instantaneous water level. These readings provided the basis for hydrologic analysis, as detailed in Chapter 1, Morphology and Hydrology. Samples for water quality analysis were taken from the water column in an open water pool, and soil samples were collected at either three or four locations as described in Chapter 2, Water Quality and Soils. Plant cover by species was determined along one or more transect lines, depending on the wetland size and complexity of water and vegetation zones. Foliar tissue was sampled for analysis of metals content, and plant standing crop was cut for measurement of biomass gravimetrically. More on the methods used in these monitoring activities is included in Chapter 3. Adult insect emergence was continuously monitored using triplicate emergence traps (see Chapter 4). Amphibian breeding success was monitored along transects (labeled Herp. A, B, and C in Figure 3). Adult amphibians as well as small mammals were live-trapped along other transects (labeled Mammal line A, B). Chapters 5 and 7 address amphibian and small mammal communities, respectively, in wetlands and elaborate on the methods used. Birds were censused at stations as described in Chapter 6.

DEFINITION OF WATERSHED AND SURROUNDING LANDSCAPE CHARACTERISTICS

Essential to understanding the relationships between urban stormwater discharge and wetlands ecology was definition of the characteristics of wetland watersheds and surrounding landscapes. Each land use includes distinctive features, such as imperviousness and vegetative cover that directly affect wetland conditions.[118] Use of geographical information in the analysis of the effects of urbanization on wetlands allows the linking of effects with specific land use changes associated with urban development.

To this end, the program used a geographical information system (GIS) to inventory land uses in the watersheds of the study wetlands (see Table 1).[118] The GIS furnished quantitative and graphical representations of land use patterns. Study sites were located on U.S. Geographical Survey 7.5-minute-series topographic maps and the maps were used to locate wetland and watershed boundaries. Aerial photographs from 1989 were digitized into a computer database and used to delineate wetland boundaries on the basis of wetland vegetation and open water. Land uses were classified according to a standard land use classification scheme. The GIS provided the areas of watersheds, wetlands, and land uses. These data were expressed in three ways:

1. Wetland and watershed areas in hectares;
2. Watershed land uses and vegetative cover as percentages of watershed areas; and
3. Ratios of the areas of watersheds, land uses, and vegetative cover to wetland areas.

The most important quantities yielded by the third method were the ratios of watershed and wetland areas (wetland areas were subtracted from their watershed areas in calculating these ratios). The method also was used to determine the ratios of impervious and forested areas to wetland areas. GIS data obtained in 1989 were updated through visual examination of 1995 aerial photographs. In addition, in 1996, the same information was developed for 1000-m bands of the landscapes surrounding the study wetlands using 1995 satellite images.

With regard to calculating watershed imperviousness, the program found that the relevant literature generally did not provide the level of detail necessary to establish the relationships between imperviousness and the land use definitions used in the GIS inventory. The program, therefore, relied on a variety of sources linking specific land uses to imperviousness levels. Estimates of imperviousness were made using values from the literature for similar land uses and adjusting them as necessary with best professional judgment.[118-120]

Effective Impervious Area (EIA) represents the impervious area that is actually connected to constructed drainage systems. This value was estimated as a proportion of Total Impervious Area (TIA) according to the formula EIA = 0.15 * TIA1.41.[119] This equation was developed in Denver and its accuracy (correlation coefficient = 0.98 and standard error = 0.075) probably varies in other areas. However, the author's estimates were compatible with those in Puget Sound lowland hydrologic models.[121] After determining EIA and TIA values for each land use, EIAs for entire watersheds were determined using the formula EIADB = Σ1 → k (EIAk * LUk), where EIADB is the percentage of watershed area that is effectively impervious, k corresponds to the land uses inventoried in the basin, EIAk is the percentage of watershed area associated with land use k, and LUk is the percentage of the watershed classified as land use k. TIAs were calculated using the same formula.

ORGANIZATION OF THE MONOGRAPH

The chapters that follow trace the major areas of progress in filling in the conceptual framework presented in Figure 2. Section II provides several chapters describing the ecology of the palustrine wetlands of the central Puget Sound lowlands, organized according to the major structural components monitored during the program. Section III presents five chapters assessing the effects of urban stormwater and other influences of urbanization observed during the study. Finally, Section IV makes recommendations for managing urban stormwater to protect wetlands in urbanizing watersheds.

REFERENCES

1. Athanas, C., Wetlands creation for stormwater treatment, in J. Zelanzny and J. S. Feierabend, Eds., *Increasing Our Wetland Resources*, National Wildlife Federation, Washington, D.C., 1988, p. 61.
2. McArthur, B. H., The use of isolated wetlands in Florida for stormwater treatment, in *Wetlands: Concerns and Successes, Proc. Am. Water Resour. Assoc. Symp.*, D. N. Fisk, Ed., Tampa, FL, Sept. 17–22, 1989, p. 185.
3. Chan, E., T. A. Bursztynsky, N. Hantzsche, and Y. J. Litwin, The Use of Wetlands for Water Pollution Control, EPA-6-/S2-82-068, NTIS PB83-107466, U.S. Environmental Protection Agency, Municipal Environmental Research Laboratory, Cincinnati, Ohio, 1981.
4. Hickok, E. A., Wetlands for the control of urban stormwater, in Proceedings, National Conference on Urban Erosion and Sediment Control: Institutions and Technology, W. L. Downing, Ed., EPA 905/9-80-002, U.S. Environmental Protection Agency, Washington, D.C., 1980, p. 79.
5. Lakatos, D. F. and L. J. McNemar, Wetlands and stormwater pollution management, in *Proceedings of a National Symposium: Wetland Hydrology*, J. A. Kusler and G. Brooks, Eds., Association of State Wetland Managers, Berne, NY, 1988, p. 214.
6. Livingston, E. H., The use of wetlands for urban stormwater management, in *Design of Urban Runoff Quality Controls*, L. A. Roesner, B. Urbonas, and M. B. Sonnen, Eds., American Society of Civil Engineers, New York, NY, 1989, p. 467.

7. U.S. Congress, Office of Technology Assessment, Wetlands: Their Use and Regulation, OTA-0-206. U.S. Government Printing Office, Washington, D.C., 1984.
8. Zedler, J. B. and M. E. Kentula, Wetlands Research Plan, EPA/600/3-86/009, NTIS PB86-158656, U.S. Environmental Protection Agency, Environmental Research Laboratory, Corvallis, OR, 1986.
9. Newton, R. B., *The Effects of Stormwater Surface Runoff on Freshwater Wetlands: A Review of the Literature and Annotated Bibliography*, Publication 90-2, University of Massachusetts, Environmental Institute, Amherst, MA, 1989.
10. Brown, R. G., Effects of Wetlands on Quality of Runoff Entering Lakes in the Twin Cities Metropolitan Area, Minnesota, Water Resources Investigations Report 85-4170, U.S. Geological Survey, Washington, D.C., 1985.
11. Canning, D. J., Urban Runoff Water Quality: Effects and Management Options, Shorelands Technical Advisory Paper No. 4, 2nd ed., Washington Department of Ecology, Shorelands and Coastal Zone Management Program, Olympia, WA, 1988.
12. ABAG (Association of Bay Area Governments), Urban Stormwater Treatment at Coyote Hills Marsh. Association of Bay Area Governments, Oakland, CA, 1986.
13. Woodward-Clyde Consultants, The Use of Wetlands for Controlling Stormwater Pollution (Draft). U.S. Environmental Protection Agency, Region 5, Water Division, Watershed Management Unit, Chicago, IL, 1991.
14. Silverman, G. S., Seasonal Fresh Water Wetlands Development and Potential for Urban Runoff Treatment in the San Francisco Bay Area, Ph.D. thesis, University of California, Los Angeles, 1983.
15. U.S. Environmental Protection Agency, Freshwater Wetlands for Wastewater Management: Environmental Assessment Handbook, EPA 904/9-85-135, U.S. Environmental Protection Agency, Region 4, Atlanta, GA, 1985.
16. Wentz, W. A., Ecological/environmental perspectives on the use of wetlands in water treatment, in *Aquatic Plants for Water Treatment and Resource Recovery*, K. R. Reddy and W. H. Smith, Eds., Magnolia Publishing, Orlando, FL, 1987, p. 17.
17. Benforado, J., Ecological considerations in wetland treatment of wastewater, in *Selected Proceedings of the Midwest Conference on Wetland Values and Management,* B. Richardson, Eds., Freshwater Society, Navarre, MN, 1981, p. 307.
18. Guntenspergen, G. and F. Stearns, Ecological limitations on wetland use for wastewater treatment, in *Selected Proceedings of the Midwest Conference on Wetland Values and Management,* B. Richardson, Ed., Freshwater Society, Navarre, MN, 1981, p. 272.
19. Sloey, W. E., F. L. Spangler, and C. W. Fetter, Jr., Management of freshwater wetlands for nutrient assimilation, in *Freshwater Wetlands: Ecological Processes and Management Potential*, R. E. Good et al., Eds., Academic Press, New York, 1978, p. 321.
20. Dawson, B., High hopes for cattails, *Civ. Eng.,* 59, 5, 1989.
21. Heliotis, F. D., *Wetland Systems for Wastewater Treatment: Operating Mechanisms and Implications for Design,* IES Report 117, Institute for Environmental Studies, University of Wisconsin, Madison, 1982.
22. Brennan, K. M., Effects of wastewater on wetland animal communities, in *Ecological Considerations in Wetlands Treatment of Wastewater*, P. J. Godfrey et al., Eds., Van Nostrand Reinhold, New York, 1985, p. 199.
23. Kadlec, R. H., Northern natural wetland water treatment systems, in *Aquatic Plants for Water Treatment and Resource Recovery*, K. R. Reddy and W. H. Smith, Eds., Magnolia Publishing, Orlando, FL, 1987, p. 83.

24. Stark, J. R. and R. G. Brown, Hydrology and water quality of a wetland used to receive wastewater effluent, St. Joseph, Minnesota, In *Proceedings of a National Symposium: Wetland Hydrology*, J. A. Kusler and G. Brooks, Eds., Association of State Wetland Managers, Berne, NY, 1988.
25. Bevis, F. B. and R. H. Kadlec, Effects of long-term discharge of wastewater on a northern Michigan wetland, in *Abstr. Conf. Freshwater Wetlands and Sanitary Wastewater Disposal,* Higgins Lake, MI, July 1979.
26. Whigham, D. F., R. L. Simpson, and K. Lee, The Effect of Sewage Effluent on the Structure and Function of a Freshwater Tidal Marsh Ecosystem, Rutgers University, Water Resources Research Institute, New Brunswick, NJ, 1980.
27. Boto, K. G. and W. H. Patrick, Jr., Role of wetlands in the removal of suspended sediments, in *Wetland Functions and Values: The State of Our Understanding*, P. E. Greeson et al., Eds., American Water Resources Association, Minneapolis, MN, 1978, p. 479.
28. Darnell, R. M., Impacts of Construction Activities in Wetlands of the United States, EPA-600/3-76-045, U.S. Environmental Protection Agency, Environmental Research Laboratory, Corvallis, OR, 1976.
29. Keddy, P. A., Freshwater wetlands human-induced changes: indirect effects must also be considered, *Environ. Manage.,* 7(4), 299, 1983.
30. Weller, M. W., Issues and approaches in assessing cumulative impacts on waterbird habitats in wetlands, *Environ. Manage.,* 12(5), 695, 1988.
31. U.S. Environmental Protection Agency, Natural Wetlands and Urban Stormwater: Potential Impacts and Management, 843-R-001, U.S. Environmental Protection Agency, Office of Water, Washington, D.C., 1993.
32. Johnson, P. G. and L. F. Dean, Stormwater Management Guidebook for Michigan Communities, Clinton River Watershed Council, Utica, MI, 1987.
33. Hemond, H. F. and J. Benoit, Cumulative impacts on water quality functions of wetlands, *Environ. Manage.,* 12(5), 639, 1988.
34. Gosselink, J. G. and R. E. Turner, The role of hydrology in freshwater wetland ecosystems, in *Freshwater Wetlands: Ecological Processes and Management Potential*, R. E. Good et al., Eds., Academic Press, New York, 1978.
35. Mitsch, W. J. and J. G. Gosselink, *Wetlands*, 2nd ed., Van Nostrand Reinhold, New York, 1993.
36. Leopold, L. B., Hydrology for Urban Land Planning: A Guidebook on the Hydrologic Effects of Urban Land Use, U.S. Geological Survey Circular 554, U.S. Department of the Interior, Geological Survey, Washington, D.C., 1968.
37. Pedersen, E. R., The Use of Benthic Invertebrate Data for Evaluating Impacts of Urban Stormwater Runoff, M.S. thesis, University of Washington, College of Engineering, Seattle, 1981.
38. PSWQA (Puget Sound Water Quality Authority), Issue Paper: Nonpoint Source Pollution, PSWQA, Seattle, 1986.
39. Stockdale, E. C., Freshwater Wetlands, Urban Stormwater, and Nonpoint Pollution Control: A Literature Review and Annotated Bibliography, Washington Department of Ecology, Olympia, 1991.
40. Loucks, O. L., Restoration of the pulse control function of wetlands and its relationship to water quality objectives, in Wetlands Creation and Restoration: The Status of the Science: Vol. II, Perspectives, J. A. Kusler and M. E. Kentula, Eds., EPA/600/3-89/038, U.S. Environmental Protection Agency, Environmental Research Laboratory, Corvallis, OR, Pp. 55-65, 1989.

41. Adamus, P. R. and L. T. Stockwell, A Method for Wetland Functional Assessment: Volume 1, Critical Review and Evaluation, FHWA-1P-82-23, U.S. Department of Transportation, Federal Highway Administration, Washington, D.C., 1983.
42. Reinelt, L. E. and R. R. Horner, Characterization of the Hydrology and Water Quality of Palustrine Wetlands Affected by Urban Stormwater, PSWSMRP, Seattle, 1990.
43. Whigham, D. F., C. Chitterling, and B. Palmer, Impacts of freshwater wetlands on water quality: a landscape perspective, *Environ. Manage.,* 12(5), 663, 1988.
44. Hopkinson, C. S. and J. W. Day, Jr., Modeling the relationship between development and stormwater and nutrient runoff, *Environ. Manage.,* 4(4), 315, 1980
45. Azous, A., An Analysis of Urbanization Effects on Wetland Biological Communities, M. S. thesis. University of Washington, Department of Civil Engineering, Environmental Engineering and Science Program, Seattle, 1991.
46. Franklin, K. T. and R. E. Frenkel, Monitoring a Wetland Wastewater Treatment System at Cannon Beach, OR, U.S. Environmental Protection Agency, Region 10, Seattle, 1987.
47. Hickok, E. A., M. C. Hannaman, and N. C. Wenck, Urban Runoff Treatment Methods: Volume 1. Non-Structural Wetlands Treatment, EPA-600/2-77-217, Minnehaha Creek Watershed District, Wayzata, MN, 1977.
48. Lynard, W. G., E. J. Finnemore, J. A. Loop, and R. M. Finn, Urban Stormwater Management and Technology: Case Histories, EPA-600/8-80-035. U.S. Environmental Protection Agency, Washington, D.C., 1980.
49. Martin, E. H, Effectiveness of an urban runoff detention pond wetlands system, *J. Environ. Eng.,* 114(4), 810, 1988.
50. Morris, F. A., M. K. Morris, T. S. Michaud, and L. R. Williams, Meadowland Natural Treatment Processes in the Lake Tahoe Basin: A Field Investigation (Final Report), EPA-600/54-81-026, NTIS PB81-185639, U.S. Environmental Protection Agency, Washington, D.C., 1981.
51. Oberts, G. L. and R. A. Osgood, The water quality effectiveness of a detention/wetland treatment system and its effect on an urban lake, presented at the 8th Annu. Int. Symp. Lake and Watershed Manage., St. Louis, MO, November 15–19, 1988.
52. Chan, E., *Treatment of Stormwater Runoff by a Marsh/Flood Basin*, Association of Bay Area Governments, Oakland, CA, 1979.
53. Ehrenfeld, J. G. and J. P. Schneider, *The Sensitivity of Cedar Swamps to the Effects of Nonpoint Source Pollution Associated with Suburbanization on the New Jersey Pine Barrens*, Rutgers University Press, New Brunswick, NJ, 1983.
54. Isabelle, P. S. et al., Effects of roadside snowmelt on wetland vegetation: an experimental study, *J. Environ. Manage.,* 25(1), 57, 1987.
55. Morgan, M. D. and K. R. Philipp, The effect of agricultural and residential development on aquatic macrophytes in the New Jersey Pine Barrens, *Biol. Conserv.,* 35, 143, 1986.
56. Mudrock, A. and J. A. Capobianco, Effects of treated effluent on a natural marsh, *J. Water Pollut. Control Fed.,* 51(9), 2243, 1979.
57. Tilton, D. L. and R. H. Kadlec, The utilization of freshwater wetland for nutrient removal from secondarily treated wastewater effluent, *J. Environ. Qual.,* 8, 328, 1979.
58. Galvin, D. V. and R. K. Moore, Toxicants in Urban Runoff, Toxicant Program, Report No. 2, Municipality of Metropolitan Seattle, Seattle, WA, 1982.
59. Harrill, R. L., Urbanization, water quality and stormwater management: a Maryland perspective, in Wetlands of the Chesapeake, Proc. Conf. held April 9–11, Easton, MD, 1985, p. 246.

60. Oberts, G. L., Water Quality Effects of Potential Urban Best Management Practices: A Literature Review, Techn. Bull. No. 97, Wisconsin Department of Natural Resources, Madison, 1977.
61. Hammer, D. A., *Creating Freshwater Wetlands*, Lewis Publishers, Chelsea, MI, 1992.
62. Fries, B. M., Fate of Phosphorus from Residential Stormwater Runoff in a Southern Hardwood Wetland, M.S. thesis. University of Central Florida, Orlando, 1986.
63. Lyon, J. G., R. D. Drobney, and C. E. Olson, Jr., Effects of Lake Michigan water levels on wetland soil chemistry and distribution of plants in the Straits of Mackinac, *J. Great Lakes Res.,* 12(3), 175, 1986.
64. Brinson, M. M., Strategies for assessing the cumulative effects of wetland alteration on water quality, *Environ. Manage.,* 12(5), 655, 1988.
65. ABAG (Association of Bay Area Governments), Treatment of Stormwater Runoff by a Marsh/Flood Basin (Interim Report), Association of Bay Area Governments, Oakland, CA, 1979.
66. Schiffer, D. M., Effects of Highway Runoff on the Quality of Water and Bed Sediments of Two Wetlands in Central Florida, Report No. 88-4200, U.S. Department of the Interior, Geological Survey, Washington, D.C., 1989.
67. Richardson, C. J., Wetlands as transformers, filters and sinks for nutrients, in *Freshwater Wetlands: Perspectives on Natural, Managed and Degraded Ecosystems,* 9th Symp., Charleston, SC, University of Georgia, Savannah River Ecology Laboratory, 1989.
68. Richardson, C. J., Mechanisms controlling phosphorus retention in freshwater wetlands, *Science,* 228, 1424, 1985.
69. Nichols, D. S., Capacity of natural wetlands to remove nutrients from wastewater, *J. Water Pollut. Control Fed.*, 55(5), 495, 1983.
70. Marshland Wastewater Treatment Evaluation for the City of Black Diamond, R. W. Beck and Associates, Seattle, WA, 1985.
71. Horner, R. R., Long-term effects of urban stormwater on wetlands, in *Design of Urban Runoff Controls*, L. A. Roesner, B. Urbonas, and M. B. Sonnen, Eds., American Society of Civil Engineers, New York, 1989, p. 451.
72. Cooke, S. S., The effects of urban stormwater on wetland vegetation and soils: a long-term ecosystem monitoring study, in Puget Sound Res. '91: Proc., January 4–5, 1991, Seattle, WA, Puget Sound Water Quality Authority, Olympia, WA, 1991, p. 43.
73. Horner, R. R., F. B. Gutermuth, L. L. Conquest, and A. W. Johnson, Urban stormwater and Puget Trough wetlands, in *Proc. First Annu. Meet. Puget Sound Res.*, Seattle, WA, March 18-19, 1988, Puget Sound Water Quality Authority, Seattle, 1988, p. 723.
74. Bedinger, M. W., Relation between forest species and flooding, in *Wetland Functions and Values: The State of Our Understanding*, P. E. Greeson et al., Eds., American Water Resources Association, Minneapolis, MN, 1979.
75. Green, W. E., Effect of water impoundment on tree mortality and growth, *J. For.,* 45(2), 118, 1947.
76. Brink, V. C., Survival of plants under flood in the lower Fraser River Valley, British Columbia, *Ecology,* 35(1), 94, 1954.
77. Ahlgren, C. E. and H. L. Hansen, Some effects of temporary flooding on coniferous trees, *J. For.*, 55, 647, 1957.
78. Rumberg, C. B. and W. A. Sawyer, Response of wet meadow vegetation to length and depth of surface water from wild flood irrigation, *Agron. J.*, 57, 245, 1965.
79. Minore, D., Effects of Artificial Flooding on Seedling Survival and Growth of Six Northwestern Tree Species, Forest Service Reserve Note PNW-92, U.S. Department of Agriculture, Forest Service, Washington, D.C., 1968.

80. Gill, C. J., The flooding tolerances of woody species: a review, *For. Abst.*, 31, 671, 1970.
81. Cochran, P. H., Tolerances of lodgepole pine and ponderosa pine seedlings to high water tables, *N. W. Sci.*, 46(4), 322, 1972.
82. Teskey, R. O. and T. M. Hinckley, Impact of Water Level Changes on Woody Riparian and Wetland Communities: Volume I — Plant and Soil Responses to Flooding, FWS/OBS-77/58, U.S. Department of the Interior, Fish and Wildlife Service, Washington, D.C., 1977.
83. Teskey, R. O. and T. M. Hinckley, Impact of Water Level Changes on Woody Riparian and Wetland Communities: Volume II — Southern Forest Region, FWS/OBS-77/59, U.S. Department of the Interior, Fish and Wildlife Service, Washington, D.C., 1977.
84. Teskey, R. O. and T. M. Hinckley, Impact of Water Level Changes on Woody Riparian and Wetland Communities: Volume III — The Central Forest Region, FWS/OBS-77/60, U.S. Department of the Interior, Fish and Wildlife Service, Washington, D.C., 1977.
85. Teskey, R. O. and T. M. Hinckley, Impact of Water Level Changes on Woody Riparian and Wetland Communities: Volume IV — Eastern Deciduous Forest Region, FWS/OBS-78/87, U.S. Department of the Interior, Fish and Wildlife Service, Washington, D.C., 1977.
86. Whitlow, T. H. and R. W. Harris, Flood Tolerance in Plants: A State of the Art Review, Techn. Rep. E-79-2, U.S. Army Corps of Engineers, Waterways Experiment Station, Vicksburg, MS, 1979.
87. Davis, G. B. and M. M. Brinson, Responses of Submerged Vascular Plant Communities to Environmental Change, FWS/OBS-79/33, U.S. Department of the Interior, Fish and Wildlife Service, Washington, D.C., 1980.
88. Walters, A. M., R. O. Teskey, and T. M. Hinckley, Impact of Water Level Changes on Woody Riparian and Wetland Communities: Volume VIII — Pacific Northwest and Rocky Mountain Regions, FWS/OBS-78/94, U.S. Department of the Interior, Fish and Wildlife Service, Washington, D.C., 1980.
89. McKnight, J. S., D. D. Hook, O. G. Langdon, and R. L. Johnson, Flood tolerance and related characteristics of trees of the bottomland forest of the southern United States, in *Wetlands of Bottomland Hardwood Forests*, J. R. Clark and J. Benforado, Eds., Elsevier, New York, 1981, p. 29.
90. Chapman, R. J., T. M. Hinckley, L. C. Lee, and R. O. Teskey, Impact of Water Level Changes on Woody Riparian and Wetland Communities: Volume X — Index and Addendum to Volumes I-VIII, U.S. Department of the Interior, Fish and Wildlife Service, Eastern Energy and Land Use Team, Kearneysville, WV, 1982.
91. Jackson, M. B. and M. C. Drew, Effects of flooding on growth and metabolism of herbaceous plants, in *Flooding and Plant Growth*, T. T. Kozlowski (ed.), Academic Press, San Diego, 1984.
92. Kozlowski, T. T., Ed., *Flooding and Plant Growth*, Academic Press, New York, 1984, p. 47.
93. Thibodeau, F. R. and N. H. Nickerson, Changes in a wetland plant association induced by impoundment and draining, *Biol. Conserv.*, 33, 269, 1985.
94. Gunderson, L. H., J. R. Stenberg, and A. K. Herndon, Tolerance of five hardwood species to flooding regimes, in *Interdisciplinary Approaches to Freshwater Wetlands Research*, D. A. Wilcox, Ed., Michigan State University Press, East Lansing, 1988, p. 119.
95. Kadlec, J. A., Effects of a drawdown on a waterfowl impoundment, *Ecology*, 43, 267, 1962.
96. Niering, W. A., Effects of Stormwater Runoff on Wetland Vegetation, unpublished report (cited in U.S. EPA 1993), 1989.

97. van der Valk, A. G., Response of wetland vegetation to a change in water level, in *Wetland Management and Restoration,* C. M. Finlayson and T. Larson, Eds., Rep. 3992, Swedish Environmental Protection Agency, Stockholm, 1991.
98. Conner, W. H., J. G. Gosselink, and R. T. Parrondo, Comparison of the vegetation of three Louisiana swamp sites with different flooding regimes, *Am. J. Bot.*, 68, 320, 1981.
99. Gomez, M. M. and F. P. Day, Litter nutrient content and production in the Great Dismal Swamp, Virginia, *Am. J. Bot.*, 69, 1314, 1982.
100. Cooke, S. S. and A. L. Azous, Effects of Urban Stormwater Runoff and Urbanization on Palustrine Wetland Vegetation, A Report to the U. S. Environmental Protection Agency Region 10, Seattle, 1993.
101. Kadlec, J. A., Nutrient dynamics in wetlands, in *Aquatic Plants for Water Treatment and Resource Recovery*, K. R. Reddy and W. H. Smith, Eds., Magnolia Publishing, Orlando, 1987, p. 393.
102. Wickcliff, C., H. J. Evans, K. K. Carter, and S. A. Russell, Cadmium effects on the nitrogen fixation system of red alder, *J. Environ. Qual.*, 9, 180, 1980.
103. Brown, B. T. and B. M. Rattigan, Toxicity of soluble copper and other metal ions in Eleodea Canadensis, *Environ. Pollut.*, 21, 303, 1979.
104. Portele, G. J., The Effects of Highway Runoff on Aquatic Biota in the Metropolitan Seattle Area, M.S. thesis, University of Washington, Department of Civil Engineering, Seattle, 1981.
105. Nordby, C. S. and J. B. Zedler, Responses of fish and macrobenthic assemblages to hydrologic disturbances in Tijuana Estuary and Los Peñasquitos Lagoon, CA, *Estuaries,* 14(1), 80, 1991.
106. Marshall, W. A., The Effect of Stormdrain Runoff on Algal Growth in Nearshore Areas of Lake Sammamish, M.S. thesis, University of Washington, Department of Civil Engineering, Seattle, 1980.
107. Minton, S. A., Jr., The fate of amphibians and reptiles in a suburban area, *J. Herpetol.,* 2(3–4), 113, 1968.
108. Lloyd-Evans, T. L., Use of Wetland for Stormwater Detention Effects on Wildlife Habitat, Manomet Bird Observatory (cited in Azous, 1991), 1989.
109. Richter, K. O., A. Azous, S. S. Cooke, R. Wisseman, and R. Horner, Effects of Stormwater Runoff on Wetland Zoology and Wetland Soils Characterization and Analysis, PSWSMRP, Seattle, WA, 1991.
110. Johnsgrad, P. A., Effects of water fluctuation and vegetation change on bird populations, particularly waterfowl, *Ecology*, 37, 689, 1956.
111. Richter, K. O. and R. Wisseman, Effects of Stormwater Runoff on Wetland Zoology, PSWSMRP, Seattle, WA, 1990.
112. Bursztynsky, T. A., Personal communication to E. C. Stockdale, King County, Resources Planning Section (cited in Stockdale, 1991), 1986.
113. Weller, M. W., The influence of hydrologic maxima and minima on wildlife habitat and production values of wetlands, in *Wetland Hydrology*, J. A. Kusler and G. Brooks, Eds., Association of State Wetland Managers, Chicago, 1987.
114. Stockdale, E. C., The Use of Wetlands for Stormwater Management and Nonpoint Pollution Control: A Review of the Literature, Washington Department of Ecology, Olympia, 1986.
115. Stockdale, E. C., Viability of Freshwater Wetlands for Urban Surface Water Management and Nonpoint Pollution Control: An Annotated Bibliography, Washington Department of Ecology, Olympia, 1986.
116. Reinelt, L. E. and R. R. Horner, Pollutant removal from stormwater runoff by palustrine wetlands based on a comprehensive budget, *Ecol. Eng.*, 4, 77, 1995.

117. Cowardin, L. M., V. Carter, F. C. Golet, and E. T. LaRoe, Classification of Wetlands and Deepwater Habitats of the United States, FWS/OBS-79/31, U.S. Fish and Wildlife Service, Washington, D.C., 1979.
118. Taylor, B. L., The Influence of Wetland and Watershed Morphological Characteristics on Wetland Hydrology and Relationships to Wetland Vegetation Communities, M.S. thesis, University of Washington, Department of Civil Engineering, Seattle, 1993.
119. Alley, W. M. and J. E. Veenhuis, Effective impervious area in urban runoff modeling, *J. Hydraul. Eng.,* 109(2), 313, 1983.
120. Prych, E. A. and J. C. Ebbert, Quantity and Quality of Storm Runoff from Three Urban Catchments in Bellevue, Washington, Water Resources Investigations Report 86-4000, U. S. Geological Survey, Tacoma, 1986.
121. PEI/Barrett Consulting Group, Snoqualmie Ridge Draft Master Drainage Plan, Bellevue, WA, 1990.
122. U.S. Environmental Protection Agency, Results of the Nationwide Urban Runoff Program: Executive Summary, NTIS PB84-185545, U.S. Environmental Protection Agency, Water Planning Division, Washington, D.C., 1983.

Section II

Descriptive Ecology of Freshwater Wetlands in the Central Puget Sound Basin

1 Morphology and Hydrology

Lorin E. Reinelt, Brian L. Taylor, and Richard R. Horner

CONTENTS

Introduction 31
Puget Sound Wetlands and Stormwater Management Research Program 32
Wetlands in Urbanizing Areas 33
Wetland Hydrologic Functions 33
Hydrology of Palustrine Wetlands 33
  Precipitation 34
  Surface Inflows 34
  Groundwater 34
  Change in Wetland Storage 35
  Evapotranspiration 35
  Surface Outflow 35
Research Methods and Wetland Descriptors 36
  Wetland Morphology 36
  Watershed Characteristics 36
  Watershed Imperviousness 37
  Watershed Soils 38
Wetland Hydrology 39
  Wetland Water Level Measurements and Fluctuation 39
  Seasonal Fluctuation in Wetland Water Levels 39
  Length of Summer Dry Period 40
Results and the Conceptual Model 40
  Water Level Fluctuation Patterns 40
  Conceptual Model of Influences on Wetland Hydroperiod 41
Conclusions 41
Acknowledgments 43
References 44

INTRODUCTION

This chapter provides an overview of the morphologic and hydrologic characteristics of palustrine (depressional freshwater) wetlands and their watersheds in the central

1-56670-386-7/00/$0.00+$.50

Puget Sound Basin. Natural and anthropogenic factors that affect wetland morphology and hydrology are discussed, with particular attention to the effects of development (typically the conversion of forested lands to urban areas) on changing watershed and wetland hydrology. It was concluded that wetland water level fluctuation (WLF) estimates, measured with staff and crest-stage gauges, provide a good overall indicator of wetland hydrologic conditions. Analysis methods and materials used in the research program are also presented.

Wetlands are ecosystems that develop at the interface of aquatic and terrestrial environments when hydrologic conditions are suitable. Wetlands are recognized as biologically productive ecosystems offering extensive, high-quality habitat for a diverse array of terrestrial and aquatic species, as well as multiple beneficial uses for humans, including flood control, groundwater recharge, and water quality treatment. However, as urbanization of natural landscapes occurs, some or all of the functions and values of wetlands may be affected. Some wetlands may be impacted by direct activities such as filling, draining, or outlet modification, while others may be affected by secondary impacts, including increased or decreased quantity and reduced quality of inflow water.

The morphology of a wetland and the wetland's position within the landscape greatly influences habitat characteristics. Morphology is used here to describe the wetland's physical shape and form. As a result of a wetland's shape, it may contain significant pooled areas with little or no flow gradient (termed an open-water system), or alternatively, it may show evidence of channelization and contain a significant flow gradient (termed a flow-through system). In some instances, a wetland may also form in a local or closed depression (termed a depressional system).

The outlet condition of a wetland, as defined by the degree of flow constriction, has a direct effect on wetland hydrology and hydroperiod. Finally, a wetland's position in the landscape is also a key factor affecting wetland hydrologic conditions. Palustrine (isolated, freshwater) wetlands usually have relatively small contributing watersheds and often occur in areas with groundwater discharge conditions.

Hydrology is probably the single most important determinant for the establishment and maintenance of specific types of wetlands and wetland processes.[1] Water depth, flow patterns, and the duration and frequency of inundation influence the biochemistry of the soils and are major factors in the selection of wetland biota. Thus, changes in wetland hydrology may influence significantly the soils, plants, and animals of particular wetland systems. Precipitation, surface water inflow and outflow, groundwater exchange and evapotranspiration, along with the physical features noted above, are the major factors that influence the hydrology of palustrine wetlands.

PUGET SOUND WETLANDS AND STORMWATER MANAGEMENT RESEARCH PROGRAM

The Puget Sound Wetlands and Stormwater Management Research Program was established to determine the effects of urban stormwater on wetlands and the effect of wetlands on the quality of urban stormwater. There are two primary components of the research program: (1) a study of the long-term effects of urban stormwater on wetlands, and (2) a study of the water quality benefits to downstream receiving waters as urban stormwater flows through wetlands. In both studies, the hydrologic and morphologic conditions of the wetlands had a direct effect on observations involving water quality, soils, and the plant and animal communities.

We present here hydrologic information gained from a broad overview of the hydrology of 19 wetlands representing a variety of watershed development conditions studied from 1988 to 1995. The discussion also covers information on the hydrology of two wetlands, one each in an urban (B3I) and undeveloped (PC12) watershed, intensively studied from 1988 to 1990. Study site locations are shown in Section I, Figure 1-1.

WETLANDS IN URBANIZING AREAS

Wetlands have received increased attention in recent years as a result of continuing wetland losses and impacts resulting from new development. In urbanizing areas, the quantity and quality of stormwater can change significantly as a result of land-use conversion in a watershed. Increases in the quantity of stormwater usually result from new impervious surfaces (e.g., roads, buildings), installation of storm sewer piping systems, and removal of trees and other vegetation in the watershed. On the other hand, decreased inflow of water to wetlands can result from modifications in surface and groundwater flows.

Wetland hydrology is often described in terms of its hydroperiod, the pattern of fluctuating water levels resulting from the balance between water inflows and outflows, topography, subsurface soil, geology, and groundwater conditions.[1] Seasonal water level changes have been described as the heartbeat of Pacific Northwest palustrine systems.[2] For cases where wetlands are the primary receiving water for urban stormwater from new developments, the effects of watershed changes will be manifested through changes in the hydrology of wetlands.

WETLAND HYDROLOGIC FUNCTIONS

Wetlands provide many important hydrologic, ecological, and water quality functions. Specific hydrologic functions include flood protection, groundwater recharge, and streamflow maintenance. Wetlands provide flood protection by holding excess runoff after storms, before slowly releasing it to surface waters. While wetlands may not prevent flooding, they can lower flood peaks by providing detention of storm flows.

Wetlands that are connected to groundwater or aquifers provide important recharge waters. Wetlands retain water, allowing time for surface waters to infiltrate into soils and replenish groundwater. During periods of low streamflow, the slow discharge of groundwater maintains instream flows. The connection of wetlands with streamflows and groundwater make them essential in the proper functioning of the hydrologic cycle.

HYDROLOGY OF PALUSTRINE WETLANDS

The hydrology of palustrine wetlands is governed by the following components: precipitation, evapotranspiration, surface inflow, surface outflow, groundwater

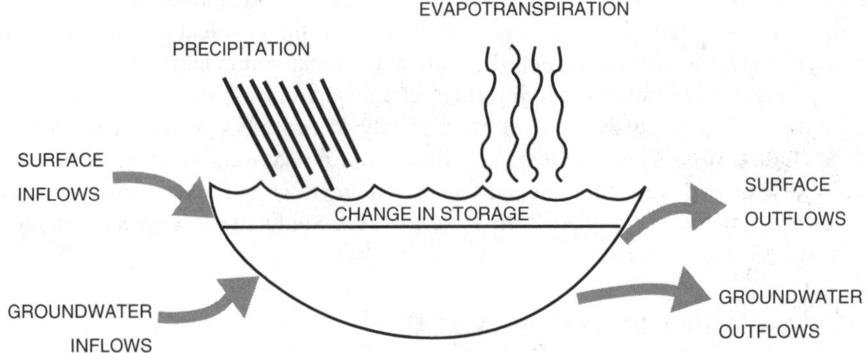

FIGURE 1-1 Wetland water budget components.

exchange, and change in wetland storage (see Figure 1-1). In a hydrologic balance, these components are represented by the following equation:

$$P + I \pm G \pm S = ET + O \tag{1-1}$$

where P = precipitation, I = surface inflow, G = groundwater exchange, S = change in wetland storage, ET = evapotranspiration, and O = surface outflow.[3]

Precipitation

Precipitation is determined by regional climate and topography. Approximately 75% of the total annual rainfall occurs from October to March during a well-defined wet season in the Puget Sound region. Generally, annual precipitation totals across central Puget Sound increase further east with increasing elevation. Rainfall tends to be more uniform geographically during the wet season, and more variable and intense during short dry-season cloudbursts.

Surface Inflows

Surface inflows result from runoff generation in the wetland's watershed. The quantity of surface inflows are determined by watershed land characteristics such as cover (e.g., impervious surface, forest), soils, slopes, as well as the wetland-to-watershed ratios. The rate of water delivery to the wetland is also affected by the predominant flow type in the watershed (e.g., overland or sheet flow, subsurface flow or interflow, concentrated flow). Generally, as a watershed becomes more developed, with more constructed storm drainage systems, the more rapid the hydrologic response in the watershed.

Groundwater

The role and influence of groundwater on wetland hydrology is highly variable. The exchange of water between the wetland and groundwater is governed by the relative elevations of surface water in the wetland and surrounding groundwater, as well as soil permeability, local geology and topography. Numerous studies have discussed the importance of groundwater in maintaining wetland hydrology.[1,4,5] Wetlands can be discharge or recharge zones for groundwater, or both, depending on the time of year. The palustrine wetlands studied in this research are predominantly groundwater discharge zones (water discharges from groundwater to the wetland).

Groundwater flow to wetlands can be quantitatively estimated using Darcy's law, an empirical law governing groundwater flow:

$$Q = -KA\ dH/dL \tag{1-2}$$

where K = hydraulic conductivity, dH/dL = the hydraulic or piezometric gradient and A = cross-sectional area or control surface across which groundwater flows.

In the detailed study of two wetlands, shallow and deep piezometers were installed at both wetlands to estimate the horizontal and vertical components, respectively, of groundwater flow to the wetlands.

Change in Wetland Storage

Wetland storage changes seasonally and in response to storm events. The water storage can be estimated as the mean water depth of the wetland multiplied by the area of the wetland.[1] Seasonal changes in wetland storage are attributable to the local patterns of precipitation and evapotranspiration.

It has been asserted that the prime factor controlling seasonal fluctuation is drainage basin topography and that wetland water levels generally coincide with regional groundwater levels.[6] Also observed is that steep slopes adjacent to a wetland can lead to increased groundwater inputs, particularly on a seasonal basis.[5]

Event changes in wetland storage result from increased surface or groundwater inputs associated with precipitation. This observation of hydrologic change is referred to as water level fluctuation (WLF). It is estimated for an occasion as the difference between the instantaneous staff-gauge measurement and a crest-stage measurement of the peak water level (since the previous sampling occasion).[7-9]

Evapotranspiration

Evapotranspiration (ET) consists of water that evaporates from wetland water or soils combined with the water that passes through vascular plants that is transpired to the atmosphere. Solar radiation, temperature, wind speed, and vapor pressure are the main factors influencing evaporation rates.[10]

The ratio of ET to evaporation varies widely depending on vegetation type and site conditions. Reported ET ratios vary between 0.67 and 1.9.[11-13] Generally, emergent wetland vegetation transpires more than woody vegetation; however, factors such as plant density also affect transpiration rates. Evapotranspiration is greatest from May to August, exceeding 100 mm per month, and least from November to March.

Surface Outflow

Surface outflows are affected by all the hydrologic factors noted above. For wetlands with relatively large watersheds, outflows are often comparable in magnitude to inflows. The physical features that affect surface outflows include outlet conditions, wetland-to-watershed ratios, and wetland morphometry.

RESEARCH METHODS AND WETLAND DESCRIPTORS

Many of the methods and materials used for morphologic, hydrologic, and watershed data collection were previously reported in PSWSMRP papers.[7,9,14] Here, we provide a summary of the methods, with additional information on data processing and analysis.

WETLAND MORPHOLOGY

Three different measures of wetland morphology that influence the hydrology and hydroperiod of wetlands were defined and include wetland type (open water, flow through, depressional), outlet condition, and wetland-to-watershed ratio.[7] Wetlands were classified as open-water systems if significant open water pools were present and surface water velocities were predominantly low (less than 5.0 cm/s). Wetlands were classified as flow-through systems if there was evidence of channelization and significant water velocities. All depressional wetlands were also open water wetlands.

Outlet conditions were defined by level of constriction as high (e.g., undersized culvert, closed depression, confined beaver dam), or low to moderate (e.g., overland flow to stream, oversized culvert, broad bulkhead or beaver dam).[9,14] Wetland-to-watershed ratios was determined by the area of wetland in relationship to the contributing watershed. Watershed areas were delineated based on U.S. Geological Survey (USGS) quadrangle map contours and wetland areas were obtained from the King County Wetlands Inventory.[15] The hydroperiod of wetlands with low wetland-to-watershed ratios (less than 0.05) tends to be dominated by surface inflows, whereas wetlands with higher ratios are more influenced by regional groundwater conditions.

WATERSHED CHARACTERISTICS

Changes in land use ultimately affect wetlands receiving water from an urbanizing drainage basin. Different land uses have unique combinations of factors that directly affect watershed hydrology, such as imperviousness and vegetative cover. By collecting information about drainage basin land use, it is possible to link wetland hydroperiod characteristics to specific land uses, as well as general changes associated with urban development.

A geographic information system (GIS) was developed to manage land use data for the watersheds of the study wetlands, and to facilitate quantitative and graphical analysis of land-use patterns. Land-use classifications, based on a national standard, were determined from 1989 aerial photographs and subsequently digitized for viewing and analysis using computer software.[14,16] For each study site, the GIS contained information about the total watershed, wetland area, and the area of watershed in each land use type (e.g., urban, agriculture, forest).

TABLE 1-1
Total Impervious (TIA) and Effective Impervious Areas (EIA) Associated with Land Uses

Code	National Standard	TIA%	EIA%	Reference Number or Source
111	Low Density SFR (<1 unit/acre)	<15	4	24
112	Med. Density SFR (1–3 unit/acre)	20	10	24
113	High Density SFR (3–7 units/acre)	40	25	24
114	Mobile Homes	70	60	22
115	Low Density MFR (>7 units/acre)	80	72	22
120	Commercial (general)	90	85	22
121	Retail sales and services	80	72	22
123	Offices and professional services	75	66	22
124	Hotels and Motels	75	66	Est.
131	Light Industrial	60	48	21
132	Heavy Industrial	80	72	Est.
144	Freeway Right-of-way	100	99	22
151	Energy Facilities	80	72	Est.
152	Water Supply Facilities	80	72	Est.
155	Utility Right-of-way	5	1.5	Est.
160	Community Facilities (general)	75	66	Est.
161	Educational Facilities	40	27	22
162	Religious Facilities	70	60	Est.
171	Golf Courses	20	10	22
172	Parks	5	1.5	22
190	Open Land (general)	2	1	Est.
192	Land being developed	50	37	Est.
193	Open space — designated	2	1	Est.
200	Agricultural Land	5	1.5	Est.
300	Grassland	2	0	Est.
400–430	Forest Lands	2	0	Est.
440	Clearcut areas	5	0	Est.

Note: Est. = Estimate based on similar land uses.

WATERSHED IMPERVIOUSNESS

The literature consistently identifies hydrologic effects of urbanization with increased impervious areas within the watershed.[18] Increases of impervious area within a watershed reduce infiltration due to forest clearing for urban conversion. This also results in a loss of vegetative storage and decreased transpiration.[19]

Imperviousness was estimated from aerial photos and empirical relationships between land uses and percent impervious cover (see Table 1-1).[20,21,24] This estimation technique was found to produce results consistent with values used in Puget Sound lowland hydrologic models.[22] Effective impervious area (impervious surfaces connected to a storm drainage system) was also estimated according to a formula based on drainage basins in the Denver area:

$$EIA = 0.15\ TIA\ 1.41 \quad (r^2 = 0.98, \text{standard error} = 7.5\%) \qquad (1\text{-}3)$$

where EIA and TIA are the percent effective and total impervious area, respectively.[21]

Watershed Soils

The Soil Conservation Service (SCS) Soil Survey for King County was used to evaluate the drainage characteristics of the soils in each of the 19 drainage basins.[23] Two soil parameters were reviewed to determine which would be an appropriate index of the soil hydrologic characteristics relevant to the analysis and included permeability and general drainage characteristics.

Soil permeability is measured as a range of infiltration rates, the units of which are distance over time. Soil permeability for the majority of the soils found in the watersheds was in the range of 2.0 to 6.3 in./hr. The drainage class is a more general description of the soil characteristics such as "Moderately well-drained" or "Somewhat excessively drained." Many soils in the Pacific Northwest (e.g., Alderwood series) are underlain by glacial till, a hardpan layer that limits the ultimate depth of percolation and plays an important role in routing subsurface flow. Drainage class was therefore thought to be a better estimator of the hydrologic role of the watershed soils than permeability because it represents the effects of the multiple soil horizons characterized as a particular soil type; whereas infiltration rate is only based on the top soil layer.

For this study, a watershed soils index (WSI) was calculated as an area-weighted mean of soil drainage classes found in the study basins. Each of seven drainage classes described by the SCS was assigned a number that ranged from 1 to 7, with lower numbers representing poorly drained soils (see Table 1-2). The range of the

TABLE 1-2
Wetland and Watershed Morphologic and Hydrologic Characteristics

Drainage Class	WSI	SCS Hydrology Group [a]	Examples
Very poorly drained	1	(D)	(Muck)
Poorly drained	2	D	Norma, Bellingham
Somewhat poorly drained	3	(D)	(Oridia, Renton)
Moderately well drained	4	C	Alderwood, Kitsap
Well drained	5	B, C	Ragnar, Beausite
Somewhat excessively drained	6	A	Everett, Indianola
Excessively drained	7	(C)	(Pilchuck) [b]

[a] Parentheses indicate soils that were not found in any of the wetland watersheds.
[b] The apparent contradiction between Hydrology Group and WSI is because this soil is found on terraces adjacent to streams.

WSI corresponded with the SCS Hydrologic Soil groups, which are used in the Curve Number method of runoff estimation. The WSI was preferred for the analysis because it describes soil drainage to a finer level than the hydrologic soil group.

WETLAND HYDROLOGY

WETLAND WATER LEVEL MEASUREMENTS AND FLUCTUATION

Water level measurements in wetlands can be made using a variety of gauges or instruments. Readings can be instantaneous, continuous, or representative of a peak or base level since the last site visit. In the research program, we utilized staff and crest stage gauges to record, respectively, instantaneous water levels and peak occasion water levels during each site visit. At two wetlands (B3I and PC12) continuous water levels were recorded over a two-year period (1988 to 1990) using automatic data recorders. Gauges were placed in open water areas or areas of channelized flow where water level measurements could be attained throughout most of the year.

The crest stage and staff gauge data were used to estimate wetland water level fluctuation. To estimate the water level fluctuation at a wetland site, two factors were considered: (1) the water level prior to the storm event, hereafter referred to as the base water level, and (2) the water level change resulting from the event. Four methods of calculating water level fluctuation were investigated in a preliminary analysis before choosing a preferred method to use in the analysis.[8,9] Methods differed primarily in how the base water level prior to the stormwater influx was estimated. The fluctuation was then calculated as the difference between the maximum and base water levels.

The selected method used the midpoint of the sampling interval to estimate the base water level:

$$\text{WLFi} = \text{Ci} - 0.5(\text{Si} + \text{Si} - 1) \tag{1-4}$$

where WLFi, Ci, and Si = the water level fluctuation, crest level, and base level, respectively, for sampling occasion i, and Si – 1 = the base level for occasion i – 1.

The water level fluctuation data were used in three ways during the analysis. The data from each sampling occasion were used when evaluating the relationship between precipitation and water level fluctuation. Mean and maximum study period WLF values were used when assessing the effects of land use and wetland characteristics on the wetland hydroperiod.

SEASONAL FLUCTUATION IN WETLAND WATER LEVELS

Seasonal fluctuation in wetland water levels is probably the most important factor governing wetland development and functioning in the Pacific Northwest.[2] A quantitative measure of seasonal WLF was developed based on an examination of the hydroperiod plots for the study wetlands.

May and October are months when the water level changes dramatically in those sites that undergo large seasonal fluctuations. Noting this, the dry season water level was estimated as the mean of staff gauge measurements collected during the months June through September. Similarly, the wet season water level was estimated as the mean of staff gauge measurements collected between November and May. Approximately equal sample sizes were used to calculate each of the seasonal mean water levels. The mean seasonal difference in water levels was calculated as the difference in these seasonal mean water levels. The data from the early study period (April 1988 to April 1991) were used to calculate seasonal WLFs.

A second measure of the seasonal WLF is the range of water levels observed. The water depth range was calculated as the difference between the maximum and minimum water levels during the study period. This measure, used with the mean seasonal water level difference described above, provided a picture of the wetland hydroperiod suitable for analysis, because both typical and extreme events were addressed.

Length of Summer Dry Period

The length of the summer dry period (defined by the absence of surface water) was also analyzed; however, this dry period estimate was subject to the following limitations:

1. Estimating the length of the dry period was affected by the flow characteristics and topography within the wetland; that in turn determined which areas dry first. Because gauges were placed in the wetland areas thought to be the last to dry out during the summer, a water level of 3 cm or less constituted "dry" in this analysis.
2. The exact length of the dry period was uncertain, because of the frequency of site visits. The approximate monthly sampling interval during the summer months did not allow for the determination of the date the water level reached "zero." To compensate for this uncertainty, the transition from "wet" to "dry" (or vice versa) was assumed to occur at the midpoint of the sampling interval.

RESULTS AND THE CONCEPTUAL MODEL

Descriptive results of the morphologic and hydrologic analysis of the study wetlands are shown in Table 1-3. The various water level fluctuation patterns observed in the wetlands and a conceptual model relating wetland and watershed characteristics to wetland hydroperiod are also presented below.

Water Level Fluctuation Patterns

Based on the water level fluctuation analysis, wetlands were classified into four distinguishable types of hydroperiods (Figure 1-2):

1. Stable base water level with low event fluctuations (SL),
2. Stable base water level with high event fluctuations (SH),
3. Fluctuating base water level with low event fluctuations (FL), and
4. Fluctuating base water level with high event fluctuations (FH).

TABLE 1-3
Soil Drainage Classes and Watershed Soils Index (WSI)

Wetland	Outlet Condition	Outlet Constriction	WLF Type	Dry in Summer?	System Type	% TIA 1989	% TIA 1995
AL3	None	High	FL	Y	OW/D	4	4
B3I	Culvert	High	SH	N	FT	55	55
BBC24	Beaver dam	Low	SL	N	OW	3	11
ELS39	Culvert	High	FH	Y	OW	25	25
ELS61	Stream	Low	FL	N	OW	5	11
ELW1	Lake	Low	SH	N	FT	20	20
FC1	Beaver dam	Moderate	S/FH	N	FT	31	31
HC13	Beaver dam	High	FL	N	OW	4	4
JC28	Stream	Low	SL	Y	FT	20	21
LCR93	None	High	FH	Y	FT	6	6
LPS9	Drain inlet	High	FH	Y	FT	22	22
MGR36	Stream	Low	SL	N	FT	3	3
NFIC12	None	High	FL	Y	OW/D	2	40
PC12	Beaver dam	High	FL	N	OW	5	7
RR5	Beaver dam	Low	FL	N	OW	3	3
SR24	Road	Low	FL	N	OW	2	2
SC4	Culvert	Low	SL	Y	FT	12	12
SC84	Stream	Low	FL	Y	OW	19	17
TC13	Drain inlet	Moderate	FL	Y	OW	2	2

The four patterns were defined quantitatively using a threshold of 20 cm. Wetlands with a base water level range less than or greater than 20 cm were considered stable or fluctuating, respectively. Similarly, wetlands with event fluctuations less than or greater than 20 cm were considered low or high, respectively. Figure 1-2 shows the WLF pattern for the 19 study wetlands.

Conceptual Model of Influences on Wetland Hydroperiod

A conceptual model was developed to characterize the relationships between watershed and wetland morphological characteristics and wetland hydroperiod (Figure 1-3).[9] This model was used to examine, through application of a multivariate regression model, which wetland and watershed hydrologic processes, and factors governing these processes had the greatest influence on wetland hydroperiod. Results from this analysis are presented in Section III of this report.

CONCLUSIONS

There are many descriptive measures of the morphologic and hydrologic characteristics of freshwater wetlands in the central Puget Sound basin. Summarized here are those that were examined and utilized by the PSWSMRP. The physical shape or type of wetland (e.g., open water, flow-through), the wetland's position within the landscape particularly as related to the wetland-to-watershed ratio, and the degree of outlet constriction were presented as key wetland characteristics affecting hydro-

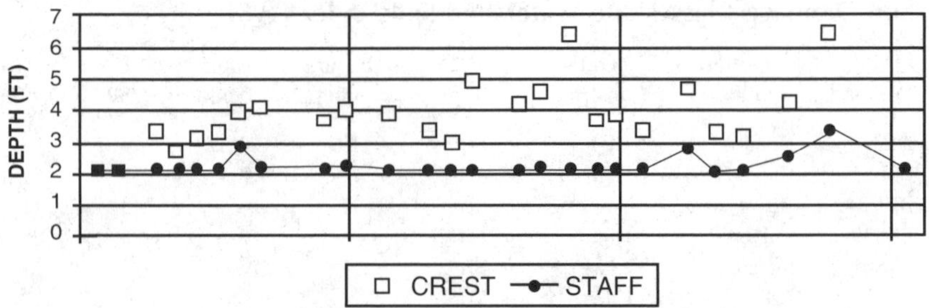

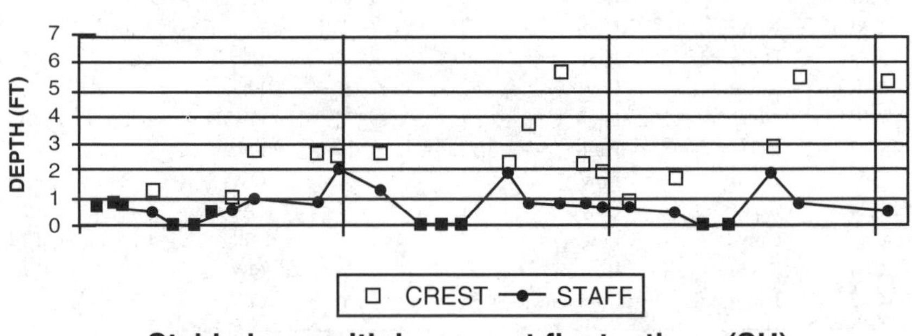

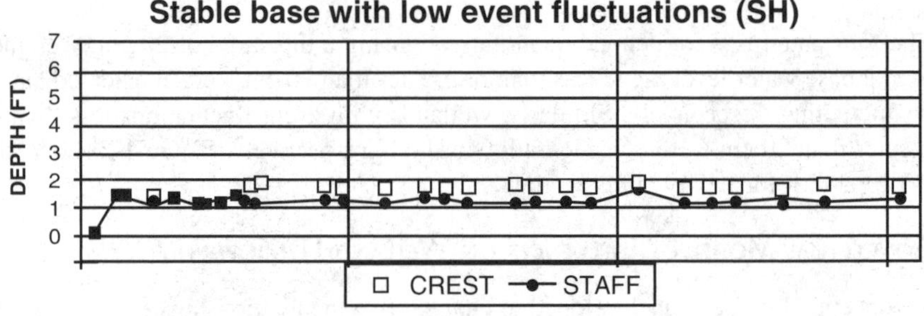

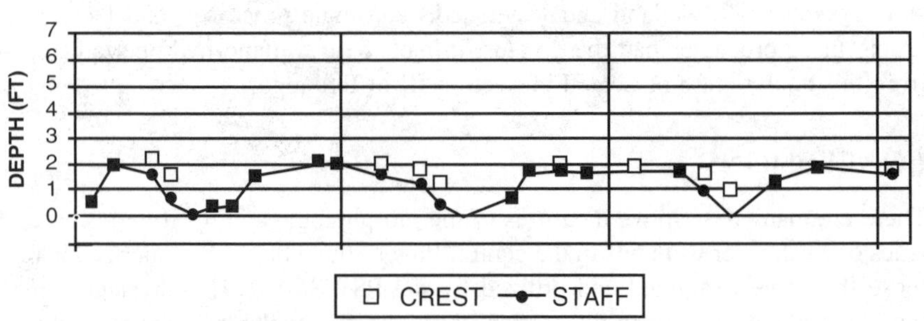

FIGURE 1-2 Four water-level fluctuation patterns.

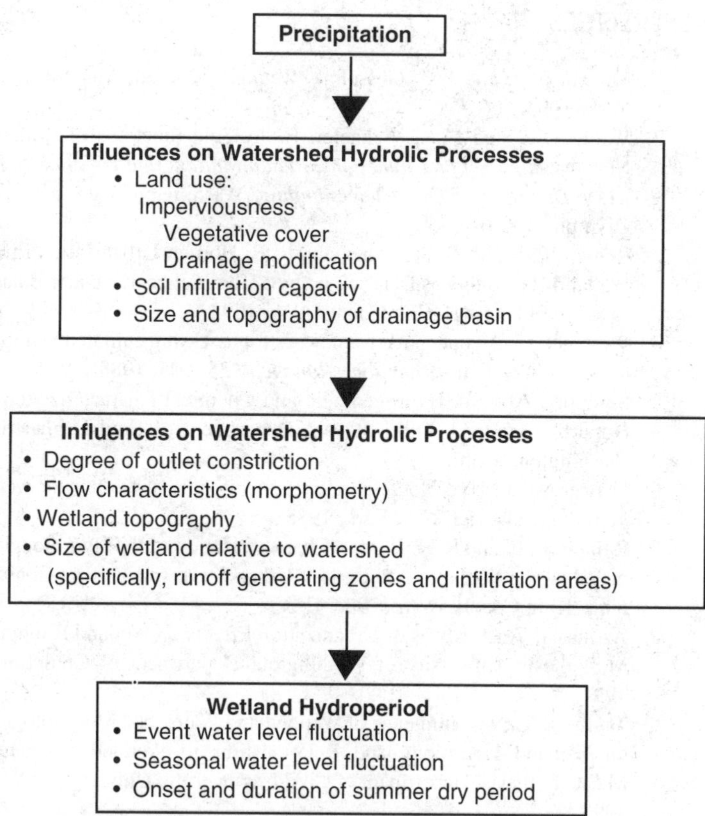

FIGURE 1-3 Conceptual model of influences on wetland hydroperiod.

periods. The imperviousness, land cover, and soils of the watershed were also found to be important characteristics affecting surface runoff and wetland hydrology.

The quantity of stormwater entering many wetlands in the central Puget Sound region has changed as a result of rapid development in urbanizing areas. These changes may affect the functions and values of wetlands by impacting the hydrology, which in turn may affect the plant and animal communities. If the relationships between watershed and wetland changes and their impacts on wetland hydroperiod can be characterized and documented, it may be possible to mitigate these effects through improved watershed controls or development regulations.

ACKNOWLEDGMENTS

The authors wish to thank Mike Surowiec for performing the hydrogeologic and groundwater components for the intensive study of two wetlands (Bellevue 3I and Patterson Creek 12).

REFERENCES

1. Mitsch, W. J. and J. G. Gosselink, *Wetlands,* 2nd ed., Van Nostrand Reinhold, New York, 1993.
2. Wald, A. R. and M. G. Schaefer, Hydrologic functions of wetlands of the Pacific Northwest, in *Wetland Functions, Rehabilitation, and Creation in the Pacific Northwest: The State of Our Understanding,* Washington State Department of Ecology, Olympia, 1986.
3. Reinelt, L. E., M. S. Surowiec, and R. R. Horner, Urbanization Effects on Palustrine Wetland Hydrology as Determined by a Comprehensive Water Balance, King County Resource Planning, King County, WA, 1993.
4. Winter, T. C., A conceptual framework for assessing cumulative impacts on hydrology of nontidal wetlands, *Environ. Manage.,* 125, 605, 1988.
5. Surowiec, M.S., A Hydrogeologic and Chemical Characterization of an Urban and Nonurban Wetland, M.S.E. thesis, Department of Civil Engineering, University of Washington, Seattle, 1989.
6. Duever, M. J., Hydrologic processes for models of freshwater wetlands, in *Wetland Modeling*, Elsevier, New York, 1988.
7. Reinelt, L. E. and R. R. Horner, Characterization of the Hydrology and Water Quality of Palustrine Wetlands Affected by Urban Stormwater, King County Resource Planning, King County, WA, 1990.
8. Azous, A., An Analysis of Urbanization Effects on Wetland Biological Communities, M.S. thesis, University of Washington, Department of Civil Engineering, Seattle, 1991.
9. Taylor, B. L., The Influence of Wetland and Watershed Morphological Characteristics on Wetland Hydrology and Relationships to Wetland Vegetation Communities, M.S.C.E. thesis, Department of Civil Engineering, University of Washington, Seattle, 1993.
10. Linsley, R. K., Jr., M. A. Kohler, and J. L. H. Paulhus, *Hydrology for Engineers*, McGraw-Hill, New York, 1982.
11. Dolan, T. J., A. J. Hermann, S. E. Bayley, and J. Soltek, Evapotranspiration of a Florida, U.S., freshwater wetland, *J. Hydrol.*, 74, 355, 1984.
12. Boyd, C. E., Evapotranspiration/evaporation (E/Eo) ratios for aquatic plants, *J. Aquat. Plant Manage.*, 25, 1, 1987.
13. Koerselman, W. and B. Beltman, Evapotranspiration from fens in relation to penman's potential free water evaporation (Eo) and pan evaporation, *Aquat. Bot.*, 31, 307, 1988.
14. Reinelt, L. E. and R. R. Horner, Urban stormwater impacts on the hydrology and water quality of palustrine wetlands in the Puget Sound region, *Puget Sound Water Qual. Auth.,* 1, 33, 1991.
15. King County Wetlands Inventory, Resource Planning, King County, WA, 1991.
16. Anderson, J. R., E. E. Hardy, J. T. Roach, and R. E. Witmer, A land use and land cover classification system for use with remote sensor data, U.S.G.S. Professional Paper 964, U.S. Geologic Survey, Washington, D.C., 1976.
17. Reinelt, L. E., J. Velikanje, and E. J. Bell, Development and application of a geographic information system for wetland/watershed analysis, *Comput. Environ. Urban Syst.,* 15, 239, 1991.
18. Schueler, T. R., The importance of imperviousness, *Watershed Protect. Tech.,* 13, 100, 1994.
19. Lazaro, T. R., *Urban Hydrology: A Multidisciplinary Perspective*, Ann Arbor Science, Ann Arbor, MI, 1979.

20. Gluck, W. R. and R. H. McCuen, Estimating land use characteristics for hydrologic models, *Water Resour. Res.,* 1(11), 177, 1975.
21. Alley, W. M. and J. E. Veenhuis, Effective impervious area in urban runoff modeling, *J. Hydraul. Eng.,* 109(2), 313, 1983.
22. PEI/Barrett Consulting Group, Snoqualmie Ridge Draft Master Drainage Plan, 1990.
23. Snyder, D. E., P. S. Gale, and R. F. Pringle, Soil Survey of King County Area, Washington, U.S. Department of Agriculture, Soil Conservation Service, Washington, D.C., 1973.
24. King County Surface Water Management, Surface Water Design Manual, King County Department of Public Works, Seattle, WA, 1990.

2 Water Quality and Soils

Richard R. Horner, Sarah S. Cooke, Lorin E. Reinelt, Kenneth A. Ludwa, and Nancy T. Chin

CONTENTS

Introduction 47
Water Quality 48
Collection and Methods 48
Research Findings: A Portrait of Puget Sound Basin Wetland Water Quality 49
Wetland Water Quality in Context 51
Seasonal Variation 54
Variation with Wetland Morphology 57
Soils 59
Collection and Methods 59
Research Findings: A Portrait of Puget Sound Basin Wetland Soils 60
General Soil Chracteristics and Nutrients 60
Metals in Soils 63
Summary of Soil Characteristics of Wetlands with Nonurbanized Watersheds 66
References 66

INTRODUCTION

This chapter discusses water and soil quality in wetlands without significant urbanization in their watersheds. Like other chapters in this section, its purpose is to characterize particular elements of Puget Sound Basin freshwater wetland ecology in a state relatively unaffected by human activity. The wetlands profiled in this group were those with less than 4% impervious surface and greater than or equal to 40% forested area in their watersheds. It is recognized that human influence is not entirely absent in these cases, but truly pristine examples do not exist in the lowlands of the Puget Sound Basin. While there are palustrine wetlands in the Pacific Northwest that are not directly affected by urbanization, it is difficult to locate wetlands that are completely unaffected by humans. The wetlands considered here are regarded as representative of the closest to a natural state attainable in the ecoregion. Chapter 9 in Section III concentrates on water and soil quality in wetlands with watersheds

1-56670-386-7/00/$0.00+$.50

that are moderately and highly urbanized, in addition to wetlands with watersheds that had new development during the years of the study.

It is important to reiterate that the research program concentrated on palustrine wetlands of the general type most prevalent in the lower elevations of the central Puget Sound Basin. The results and conclusions presented here are probably applicable to similar wetlands somewhat to the north and south of the study area, but may not be representative of higher, drier, or more specialized systems, like true bogs and poor (low nutrition) fens.

WATER QUALITY

COLLECTION AND METHODS

Collection of samples for water quality analysis was performed in 1988–1990, 1993, and 1995. Sampling was concentrated during the wet and dry seasons, with fewer samples taken in the transition seasons between those periods. This scheduling was to concentrate effort during the times most pollutants enter wetlands: the wet season when runoff is high, and during the dry season when the decrease in surface water due to relatively low inflow and high evapotranspiration tends to concentrate pollutants most.

In the last four years samples were collected in 19 wetlands on the following schedule: November 1 to March 31 — 4 samples, April 1 to May 31 — 1 sample, June 1 to August 31 — 2 samples, and September 1 to October 31 — 1 sample. Sampling occurred at about the same times each year in order to get a consistent view of seasonal water quality variation. The same general pattern was observed in 1988; but there were only 14 wetlands in the program at that time, sampling did not begin until May, and a total of 7 instead of 8 samples was taken. Some of the wetlands, in most years 9 of the 19, had no surface water for varying lengths of time between late spring and early fall and could not be sampled during those times.

Samples were taken from the largest open water pool in each wetland, if there was one. If no pool was present, samples were collected near the outlet if there was surface water, otherwise downstream of the inlet. The standard grab method was generally used to collect the samples manually. A hand-operated pump device was employed to take samples intended for dissolved oxygen analysis and in cases where shallow water prevented conventional grab sampling without entraining material from the bottom.[1]

Temperature and pH were measured in the field, temperature either by mercury thermometer or electronic meter. The pH was determined with the electronic meter, in latter years a Beckman Model ϕ 11 instrument. Dissolved oxygen samples were stabilized in the field and transported on ice, along with samples for other analyses, to one of several laboratories used in the different years.

Water quality analyses varied somewhat from the beginning to the end of the program. Some analyses that did not produce much usable information in the early years were dropped. Analyses that were performed in all years are the focus of this chapter and include:

- Temperature
- PH
- Soluble reactive phosphorus (SRP)
- Total phosphorus (TP)

- Dissolved oxygen (DO)
- Conductivity (Cond)
- Total suspended solids (TSS)
- Ammonia-nitrogen (NH_3-N)
- Nitrate + nitrite-nitrogen (NO_3 + NO_2-N)
- Fecal coliforms (FC)
- Total lead (Pb)
- Total copper (Cu)
- Total zinc (Zn)

Among the analyses deleted after the early years were dissolved metals, which were usually below detection limits. It is probable that the use of exceptional methods would detect these constituents, but doing so was outside the objectives of this research. Enterococcus was dropped as a bacteriological measure because it did not yield the hoped-for reduced variability often prevalent with fecal coliforms, and was never widely adopted as a standard analyte as had been anticipated 10 years ago. Oil and grease and total petroleum hydrocarbons were measured in a relatively small number of samples but were present in very small concentrations, with the exception of an isolated incident when an oil spill was suspected. In the final two years of sampling data became available on a number of metals in addition to the three of most interest since they were run routinely on the inductively coupled plasma–mass spectrometer (ICP-MS) used by the laboratory handling those samples.

A monitoring and quality assurance/quality control (QA/QC) plan specifies in detail the sampling and analytical methods and QA/QC provisions for the last two years of the program, which were typical of all years.[2] A report by Reinelt and Horner is the best source of detail on methods for the initial years.[3] Water quality methods and results are also reported in several technical reports.[4-10]

Research Findings: A Portrait of Puget Sound Basin Wetland Water Quality

The main objective of this section is to develop a water quality profile of the least developed wetlands in the data set as presumably representative of the best attainable condition in the Puget Sound Basin lowlands. In developing this profile companion data are also presented for more urbanized cases, in part to allow some comparisons now and also for more extensive discussion of those cases in Chapter 9. Later in this chapter, wetlands in the data set are classified according to morphological characteristics and again compared. These comparisons are performed with the use of basic summary statistics (primarily: means, standard deviations, and medians). For the most part, tests for statistical significance of differences and analyses of variance were not performed because of lack of replication of conditions with any exactness, large natural variability, and relatively small sample sizes under any given set of conditions.

Table 2-1 gives a statistical summary of the water quality data gathered over the full project from wetlands whose watersheds did not experience significant urbanization change during that period (control wetlands) grouped by urbanization status. Chapter 9 takes up wetlands with watersheds that did change. Nonurban watersheds (N) were classed as those with both less than 4% impervious land cover and greater than or equal to 40% forest; highly urbanized watersheds (H) were considered to be those being both greater than or equal to 20% impervious and less than or equal

TABLE 2-1
Water Quality Statistics for Wetlands Not Experiencing Significant Urbanization Change (1988–1995)

Urban Status	Statistic	pH	DO (mg/L)	Cond. (μ S/cm)	TSS (mg/l)	NH_3-N (μ g/l)	NO_3+NO_2-N (μ g/l)	SRP (μ g/l)	TP (μ g/l)	FC (CFU/100 ml)	Cu (μ g/l)	Pb (μ g/l)	Zn (μ g/l)
N	Mean	6.38	5.7	72.5	<4.6	<59.9	<368.2	<17.6	52.3	>271.3	<3.3	<2.7	<8.4
	Maximum	7.65	11.3	230.0	73.0	1373.0	3200.0	414.0	850.0	6240.0	15.0	21.0	49.0
	Std. Dev.	0.53	2.6	63.8	>8.5	>129.3	>484.6	>47.6	86.6	>1000.4	>2.7	>2.8	>8.3
	CV	8%	45%	88%	>185%	>216%	>132%	>271%	166%	>369%	>80%	>105%	>99%
	Median	6.36	5.9	46.0	2.0	21.0	111.5	6.0	29.0	9.0	<5.0	1.0	5.0
	n	162	205	190	204	205	206	200	206	206	93	136	136.0
M	Mean	6.54	<5.5	142.4	<9.2	<125.7	<598.2	<31.5	92.5	>2664.8	<3.7	<3.4	<9.8
	Maximum	7.88	14.8	275.0	180.0	2270.0	7210.0	280.0	780.0	359550.0	7.0	13.0	33.0
	Std. Dev.	0.82	>3.6	72.8	>21.6	>266.8	>847.2	>37.9	91.8	>27341.7	>1.9	>2.7	>7.2
	CV	13%	>66%	51%	>235%	>212%	>142%	>120%	99%	>1026%	>51%	>79%	>73%
	Median	6.72	5.1	160.0	2.8	43.0	304.0	16.0	70.0	46.0	<5.0	3.0	8.0
	n	132	173	161	175	177	177	172	177	173	78	122	122.0
H	Mean	6.73	<5.4	150.9	<9.2	<68.3	<395.4	31.2	109.5	>968.6	<4.1	<4.5	<20.2
	Maximum	7.51	10.5	271.0	87.0	516.8	1100.0	79.0	1940.0	38000.0	12.0	22.0	73.0
	Std. Dev.	0.57	>2.9	85.5	>15.1	>104.4	>239.4	15.7	233.5	>4752.8	>2.5	>4.0	>16.7
	CV	9%	>53%	57%	>164%	>153%	>61%	50%	213%	>491%	>62%	>89%	>83%
	Median	6.88	6.3	132.2	4.0	32.0	376.0	28.2	69.0	61.0	<5.0	5.0	20.0
	n	52	67	61	66	67	67	65	67	66	29	44	44.0

Note: Nonurban watersheds (N) = less than 4% impervious land cover and greater than or equal to 40% forest. Highly urbanized watersheds (H) = greater than or equal to 20% impervious and less than or equal to 7% forest. Moderately urbanized watersheds (M) = wetlands not fitting either of the other categories.

to 7% forest. Those wetlands not fitting either of the other categories were classified as moderately urbanized watersheds (M).[11] This classification scheme was developed for analysis of the probable origin of soil metals, and was used for water quality as well. Characteristics of the individual watersheds can be found in Section I, Table 1. Indeterminate statistics (greater than or less than a given value) are the result of some measurements being below detection or, in the case of FC, bacterial colonies too numerous to count in some very concentrated samples.

Examination of Table 2-1 reveals several general points about wetland water quality. First, excepting pH, concentrations varied greatly, as indicated by the relatively high coefficients of variation (CV). The principal sources of water quality variability are examined later in the chapter. Fecal coliform was the most variable of the analyses overall, followed by TSS and NH_3-N. Other than for pH, DO, and conductivity, medians were usually lower than arithmetic means, signifying the influence on means, but not on medians, exerted by a relatively few high values. This trend is consistent with a lognormal probability distribution of values, a distribution frequently observed in environmental data.[12]

In the nonurban wetlands, a cursory comparison of the Table 2-1 medians, shows that pH rose slightly and DO marginally declined with increasing urbanization. Conductivity and NH_3-N increased substantially from nonurban to moderately urban wetlands but actually were a bit lower in highly urbanized cases. NO_3+NO_2-N and TP increased from N to M status, but not further, with H status. Cu showed little difference among categories, but many values were below detection. The remaining variables (TSS, SRP, FC, Pb, and Zn) all increased with each step up in urbanization level.

A water quality portrait of Puget Sound Basin lowland palustrine wetlands relatively unaffected by humans, then, shows slightly acidic (median pH = 6.4) systems with DO often well below saturation, and in fact sometimes quite low (less than 4 mg/l). Dissolved substances are relatively low (most conductivity readings less than 50 μS/cm) but somewhat variable. Suspended solids are routinely low but quite variable, reflecting the strong influence of storm runoff events on TSS. Median total dissolved nitrogen concentrations (the sum of ammonia, nitrate, and nitrite) are more than 20 times as high as dissolved phosphorus, suggesting general plant and algal growth is generally limited by P. Some of the fairly abundant TP would become available over time to support photosynthesis, but probably not enough to modify the general picture. The low median fecal coliform indicates that most readings are very low (less than 10 CFU/100 ml), but a small number is so high that the mean is 30 times the median. Both mean and median heavy metals concentrations are in the low parts per billion range, with standard deviations just about identical to the means.

Wetland Water Quality in Context

To proceed with a descriptive picture of regional wetland water quality, it is useful to provide some context for the quantitative information. This portion of the chapter discusses the statistical data with respect to informal criteria for separating the data into groups that can be associated with various factors that may influence the magnitudes. The account also gives a sense of how water quality compares in regional wetlands versus streams.

The informal criteria are based on several considerations and were slightly modified from past studies for this monograph.[3] Some are regulatory standards applied to other water body types (water quality standards have not yet been adopted for wetlands in Washington). Others have generally recognized biological relevance, but some are simply arbitrary breakpoints in the data distributions. In all cases professional judgment was applied in adopting a numerical informal criterion. Table 2-2 gives the distribution of wetlands, using median values, among the three urbanization categories relative to the informal criteria. It also repeats the medians and means for each category from Table 2-1.

TABLE 2-2
Comparison of Medians of Water Quality Variables for Wetlands Not Experiencing Significant Urbanization Change (1988–1995) with Informal Criteria

		Nonurbanized (N)			Moderately Urbanized (M)			Highly Urbanized (H)		
Variable	Criterion	median	mean	%[a]	median	mean	%	median	mean	%
pH	5–6	6.4	6.4	16.7	6.7	6.5	14.3	6.9	6.7	50
	6–7			67.7			57.1			0
	7–8			16.7			28.6			50
DO (mg/l)	<4	5.9	5.7	33.3	5.1	<5.5	57.1	6.3	<5.4	50
	4–6			33.3			14.3			0
	>6			33.3			28.6			50
Cond. (μS/cm)	<100	46	72	83.3	160	142	28.6	132	151	50
	100–200			16.7			42.8			0
	>200			0			28.6			50
TSS (mg/l)	<2	2	<4.6	33.3	2.8	<9.2	14.3	4	<9.2	50
	2–5			67.7			71.4			0
	>5			0			14.3			50
NH3-N (μg/l)	<50	21	<60	83.3	43	<126	71.4	32	<68	100
	50–100			16.7			14.3			0
	>100			0			14.3			0
$NO_3 + NO_2$-N (μg/l)	<100	112	<368	67.7	304	<598	57.1	376	<395	0
	100–500			16.7			28.6			50
	>500			16.7			14.3			50
TP (μg/l)	<20	29	52	0	70	92	0	69	110	0
	20–50			83.3			42.8			0
	>50			16.7			57.1			100
FC (CFU/100 ml)	<50	9	>271	83.3	46	>266	71.4	61	>969	50
	50–100			16.7			0			0
	>100			0	>200		28.6	>200		50
Zn (μg/l)	<10	5	<8.4	83.3	8	<9.8	85.7	20	<20.2	50
	>10[b]			16.7			14.3			50

[a] Percent of sites with this urbanization status (see Table 1 in Section 1 for definitions) fitting the criterion.
[b] Highest median is 21 μg/l, in comparison to the 59 μg/l chronic criterion for the protection of aquatic life with a water hardness of 50 mg/l as $CaCO_3$.

Some water quality variables did not appear to depend on urbanization. One site in each category had median pH less than 6, apparently as a consequence of some presence of peat in soils and peat-forming vegetation. Each group also had DO distributed among the three criteria ranges. As discussed later, it seems that DO depends more heavily on wetland morphology than on urbanization.

Several variables exhibited rising medians with urbanization; but when viewed in terms of the criteria, low concentrations predominated, suggesting relatively light pollutant loading from stormwater runoff. Most NH_3-N median values were in the lowest range in all categories. Wetlands produce ammonia in decomposing the abundant organic matter internally produced; and, absent an elevated source, concentrations would not necessarily be expected to follow urbanization.[13] Most NO_3+NO_2-N medians were also in the lowest range in the N and M wetlands but not in the most highly urbanized. For zinc, the most frequently detected metal, no median in any urbanization class approached the chronic criterion for the protection of aquatic life. In fact, the chronic criterion was violated in only one of these wetlands, a highly urbanized one, during the entire program. Although not shown in the table, the same general situation prevailed for copper but not for lead, which has a very low chronic criterion in these generally soft waters (3.2 μg/l). As can be seen in Table 2-1, H wetlands had Pb medians above that concentration, and M wetlands fell close to it.

TSS, conductivity, TP, and fecal coliforms exhibited a general tendency toward more sites in the higher criteria ranges with increasing urbanization. Still, TSS medians were very low. A total phosphorus concentration greater than 20 μg/l is often recognized as one sign that a lake is eutrophic, and greater than 50 μg/l as an indication of a hypertrophic state.[14] No wetland had a median below 20 μg/l, and the majority of M and H wetlands fell above 50 μg/l. Wetlands are recognized as systems more prone to eutrophication than lakes for a number of reasons (e.g., rapid nutrient cycling, often having the entire water column in the photic zone). Even those subject to little or no urbanization appear to have a rather high trophic state, and more urbanized systems are even higher. However, since wetlands flush more rapidly than lakes, these elevated TP concentrations may be a lesser concern in wetlands than they would be in lakes.

All but three wetlands would meet the 50 CFU/100 ml fecal coliform standard that applies to lakes and the highest-class streams in Washington on the basis of their means. Two moderately and one highly urbanized site could not meet even the least stringent standard. Of course, a number of individual values were far higher.

For the least urbanized wetlands, the following general statements can be made to characterize the water quality of Puget Sound Basin lowland palustrine wetlands in a fairly natural state:

- These wetlands are highly likely (83% of cases observed) to have median conductivity less than 100 μS/cm, NH_3-N less than 50 μg/l, TP in the range 20–50 μg/l, fecal coliforms less than 50 CFU/100 ml, and total Zn less than 10 μg/l.
- These wetlands are also likely (68% of cases observed) to have median TSS in the range 2–5 mg/l and NO_3+NO_2-N less than 100 μg/l.

- The pH and DO in these wetlands are unpredictable from consideration of urbanization status alone, being dependent on other factors.

Table 2-4 statistically summarizes water quality data from 50 locations on western King County streams collected by the Municipality of Metropolitan Seattle during 1990 to 1993. These data represent grab samples taken on a regular schedule, by chance most often under baseflow conditions. In these ways they are comparable to the wetland data produced by this research program. Unlike Tables 2-1 and 2-2, though, Table 2-3 mixes results from streams with very different influences. Nevertheless, it is useful to show how regional wetland and stream water quality compare.

While most wetlands tended strongly to be slightly acidic, and some were rather more so, streams tended just as strongly to be slightly alkaline. This difference is very likely the result of organic acid production by plants that are virtually absent in lotic systems. As expected, flowing streams were observed to be better oxygenated than wetlands, with median DO about twice as high. Streams at the median level were similar to moderately and highly urbanized wetlands in conductivity, but the nonurbanized wetlands had a central tendency below even the minimum measured stream value. TSS median concentrations were generally similar in the two types of water bodies. NH_3-N was generally higher in wetlands, reflecting the relatively high production rate of this species accompanying organic matter decomposition. On the other hand, NO_3+NO_2-N was for the most part lower in the wetlands, perhaps because of slower nitrification in the more oxygen-depleted environment. Median stream TP fell between the levels in the nonurbanized and more highly urbanized wetlands. Stream median fecal coliforms were higher than in any wetland category, but there were no extremely high values such as were measured in the wetlands. All in all, the two sets of results exhibit rough comparability, with most deviations mirroring the physical and biological differences in the two systems.

Seasonal Variation

Wetlands are highly variable systems with annual, seasonal, and diurnal variability in water chemistry. They often have several sources of water supply, each possessing a distinctive chemical blend that varies from year to year. Many water quality parameters exhibited clear seasonal fluctuations in the wetlands studied. DO concentrations were generally higher from mid-November to mid-May than during the remainder of the year.[3] This pattern is not surprising considering that most precipitation and runoff and the coolest temperatures in the Pacific Northwest occur during this period, and cooler, more turbulent water absorbs more oxygen.

Conductivity and pH did not exhibit such variation in most wetlands monitored. However, some had higher conductivity from May to November, when wetland water levels drop and dissolved substances become more concentrated.[3] Many wetlands had substantially higher TSS concentrations during the winter and early spring, the period of greatest runoff and erosion. However, colonial algae can cause high TSS readings in the late summer as well.[3]

While many wetlands monitored by the program had lower concentrations of NH_3-N, SRP, and TP from November to May, they had higher nutrient concentrations

TABLE 2-3
Water Quality Statistics for Wetlands Not Experiencing Significant Urbanization Change (1988–1995) Grouped by Urbanization and Morphological Status

Urban Status[a] & Morph.	Statistic	pH	DO (mg/l)	Cond. (µS/cm)	TSS (mg/l)	NH_3-N (µg/l)	NO_3+NO_2-N (µg/l)	SRP (µg/l)	TP (µg/l)	FC (CFU/100 ml)	Tot Cu (µg/l)	Tot Pb (µg/l)	Tot Zn (µg/l)
N/OW	Mean	6.14	5.3	36.8	<4.3	<66.8	<211.6	<21.2	63.2	>254.4	<3.4	<2.6	<9.0
Open Water	Maximum	7.46	11.3	94.0	49.6	1373.0	3200.0	414.0	850.0	6240.0	15.0	14.0	49.0
	St. Dev.	0.45	2.6	16.1	>7.2	>138.0	>441.7	>56.9	103.2	>1080.9	>2.8	>2.4	>8.6
	CV	7%	50%	44%	>169%	>207%	>209%	>268%	163%	>425%	>84%	>93%	>95%
	Median	6.18	5.3	33.2	2.0	23.0	49.5	5.1	31.0	4.0	5.0	1.0	6.0
	n	107	135	126	135	135	136	132	136	136	62	90	90
N/FT	Mean	6.84	6.6	142.7	<5.3	<46.6	672.3	<10.5	31.1	<304.1	<3.2	<2.8	<7.2
Flow Through	Maximum	7.65	11.2	230.0	73.0	872.0	1940.0	144.0	188.0	5400.0	8.0	21.0	30.0
	St. Dev.	0.31	2.3	64.5	>10.6	>110.4	416.9	>18.2	27.4	>828.1	>2.4	>0.0	>7.7
	CV	5%	35%	45%	>201%	>237%	62%	>173%	88%	>272%	>74%	>0%	>107%
	Median	6.92	6.5	171.4	2.0	20.0	612.5	6.1	24.2	47.0	<MDL	<MDL	2.5
	n	55	70	64	69	70	70	68	70	70	31	46	46
M/OW	Mean	5.94	<3.6	88.6	<5.7	<129.7	<213.1	<22.2	68.7	>273.4	<3.3	<2.6	<8.7
Open Water	Maximum	7.15	14.8	220.0	84.0	2270.0	1760.0	280.0	310.0	>6000.0	7.0	8.0	31.0
	St. Dev.	0.76	>2.7	57.9	>13.5	>337.6	>356.1	>40.1	68.8	>974.3	>2.0	>2.2	>6.6
	CV	13%	>74%	65%	>238%	>260%	>167%	>180%	100%	>356%	>62%	>82%	>75%
	Median	6.14	3.2	65.0	2.0	37.0	<MDL	10.0	44.0	10.0	2.6	<MDL	7.9
	n	56	75	69	76	77	77	75	77	74	33	54	54

TABLE 2-3 *(continued)*
Water Quality Statistics for Wetlands Not Experiencing Significant Urbanization Change (1988–1995) Grouped by Urbanization and Morphological Status

Urban Status[a] & Morph.	Statistic	pH	DO (mg/l)	Cond. (μS/cm)	TSS (mg/l)	NH_3-N (μg/l)	NO_3+NO_2-N (μg/l)	SRP (μg/l)	TP (μg/l)	FC (CFU/100 ml)	Tot Cu (μg/l)	Tot Pb (μg/l)	Tot Zn (μg/l)
M/FT	Mean	6.99	<6.9	182.7	<11.9	<122.6	<894.7	38.6	110.8	>4452.2	<3.9	<4.0	<10.6
Flow Through	Maximum	7.88	12.0	275.0	180.0	1140.0	7210.0	169.0	780.0	359550.0	7.0	13.0	33.0
	St. Dev.	0.53	>3.7	54.5	>25.9	>197.7	>987.1	34.6	102.9	>36108.2	>1.7	>2.9	>7.6
	CV	8%	>53%	30%	>218%	>161%	>110%	90%	93%	>811%	>43%	>72%	>71%
	Median	7.15	7.6	199.5	4.4	47.5	662.5	29.0	85.0	215.0	5.0	5.0	8.0
	n	76	98	92	99	100	100	97	100	99	45	68	68
H/FT	Mean	6.73	<5.4	150.9	<9.2	<68.3	<395.4	31.2	109.5	>968.6	<4.1	<4.5	<20.2
Flow Through	Maximum	7.51	10.5	271.0	87.0	516.8	1100.0	79.0	1940.0	38000.0	12.0	22.0	73.0
	St. Dev.	0.57	>2.9	85.5	>15.1	>104.4	>239.4	15.7	233.5	>4752.8	>2.5	>4.0	>16.7
	CV	9%	>53%	57%	>164%	>153%	>61%	50%	213%	>491%	>62%	>89%	>83%
	Median	6.88	6.3	132.2	4.0	32.0	376.0	28.2	69.0	61.0	5.0	5.0	20.0
	n	52	67	61	66	67	67	65	67	66	29	44	44

[a] See Table 2-1 for definitions of Urban Status abbreviations.

in the rest of the year possibly as a result of greater fertilizer applications and lower water levels that concentrate nutrients. NO_3+NO_2-N values fluctuated greatly in the program wetlands, and tended to vary directly with DO.[3] This association is another sign that nitrification moderated by the degree of aerobiosis has a strong influence on how much NO_3+NO_2-N will be found in a wetland water column.

Medians and geometric means of fecal coliform (FC) and enterococcus bacteria were highly variable. Peak counts occurred most frequently in late August and September, and least often from mid-November through February.[3] The monitoring program found that while most water quality parameters varied seasonally, NH_3-N, SRP, TP, FC, and enterococcus were especially changeable.[3]

Variation with Wetland Morphology

Wetland morphology refers to its form and physical structure and includes its shape; perimeter length; internal horizontal dimensions; topography (also termed bathymetry), which is the pattern of elevation gradients; water inlet and outlet configurations; and water pooling and flow patterns. These factors establish zonation at early successional stages by determining the extent of inundation from place to place and the hydrodynamic characteristics of flow. From these structural zones stem vegetation composition, distribution, and productivity, and ultimately, the character of the animal communities. Of course, these biota in turn influence morphological development over time through detrital and sediment accretion and animal activities like burrowing and dam building by beavers. The various morphological characteristics entirely determine the flood-flow storage and alteration function of wetlands. Along with the friction produced by vegetation, they set the residence time of water within the wetland, which is a key regulator of sediment trapping, nutrient processing, and other water quality functions.

Early work in the program determined that one aspect of morphology in particular, water pooling and flow patterns, had a substantial influence on wetland water quality.[3,6] The wetlands in the study were classified as either open water (OW) or flow-through (FT) types. The OW systems contain significant pooled areas and possess little or no flow gradient, while the FT wetlands are often channelized and have a clear flow gradient.

Using the first three years of data, it was found, unsurprisingly, that temperatures ranged higher in wetlands characterized by relatively large open pools, especially from May to September.[3] On an annual basis, the photosynthetic pigments chlorophyll a and phaeophytin a attained higher concentrations in wetlands characterized by large open pools, which have greater light exposure and longer residence times, and ranged much higher than in flow-through wetlands during the growing season. Dissolved oxygen tended to be significantly lower than in flow-through wetlands during these periods.

Table 2-4 summarizes statistics for the wetlands whose watersheds stayed relatively stable during the program broken down by urbanization and morphological status. Comparing open water vs. flow-through wetlands in the N and M categories, it can be seen that medians were higher, often substantially so, for flow-through than for open water wetlands in both urbanization categories for pH, DO, Cond,

TABLE 2-4
Distribution of Water Quality Data for Baseflow Samples from 50 Stream Sites[15]

	Maximum	75th Percentile	50th Percentile	25th Percentile	Minimum
Conductivity (μmho/cm)	30,900	203	130	104	53
Suspended Solids (mg/l)	12.6	4.8	3.4	2.8	1.6
Ammonia (μg/l)	190	24	15	13	5
Nitrate+nitrite (μg/l)	3000	1100	630	320	73
Temperature (C)	13.5	11.1	10.6	10	8
Dissolved Oxygen (mg/l)	11.4	11	10.4	9.6	5.8
Turbidity (NTU)	16.5	2.7	1.8	1.4	0.7
Fecal coliform (org/100 ml)	900	220	100	49	7
Enterococcus (org/100 ml)	410	170	53	22	5
pH	8.2	7.6	7.5	7.3	6.9
Total phosphorus (μg/l)	150	66	48	32	13

NO_3+NO_2-N, SRP, FC, and Pb. In addition, the flow-through means were higher in moderately urbanized wetlands for TSS, NH_3-N, TP, Cu, and Zn. Over all levels of urbanization flow-through wetland medians were higher for all water quality variables reported in the table.

It is clear from these results that flow-through wetlands strongly tend to be less acidic and better oxygenated than open water sites, as would be expected. Humic acid-producing vegetation thrives in an environment with low inflow, and attendant nutrient income. In these ponded systems oxygen renewal from the atmosphere is not as efficient as in flowing water, and they are warmer and hence have lower oxygen solubility. Also, more primary production and more oxygen-consuming organic decomposition occurs in the relatively long period of water residence. It is also clear that flow-through wetlands generally have higher pollutant concentrations, probably due to the greater loading of pollutants by the flow, and reduced pollutant removal from the water column with the shorter hydraulic residence times.

Concentrations of NO_3+NO_2-N exhibited one of the greatest disparities between open water and flow-through wetlands. In addition to greater loading introduced by the flow, this phenomenon is probably partially due to higher oxygen levels in flow-through cases, which promote nitrification that converts ammonia to nitrite and then nitrate forms. In fact, ammonia differed between the two types of morphology much less than did NO_3+NO_2-N, suggesting that ammonia discharged to wetlands may be more effectively nitrified in flowing systems. Of course, these systems also support less decomposition by microorganisms and, thus, likely produce less ammonia internally than do open water wetlands.

It must be noted that a preponderance of flow-through wetlands are in more urbanized areas, which certainly affects pollutant loading and may affect the strength of conclusions, although probably not the overall trends. It is possible that this skewed distribution is not just a coincidence but reflects the urban situation, in which higher peak runoff flows, wetland filling, and stream channelization favor flow-through over open water wetland conditions.

SOILS

Collection and Methods

Soil samples were collected once from each wetland during the months July–September in 1988 to 1990, 1993, and 1995. Soil sampling areas were selected 3 m to the side of vegetation transect lines at every point where the soil type appeared either to be transitional or completely different. Small soil cores or signs of vegetation change were the basis for judgment. Two to five samples, most commonly four, were collected from each wetland. The number had a relationship to the size and zonal complexity of the wetlands. This coverage was considered to be adequate because a synoptic study of 73 urban and rural wetlands early in the program found that there were no significant differences among wetland zones (e.g., open pool, inlet, scrub-shrub, and emergent) with respect to soil texture, organic content, pH, phosphorus, and nitrogen.[15] Because oxidation-reduction potential and one metal were significantly different near inlets as compared to other locations, the inlet zone was emphasized in choosing sampling areas.

Soil samples were collected with a corer consisting of a 10-cm (4-in.) diameter ABS plastic pipe section ground to a sharp tip. The corer was twisted into the soil with a wooden rod inserted horizontally through two holes near the top. Coring depth was 15 cm (6 in.). Samples were inserted immediately into plastic bags, air was extruded, and the bags were sealed with tape. They were then transported to one of several laboratories used in the different years.

A standard 60-cm (2-ft) deep soil pit was dug at each sampling point not inundated above the surface. The pit was observed and notes were recorded for depth to water table (if within 60 cm of the surface), horizon definition (thickness of each layer and boundary type between), color (using Munsell notations), structure (grade, size, form, consistency, and moisture), and presence of roots and pores.

Soil core samples were analyzed for the following constituents:

General Characteristics	Nutrients	Metals
• Particle size distribution (PSD)	• Total phosphorus (TP)	• Arsenic (As)
• Percent organics as loss on ignition (LOI)	• Total Kjeldahl nitrogen (TKN)	• Cadmium (Cd)
• pH		• Copper (Cu)
• Oxidation-reduction potential (redox)		• Lead (Pb)

Details on the general analytical methods, as well as the sampling program design, are reported elsewhere.[4] A method for PSD, also termed soil texture, was developed during this program for soils with more than 5% organics, as most wetland soils have. Texture is the measurement of the proportions of the various sizes of mineral particles in a soil, classified from largest to smallest as sand, silt, and clay (gravel, when significant, is also recognized in the texture classification). The analysis of any soil with more than 5% organic content must include a step that removes the organic material. Failure to remove the organic component may cause clumping of particles and render the results inaccurate. The revised PSD method is considered to be accurate for soils with up to 25% organics. At higher levels it is not accurate because of sample loss during organic removal preparation, especially in the clay component.

Additional sources of detail on methods and findings are reported from the initial years.[17-21] Soils methods and results have also been reported elsewhere.[5,7,19]

Research Findings: A Portrait of Puget Sound Basin Wetland Soils

General Soil Characteristics and Nutrients

Like water quality, soil quality exhibited extensive variability. Table 2-5 statistically summarizes the soils data, excluding PSD, for wetlands that did not experience significant urbanization change during the program. This set of wetlands had watersheds that ranged from low to moderate to highly urbanized.

Coefficients of variation for the majority of the soil variables were generally in the approximate range of 75 to 150%, although some were much higher. CVs for pH were considerably lower than for other analytes for both soils and water, usually about 10 ± 3%. Similar to water quality variables, the soil variables in Table 2-5 usually exhibited medians lower than the means, except for redox and sometimes pH. Therefore, most of these data also are far from normally distributed, and are probably lognormal.

Most wetlands had at least some pockets of peat of mainly sedge and grass origins, and their soils accordingly tended to be acidic.[19] Among the different groupings of data in Table 2-5, median pH values were less than or equal to 6.1, except for highly urbanized cases, which were also flow-through systems. Overall, wetlands in highly urbanized watersheds had the highest median pH, followed by wetlands in moderately urbanized and then nonurbanized watersheds. Flow-through wetlands overall and in the N and M categories had higher median pH than open water types.

These wetlands frequently had anaerobic soils, as indicated by median redox values often less than 250 mV, the approximate point at which oxygen is fully depleted. The median for the most highly urbanized case was lower than for the N wetlands, which themselves had lower median redox than the M cases. Open water wetlands overall had higher redox readings than flow-through ones. This result is somewhat surprising, in that open water wetlands are thought to host more oxygen-consuming decomposition and to have oxygen replenished from the atmosphere less efficiently than flow-through cases. It was found in the synoptic study of 73 wetlands at the beginning of the program that open water zones had the lowest redox readings (less than 100 mV).[16] Redox was below the level at which oxygen is generally depleted in the inlet, open water, and emergent zones, but not in the scrub-shrub

TABLE 2-5
Soil Quality Statistics for Wetlands Not Experiencing Significant Urbanization Change (1988–1995) Grouped by Urbanization and Morphology

Urban Status[a] & Morph	Statistic	pH	Redox (mV).	TP (mg/kg)	TKN (mg/kg)	Org. (%)	Cu (mg/kg)	Pb (mg/kg)	Zn (mg/kg)	As (mg/kg)
N/OW	Mean	5.43	143	743	5261	45.5	58	65	27	6.5
	Maximum	6.67	649	6579	27369	97.3	2221	418	154	20
	St. Dev.	0.64	322	1087	5983	34.2	300	96	29	5.4
	CV	12%	225%	146%	114%	75%	515%	147%	106%	84%
	Median	5.40	153	251	2866	47.0	16	24	18	4.4
	n	47	46	55	58	58	54	54	54	54
N/FT	Mean	5.73	113	839	3899	36.1	21	33	31	15
	Maximum	7.12	629	8882	14223	97.3	40	322	103	225
	St. Dev.	0.48	296	1698	3297	28.8	13	60	30	42
	CV	8%	262%	202%	85%	80%	61%	184%	97%	284%
	Median	5.70	184	342	3799	34.1	17	18	17	6.6
	n	27	27	27	29	29	27	27	27	27
All N	Mean	5.54	132	775	4807	42.4	46	54	28	9.3
	Maximum	7.12	649	8882	27369	97.3	2221	418	154	225
	Std. Dev.	0.60	311	1310	5261	32.6	245	87	29	25
	CV	11%	236%	169%	109%	77%	536%	159%	103%	267%
	Median	5.60	184	283	2885	35.8	16	20	17	6.0
	n	74	73	82	87	87	81	81	81	81
M/OW	Mean	5.31	317	1114	6354	35.4	12	32	22	6.2
	Maximum	6.72	656	3827	22517	92.3	24	101	92	16
	St. Dev.	0.90	227	1019	5670	29.9	7	30	24	4.4
	CV	17%	72%	91%	89%	85%	60%	95%	107%	70%
	Median	5.11	362	945	5783	23.7	14	21	15	4.8
	n	26	26	25	25	25	28	28	28	28
M/FT	Mean	5.95	165	756	3999	25.7	18	77	60	8.1
	Maximum	6.96	611	2743	27967	92.3	63	530	334	25
	St. Dev.	0.51	279	819	5942	28.1	12	89	62	5.2

TABLE 2-5 *(continued)*
Soil Quality Statistics for Wetlands Not Experiencing Significant Urbanization Change (1988–1995) Grouped by Urbanization and Morphology

Urban Status[a] & Morph	Statistic	pH	Redox (mV).	TP (mg/kg)	TKN (mg/kg)	Org. (%)	Cu (mg/kg)	Pb (mg/kg)	Zn (mg/kg)	As (mg/kg)
	CV	9%	169%	108%	149%	109%	68%	115%	104%	65%
	Median	6.09	226	481	2021	12.6	17	57	48	7.7
	n	45	43	41	43	47	43	43	43	43
All M	Mean	5.71	222	892	4865	29.1	16	59	45	7.3
	Maximum	6.96	656	3827	27967	92.3	63	530	334	25
	Std. Dev.	0.74	269	909	5912	28.9	11	75	53	5.0
	CV	13%	121%	102%	122%	99%	69%	126%	119%	68%
	Median	5.78	306	537	2764	15.1	15	34	33	6.2
	n	71	69	66	68	72	71	71	71	71
H/FT	Mean	5.88	87	654	2703	31.6	31	89	103	13
	Maximum	6.97	640	2995	11282	80.9	63	273	456	40
	St. Dev.	1.01	291	784	2892	26.3	17	68	112	9.4
	CV	17%	335%	120%	107%	83%	56%	76%	109%	74%
	Median	6.46	91	314	2033	22.8	30	64	65	10
	n	32	32	34	36	35	32	32	32	32
All OW	Mean	5.39	206	859	5590	42.4	43	54	25	6.4
	Maximum	6.72	656	6579	27369	97.3	2221	418	154	20
	St. Dev.	0.74	302	1074	5877	33.1	244	81	27	5.1
	CV	14%	147%	125%	105%	78%	572%	151%	107%	80%
	Median	5.35	305	414	3141	34.2	15	24	16	4.5
	n	73	72	80	83	83	82	82	82	82
All FT	Mean	5.87	127	744	3540	30.3	23	69	66	11
	Maximum	7.12	640	8882	27967	97.3	63	530	456	225
	St. Dev.	0.70	286	1102	4449	27.8	15	78	80	22
	CV	12%	226%	148%	126%	92%	67%	113%	123%	199%
	Median	5.92	162	378	2274	16.1	20	46	45	7.70
	n	104	102	102	108	111	102	102	102	102

[a] See Table 2-1 for definitions of Urban Status abbreviations.

and forested zones. In another contrast with the more recent results, soils in the inlet zones of wetlands in nonurban wetlands had significantly (p less than .05) lower redox than in urban wetlands.

The highest median TP occurred in moderately urbanized wetlands, as did the highest median for TKN. The lowest median for TP was in the nonurban open water category whereas the lowest median for TKN was found among the highly urbanized flow through wetlands.

The N urbanization category had median organic content over 30%, although with extensive variation. The M and H sites overall had about half to two-thirds of that level. On the whole, open water wetlands exceeded flow-through ones in organics. This finding is as expected, since ponded systems have more primary productivity and capability of settling solids.

Soil texture is important to the nutrition, structure, drainage, and erosion prevention characteristics of a soil. Nutrients are found in a soil attached to organic matter, clay particles, and metal oxides (especially iron oxides). Soils with a high portion of clay, organic material, or both adsorb water and nutrients much more readily than soils low in these components. Fine-textured soils have a more compact structure, which may impede aeration of the soil. Clays adsorb water and if positioned lower down in the soil profile, can impede drainage, causing an impervious layer and creating a wetland. Sandy soils have very little cohesion and erode much more easily than silt- or clay-rich soils. One of the influences of urbanization on wetland ecosystems is deposition of sediments from development activities (clearing and grading).

Table 2-6 presents a comparison of soil textures in 1989 and 1995 for the wetlands that did not experience significant watershed change. As was hypothesized for these cases, little change occurred over these years regardless of urbanization or morphological status. The soils of the majority of these wetlands were dominated by silt-range particles, again irrespective of status. One N/OW and one M/FT site, located in different parts of the study area, had predominately sand. With two exceptions, the wetlands were found to have relatively little clay (less than or equal to 20%). However, a wetland in north King County and one in south King County had about 30% and 50%, respectively. It bears noting that some of the samples contributing to these statistics had greater than 25% organic content, in the range where the analytical method is less accurate.

Analysis of the PSD measurements within individual wetlands indicates that PSD often varied substantially across wetlands and showed no trends with the amount of organic matter in the soil or the soil series. No association was seen between the total suspended solids in the surface water and changes in soil texture. However, soils located near the inlets of M and H wetlands were significantly (p less than .05) more likely to have more sand than silt as compared to other locations.

Metals in Soils

Cadmium, lead, and zinc in wetland soils were observed to be highly variable from year to year, but copper and arsenic varied less. Overall, there was a declining trend in soil metal content over the years of the study. These results are somewhat sur-

TABLE 2-6
Comparison Changes in Average Particle Size Distributions from 1989 to 1995 in Wetlands that Experienced Little Urbanization Change Grouped by Urbanization and Morphological Classifications

Site	Wetland Area (ha)	PSD 1989 (% sand/silt/clay)	PSD 1995 (% sand/silt/clay)
	Nonurbanized Open Water (N/OW)		
AL3	0.81	26/54/20	No data
HC13	1.62	47/47/6	45/37/18
RR5	10.52	74/15/11	68/21/11
SR24	10.12	1/89/10	6/75/19
	Nonurbanized Flow Through (N/FT)		
LCR93	10.93	No data	30/50/20
MGR36	2.23	13/76/11	20/70/10
	Moderately Urbanized Open Water (M/OW)		
ELS39	2.02	35/49/16	15/69/16
SC84	2.83	4/81/15	11/73/16
TC13	2.06	30/41/29	38/32/30
	Moderately Urbanized Flow Through (M/FT)		
ELW1	3.84	83/13/4	75/18/7
FC1	7.28	13/71/16	10/75/15
SC4	1.62	No data	No data
	Highly Urbanized Flow Through (H/FT)		
B3I	1.98	31/62/7	24/61/15
LPS9	7.69	30/16/54	32/20/48

prising since the soil cores were 15 cm deep, representing soil horizons that would be expected to maintain fairly stable metals concentrations from year to year. Figure 2-1 shows that median concentrations of As, Cu, Pb, and Zn for all of the program wetlands generally declined each year. It is possible that metals enter and depart from wetland soils more easily than previously believed, permitting a rapid change in results in response to changes in inputs from the watershed. Declining metals pollution from vehicles and dissipating pollutants from industrial air pollution point sources, such as the closed ASARCO smelter in Tacoma, could explain the general decline of soil metals since the start of the program.

Cadmium was undetectable in the soils of most monitored wetlands, except in three that also had relatively high Pb. This result is consistent with the observation that metals often increase in tandem. Although the program detected substantial increases in Cd, Pb, and Zn at several wetlands between 1989 and 1990, it is significant that there are no apparent common characteristics among these wetlands. They represent differing hydrology, ecology, and levels of watershed development.[18]

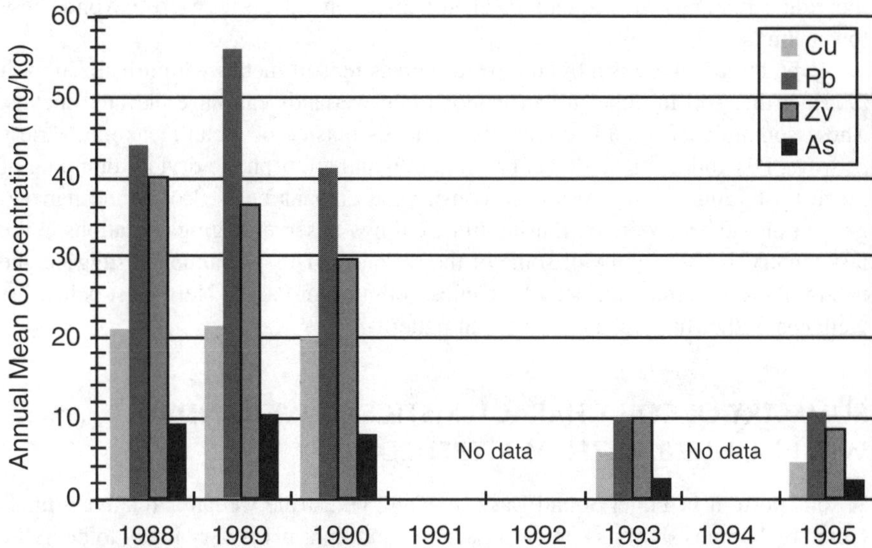

FIGURE 2-1 Annual mean metal concentrations for all wetland samples in each year.

It can be seen in reviewing the metals data in Table 2-5 that, like water quality and general soil characteristics, soil metals exhibited extensive variability. Again, too, medians were normally considerably less than means. It is further apparent in Table 2-5 that median metals concentrations increased from nonurban to moderately urbanized and increased again in highly urbanized wetlands, except for a small drop in Cu from N to M. Flow-through wetlands overall had higher median concentrations of all metals than did open water ones, although very marginally so for Cu. This tendency was again stronger for the moderately urbanized than the nonurbanized wetlands. For the most part, soils exhibited the same trend as water quality, with quantities considered to be pollutants higher in FT than in the OW wetlands. It was thought that water column contaminants might be lower in open water wetlands because of losses to the soil. However, this supposition was not borne out by the soil results. Having the two most developed sites in the FT group may be skewing the results.

The Washington Department of Ecology set metals criteria for freshwater sediments in terms of lowest effect and severe effect thresholds.[21] The criteria are

	Lowest Effect Threshold (mg/kg dry soil)	Severe Effect Threshold (mg/kg dry soil)
Copper	16	110
Lead	31	250
Zinc	120	820

No mean or median value exceeded the severe effect criteria, and very few individual readings surpassed them at any time during the program. However, some Cu and Pb means and even medians exceeded lowest effect thresholds. Many individual readings in wetlands in all urbanization categories were beyond these lower limits.

Even though there is a trend toward increasing soil metals with urbanization, it is a fact that soil in either urban or nonurban wetlands can have elevated metals. These contaminants could be entering wetlands outside of urban areas in a variety of ways. Possibilities include via precipitation and atmospheric dryfall, dumping of metal trash, and leaching from old constructed embankments. Roads and narrow-gauge railroad beds were built using mine tailings to serve logging operations in the last century, in the vicinity of some of the wetlands. This phenomenon suggests the need for site-specific inquiries into metals pollution in Pacific Northwest palustrine wetlands, rather than reliance on broad patterns.

SUMMARY OF SOIL CHARACTERISTICS OF WETLANDS WITH NONURBANIZED WATERSHEDS

A soils portrait of Puget Sound Basin lowland palustrine wetlands relatively unaffected by humans shows a somewhat acidic condition; pH is very likely to be in the range of 5 to 6. With redox as a basis, soils at many times and places will be anaerobic, but with great variability. Phosphorus is likely to be somewhere in the vicinity of 300 mg/kg, with nitrogen (TKN) approximately an order of magnitude higher. Based on these results, most soil samples from nonurban wetlands can be expected to have greater than 25% organics, and greater than 10% is extremely likely. Texture appears to be more a function of local conditions than a function of urbanization, or lack of it.

The metals As, Cu, Pb, and Zn can range over two orders of magnitude from a minimum in the low parts per million (mg/kg) region, in the soils of these nonurban wetlands. Most commonly, they appear to have approximately equal amounts of Cu, Pb, and Zn, around 20 mg/kg, and about one-quarter to one-third as much as As. This level and the observed variation around it are sufficiently high to exceed lowest threshold effect freshwater sediment criteria for Cu often and for Pb occasionally, but very rarely for Zn.

REFERENCES

1. Horner, R. R. and K. J. Raedeke, Guide for Wetland Mitigation Project Monitoring, Washington State Department of Transportation, Olympia, 1989.
2. Horner, R. R. and K. A. Ludwa, Monitoring and Quality Assurance/Quality Control Plan for Wetland Water Quality and Hydrology Monitoring, Center for Urban Water Resources Management, University of Washington, Seattle, 1993.
3. Reinelt, L. E. and R. R. Horner, Characterization of the Hydrology and Water Quality of Palustrine Wetlands Affected by Urban Stormwater, Engineering Professional Programs, University of Washington, Seattle, 1990.
4. King County Resource Planning Section, Puget Sound Wetlands and Stormwater Management Research Program: Initial Year of Comprehensive Research, Engineering Professional Programs, University of Washington, Seattle, 1988.

5. Azous, A., An Analysis of Urbanization Effects on Wetland Biological Communities, M.S. thesis, University of Washington, Department of Civil Engineering, Seattle, 1991.
6. Reinelt, L. E. and R. R. Horner, Urban stormwater impacts on the hydrology and water quality of palustrine wetlands in the Puget Sound region, in Puget Sound Research '91 Proc., Seattle, WA, January 4–5, 1991, Puget Sound Water Quality Authority, Olympia, 1991, p. 33.
7. Platin, T. J., Wetland Amphibians *Ambystoma gracile* and *Rana aurora* as Bioindicators of Stress Associated with Watershed Urbanization, M. S. C. E. thesis, University of Washington, Department of Civil Engineering, Seattle, 1994.
8. Ludwa, K. A., Urbanization Effects on Palustrine Wetlands: Empirical Water Quality Models and Development of a Macroinvertebrate Community-Based Biological Index, M. S. C. E. thesis, University of Washington, Department of Civil Engineering, Seattle, 1994.
9. Taylor, B., K. A. Ludwa, and R. R. Horner, Urbanization effects on wetland hydrology and water quality, in Puget Sound Research '95 Proc., Seattle, January 12–14, 1995, Puget Sound Water Quality Authority, Olympia, 1995, p. 146.
10. Chin, N. T., Watershed Urbanization Effects on Palustrine Wetlands: A Study of the Hydrologic, Vegetative, and Amphibian Community Response During Eight Years, M. S. C. E. thesis, University of Washington, Department of Civil Engineering, Seattle, 1996.
11. Valentine, M., Assessing Trace Metal Enrichment in Puget Lowland Wetlands, Nonthesis paper for the M. S. E. degree, University of Washington, Department of Civil Engineering, Seattle, 1994.
12. Gilbert, R. O., *Statistical Methods for Environmental Pollution Monitoring,* Van Nostrand Reinhold, New York, 1987.
13. Mitsch, W. J. and J. G. Gosselink, *Wetlands*, 2nd ed., Van Nostrand Reinhold, New York, 1993.
14. Welch, E. B., *Ecological Effects of Waste Water*, Cambridge University Press, Cambridge, U.K., 1980.
15. Municipality of Metropolitan Seattle, Water Quality of Small Lakes and Streams, Western King County, 1990–1993, Seattle, 1994.
16. Horner, R. R., F. B. Gutermuth, L. L. Conquest, and A. W. Johnson, Urban stormwater and Puget Trough wetlands, in *Proc., First Annu. Meet. Puget Sound Research,* Seattle, March 18–19, 1988, Puget Sound Water Quality Authority, Seattle, 1988, p. 723.
17. Cooke, S. S., K. O. Richter, and R. R. Horner, Puget Sound Wetlands and Stormwater Management Research Program: Second Year of Comprehensive Research, Engineering Professional Programs, University of Washington, Seattle, 1989.
18. Richter, K. O., A. Azous, S. S. Cooke, R. Wisseman, and R. Horner, Effects of Stormwater Runoff on Wetland Zoology and Wetland Soils Characterization and Analysis, PSWSMRP, Seattle, 1991.
19. Cooke, S. S., The effects of urban stormwater on wetland vegetation and soils: a long-term ecosystem monitoring study, in Puget Sound Research '91: Proc., January 4–5, 1991, Seattle, WA, Puget Sound Water Quality Authority, Olympia, 1991, p. 43.
20. Cooke, S. S. and A. Azous, Characterization of Soils Found in Puget Lowland Wetlands, Engineering Professional Programs, University of Washington, Seattle, 1993.
21. Washington Department of Ecology, Summary Criteria and Guidelines for Contaminated Freshwater Sediments, Olympia, 1991.

3 Characterization of Central Puget Sound Basin Palustrine Wetland Vegetation

Sarah S. Cooke and Amanda L. Azous

CONTENTS

Introduction....69
Methods....70
Results....72
Community Structure and Composition....72
Habitat Character....72
Wetland Plant Associations....72
Abundance and Distribution of Invasive Plant Species....72
Community Richness....75
Hydrologic Regimes by Habitat Type....78
Hydrologic Regimes of Some Species....79
Discussion....86
Acknowledgments....88
References....88
Appendix A: List of Plant Species and Frequency Found Among 19 Central Puget Sound Basin Palustrine Wetlands....90

INTRODUCTION

Nineteen wetlands in the Central Puget Sound Basin in King County, Washington were studied for five years between 1988 and 1995. An additional seven wetlands were involved in special studies within the same period. The wetlands surveyed were inland palustrine wetlands ranging in elevation from 50 to 100 m above mean sea level and characterized by a mix of aquatic bed, emergent, scrub-shrub, and forested wetland habitat classes. Our purpose was to understand the richness, composition, and structure of wetland plant communities. In later years, defining hydrologic conditions for habitats and species also became a study focus. Changes to plant communities observed during the study are explored later, in Chapter 10.

1-56670-386-7/00/$0.00+$.50

Plant community structure for all the years was examined through analysis of species richness, composition, and percent cover. Ordination and classification analyses were used to identify distinct plant communities and to examine relationships between presence, abundance, and distribution of invasive species. Relationships between species presence, habitat structure, and hydrologic conditions are discussed.

METHODS

Wetland sizes were estimated through analysis of U.S. Geographic System (USGS) 7.5-minute-series topographic maps and ranged from 0.4 to 12.4 ha. Plant communities in each wetland were characterized during a two- to three-week period in the growing season between July and August, for the years 1988, 1989, 1990, 1993, and 1995. Plant community composition and percentage cover were sampled in permanent plots adjacent to linear transects established across the hydrologic gradients of each wetland. Species cover was recorded using a cover class system based on the Octave Scale.[1,2] Detailed protocols for the vegetation fieldwork are documented elsewhere.[3] The data set also includes seven additional wetlands that were surveyed during the years 1993, 1994, and 1995 as part of related studies.

Species were identified using the method of Hitchcock et al. and were verified with specimens from the University of Washington Herbarium.[4] The Cowardin classification system was used to assign each sample plot a habitat category based on the dominant structure of the vegetation community, such as aquatic bed (PAB), emergent (PEM), scrub-shrub (PSS), forested (PFO), upland, or some transition zone between them (e.g., PEM/PSS).[5] The Cowardin classification system was selected because it is widely used in functional assessments, wetland protection regulations, and mitigation criteria.[6] In some cases the vegetation community class changed over time within sample plots and plots were recategorized as required. Sample plots in upland habitats were included in the documentation of plant species observed but were not included in analyses of wetland richness.

Plants species in the sample plots were surveyed and the data used to calculate the frequency of species presence among the wetlands studied. Plant species presence was sampled in permanent plots established every 50 m along a gradient from the upland through the transition zones and, at intervals, crossing the different wetland vegetation communities. Total plant species richness was calculated for individual wetlands and for different habitat categories. Species were also tabulated according to wetland indicator status (whether obligate [OBL], facultative wetland [FACW], facultative [FAC], or facultative upland [FACU]) and according to form (e.g. grass, herb, shrub, or tree).[7, 8]

Vegetation community types were defined and described using ordination (DECORANA) and classification (TWINSPAN) comparisons.[9,10] Plant community data were tabulated in a two-way matrix (species by cover) and were classified by grouping similar vegetation units into categories.[11,12] All species in the sample stations were included. Ordination was used to graphically display the species groups showing similar communities plotted close together and dissimilar communities plotted further apart.[2,10] The frequency of species and the relative dominance of species were both described by the proportion of vegetation sampling plots in which the species were found.

The year was divided into four seasonal periods important to plant growth, an early growing season, defined to be from March 1 through May 15, an intermediate growing season which lasts from May 16 to August 31, senescence lasting from September 1 to November 15th, and dormancy and decay, November 16 to February 28. The seasonal hydrologic regime was calculated for each vegetation sample station from 1988 to 1995. Species found in each sample station were associated with the seasonal hydrologic regime observed at the station. These data were used to describe a hydrograph for many commonly found wetland plants showing conditions for instantaneous (labeled "instant" on the graphs) and maximum water depths. The association is based on plant species presence indicating conditions favorable to species survival.

Hydrologic measurements included instantaneous water levels from staff gauges for measuring typical water depths found during monitoring visits, and peak levels from crest gauges to measure depths from storm events occurring between monitoring visits. These were recorded at least eight times annually (measured every four to six weeks) while water was present in the wetlands.[13] Typical (instantaneous) and maximum water depths were averaged for each season and year. Water level fluctuation (WLF) was calculated as the difference between the maximum depth and the average of the current and previous instantaneous water depths for each four- to six-week monitoring period. Mean WLF was calculated by averaging all WLFs for a given season or year (see Chapter 1, for the equation used).

Habitats within individual wetlands were surveyed to determine the elevation of the plant community in relation to the hydrologic monitoring location. The elevation of all vegetation-sampling stations relative to the water depth measured at the wetland gauge was used to determine the likely water depths at the vegetation sample stations during each season. The calculation was based on the assumption that water depth measured at the monitoring station would, when corrected for elevation, accurately reflect the hydrologic conditions found throughout the wetland. This was not always true as hydrologic conditions in and around the vegetation stations sometimes varied from the conditions predicted through elevation change alone. Sometimes intervening topography or large woody debris would produce localized impoundments or dry hummocks unaccounted for in the calculation. Whenever such plots were identified more accurate surveys were made or plots were eliminated from the analysis. In addition, hydrologic measurements did not tell us whether soils were dry or saturated so field observations augmented the analysis.

Negative numbers in the bar chart hydrographs are to be read as below the water surface, or inundated. Positive numbers represent the distance, in elevation, above the water surface and indicate the plant or community being examined is dry during that period. Negative numbers are interpreted as depth of inundation. Bar charts show the median and the central 80% of the range of observations for the condition being evaluated. The solid portion of the bar represents the central 50% of observations. Outliers are not shown in order to illustrate the most likely range of wettest to driest conditions where particular wetland habitats or species would be found.

RESULTS

COMMUNITY STRUCTURE AND COMPOSITION

We identified 242 plant species in 26 wetlands over the study period (a complete list of species and their occurrence is provided in Appendix A to this chapter). Of these 242 different species, most were herbs (46%), followed by shrubs (23%), grasses (10%), and trees (7%). Other species identified included ferns and horsetails (6%), sedges (6%), and rushes (2%); 65 species were obligate (27%), 56 were FACU (23%), 55 FAC (23%), and 38 FACW (16%) species; 29 (12%) species had no indicator status at the time of the study.

There were 45 species (19%) found in only one (4%) of the 26 wetlands surveyed. Over 38% of plant species were found in less than three wetlands (12%). The distribution of plants by indicator status was similar to the distribution of species: 40% of obligate, 35% of FAC, and 39% of FACU species were also found in three or fewer wetlands. FACW species, on the other hand, were generally more widely dispersed among wetlands, with all species observed in at least eight wetlands (31%).

Polystichum munitum, Rubus spectabilis, and *Rubus ursinus* were observed in all 26 wetlands but were usually low in number. *Spirea douglasii* was among the most dominant species, occurring in 96% of wetlands (25 of 26) and dominating coverage in 21% of locations where it was observed. *Alnus rubra*, *Athyrium filix-femina*, and *Salix scouleriana* were also found in 25 of 26 wetlands but were less dominant in coverage. Most of the widely distributed species were facultative and facultative upland species. Widely dispersed wetland species in addition to Douglas spirea included *Epilobium ciliatum* (Watson willowherb) (FACW) and *Lysichitum americanum* (skunk cabbage) (OBL). *Phalaris arundinacea* (reed canarygrass), a locally invasive noxious weed, was observed in 69% of wetlands (18 out of 26) and dominated 19% of observed locations. Other locally common and dominant wetland species included *Ranunculus repens* (creeping buttercup) found in 65% (17) of wetlands, and *Juncus effusus* (soft rush), observed in 58% (15) of the wetlands. Fortunately, *Lythrum salicaria* (purple loosestrife), a formidable noxious weed, was observed in only one wetland. Table 3-1 shows some of the most common and least common plants categorized by occurrence and cover dominance.

HABITAT CHARACTER

Habitats identified in the 26 wetlands include the four Cowardin categories of aquatic bed (PAB), emergent (PEM), scrub-shrub (PSS), and forested (PFO), an additional two habitats called BOG and UPL (upland transition), and combinations indicating transitions between habitats. Plant communities of different habitats were evaluated with respect to the dominant plants and their associated wetland indicator status. These distributions are shown in Table 3-2. The table shows that obligate and FACW species dominate PAB and PEM habitats. PSS habitats were more evenly distributed and PFO and upland habitats were dominated by FAC and FACU species. A limited number of bog habitats were also sampled and were predominantly obligate and FACW species.

TABLE 3-1
Species Occurrence for Different Categories of Plant Type and Cover Dominance

Cover Dominance Category	High Occurrence (>80% wetlands)	Low Occurrence (<10% wetlands)
Usually dominant. Greater than 64% coverage in more than 19% of observations.	*Phalaris arundinacea* *Spirea douglasii*	*Juncus supiniformis* *Menyanthes trifoliata*
Dominance in plots varies	*Alnus rubra* *Athyrium filix-femina* *Kalmia microphylla* *Lonicera involucrata* *Polystichum munitum* *Pteridium aquilinum* *Ranunculus repens* *Rhamnus purshiana* *Rubus laciniatus* *Rubus spectabilis* *Rubus ursinus* *Salix pedicellaris* *Salix scoulerleriana* *Salix sitchensis* *Vaccinium parvifolium*	*Azolla mexicana* *Brasenia schribneri* *Eriophorum chamissonis* *Hippurus vulgaris* *Hydrocotyl ranunculoides* *Hydrophyllum tenuipes* *Nymphaea odorata* *Polygonum amphibium* *Potentilla gramineus* *Rhynchospora alba* *Sparganium eurycarpum* *Sagittaria latifolia* *Scirpus acutus* *Veronica americana*
Always less than 1% coverage	no species	*Mimulus guttatus* *Myosotis laxa* *Potamogeton diversifolius* *Ranunculus acris* *Rorippa curvisiliqua* *Rumex obtusifolius* *Trillium ovatum* *Vaccinium ovatum* *Vaccinium uliginosum* *Vicia sativa*

Plant form was described within each of the habitats and is shown in Table 3-3. The table shows that plant species of many forms may be found within these habitats but some forms dominate more than others. Herbs are the dominant form of structure within PAB and PEM habitats, but also have a presence in the structure of scrub-shrub, wetland forested habitats, and bogs, too. Shrubs are an important component of a structure in bog and forested habitats as well as scrub-shrub, where it would be expected. Shrubs dominate species presence in all habitats with the exceptions of aquatic bed and emergent habitats, which are dominated by herbs. Shrubs and herbs are important structural forms in most habitat types.

TABLE 3-2
Distribution of Indicator Species Among Habitat Types

Habitat	Percent of Plant Type: Obligate	FACW	FAC	FACU
PAB	73%	21%	6%	0%
PEM	39%	28%	19%	14%
PSS	19%	29%	31%	21%
PFO	11%	15%	40%	34%
UPL	3%	6%	39%	52%
BOG	48%	32%	11%	9%

Wetland Plant Associations

Wetland vegetation sample plots were classified into 11 community types using TWINSPAN on the basis of species composition and percent cover (Figure 3-1).[9] The communities are further described in Table 3-4. These 11 basic community types were repeatedly observed in the study wetlands. Subdominant species changed and uncommon species were sometimes present, but dominant plant associations could often be described by one of the 11 types. These community associations may be used as a guide for understanding species composition and community structure in wetlands and are relevant to developing reference plant communities for palustrine wetlands in the Central Puget Sound Basin.

Abundance and Distribution of Invasive Plant Species

In an associated study, patterns of invasive plant species distribution, dominance, and abundance were compared among and within the wetlands.[14] The frequency of invasive species was found to be highly dependent on the conditions present, which varied for different species. For example, *Phalaris arundinacea*, *Rubus procerus*,

TABLE 3-3
Distribution of Plant Forms of Species Among Habitats (Presence Not Coverage)

Habitat	Ferns, Horsetails, Mosses & Allies	Grasses	Herbs	Rushes	Sedges	Shrubs	Trees
PAB	0%	0%	77%	4%		12%	8%
PEM	12%	4%	43%	5%	6%	20%	10%
PSS	18%	3%	23%	1%	3%	37%	14%
PFO	18%	3%	19%	0%	3%	34%	18%
Upland	20%	1%	14%	0%	2%	38%	25%
BOG	13%	8%	19%	3%	3%	46%	9%

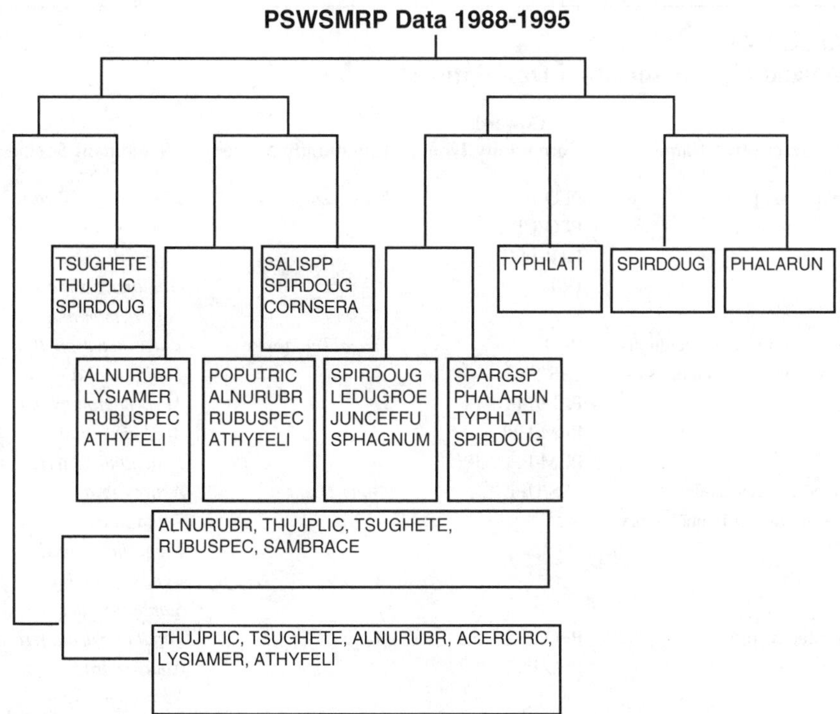

FIGURE 3-1 Classification of wetland community types present in the PSWSMRP study sites using TWINSPAN. (Four-letter codes are used for genus and species.)[9,10]

and *Solanum dulcamara* were more abundant in urbanized watersheds, while *Typha latifolia* and *Juncus effusus* were generally more abundant in less urbanized watersheds.[14] Water level fluctuation, depth of flooding, and duration of inundation were examined, but only duration of flooding was associated with the abundance of some invasive species.[14] *Typha latifolia* and *Juncus effusus* were generally more abundant in permanently flooded conditions, while *Rubus procerus* was found in sites where flooding occurred far less frequently.

The coverage of invasive species in different community types is shown in Figure 3-2. Invasive species were most abundant in aquatic bed and emergent marsh communities with *Phalaris arundinacea* and *Typha latifolia* the most prevalent species. Very few invasive species were found in coniferous-forested communities in either the wetland or upland habitats.

Community Richness

At the completion of the study, the total count ranged from 35 to 109 plant species per wetland. Seven wetlands had less than 60 species: 12 wetlands had between 60 and 84 species; and 7 had between 85 and 109 species, including 4 with 100 or more. A total of 46 species were found in bog habitats, 17 in aquatic beds, 155 in

TABLE 3-4
Wetland Plant Association Descriptions[18]

Descriptive Name	Cowardin Community Type[a]	Community Name[b]	Dominant Species
Coniferous forest	PFO PFO/UPL PAB/PFO/UPL UPL	*Tsuga-Thuja*	*Tsuga heterophylla* *Thuja plicata* *Spirea douglasii* *Gaultheria shallon* *Polystichum munitum*
Mixed coniferous-deciduous forest with shrub understory	PFO PSS/PFO PEM/PFO PEM/UPL PEM/PFO/UPL	*Tsuga-Thuja-wet*	*Tsuga heterophylla* *Thuja plicata* *Acer macrophyllum* *Acer circinatum* *Lysichitum americanum*
Mixed coniferous-deciduous forest with little understory	PSS/UPL	*Alnus-Thuja*	*Alnus rubra* *Thuja plicata* *Tsuga heterophylla* *Rubus spectabilis* *Sambucus racemosa*
Deciduous forest	PFO/PSS/UPL PFO/PSS	*Populus*	*Populus balsamifera* *Alnus rubra* *Rubus spectabilis* *Athyrium filix-femina*
Deciduous forest	PEM/PSS/PFO PFO/PAB/PEM	*Alnus*	*Alnus rubra* *Rubus spectabilis* *Cornus sericea* *Lysichitum americanum* *Athyrium filix-femina*
Mixed shrub-scrub	PAB/PSS PAB/PFO	*Salix-Spirea*	*Salix* spp. *Spirea douglasii* *Cornus sericea* *Lonicera involucrata*
Bog	BOG, BOG/PSS	*poor fen-shrub*	*Rhododendron groenlandicum (Ledum g.)* *Sphagnum* spp. *Spirea douglasii*
Mixed emergent	PAB/PEM/PSS BOG/PEM	*poor fen-marsh*	*Phalaris arundinacea* *Typha latifolia* *Rhododendron groenlandicum* *Sparganium* spp. *Spirea douglasii*
Emergent	PAB, PAB/PEM	*Typha*	*Typha latifolia* *Solanum dulcmara* *Lemna minor*

TABLE 3-4 *(continued)*
Wetland Plant Association Descriptions[18]

Descriptive Name	Cowardin Community Type[a]	Community Name[b]	Dominant Species
Emergent	PEM	*Phalaris*	*Phalaris arundinacea* *Solanum dulcmara* *Urtica dioica*
Scrub-shrub	PSS	*Spirea*	*Spirea douglasii* *Salix sitchensis* *S. Alba*

[a] Habitat type.
[b] Association name used in Figure 3-2.

emergent zones, 170 in scrub-shrub habitats, 159 in wetland-forested zones, and 119 species were found in upland habitats.

Within individual wetlands plant richness varied widely between and among the Cowardin vegetation types. Aquatic bed habitats had the fewest species, from one to eight, and averaging about four among all sampled. Emergent habitat richness varied considerably among wetlands, from 2 species to a high of 56 in one wetland, and averaged 21 species overall. Scrub-shrub habitats ranged from 4 to 39 species, with an average of 22. Forested habitats had from 5 to 61 species and averaged 25. Bog habitats varied from 15 to 24 species within individual wetlands and averaged 20. The highest total plant richness was found in wetlands with the largest number of Cowardin community types (Fisher's r to z, r = .41, p = .0001). Wetland area was found unrelated to plant richness in all years (Fisher's r to z, r <.42, p >.2).

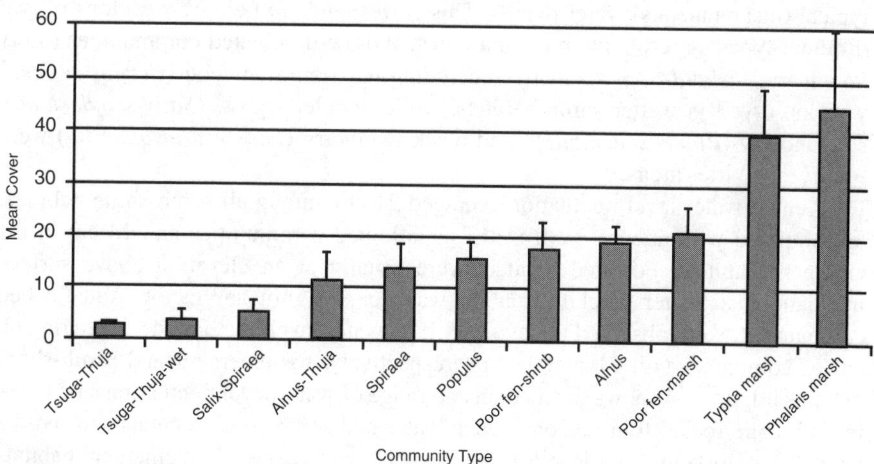

FIGURE 3-2 Portion of coverage within the community types found in the PSWSMRP study sites occupied by invasive species. Error bars are one standard error. See Table 3-4 for a description of community types.

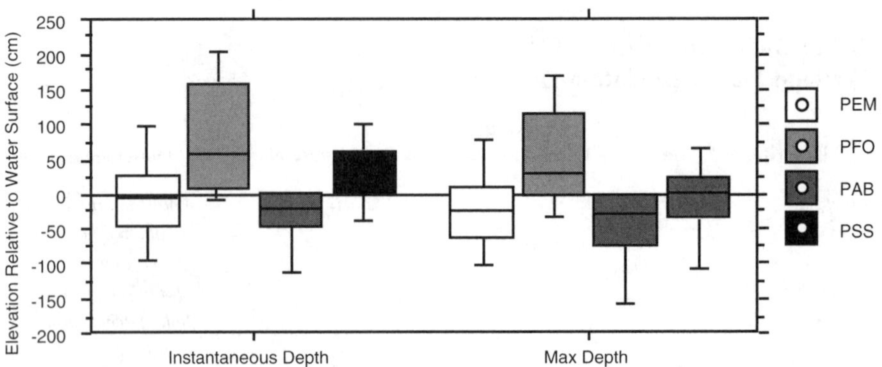

FIGURE 3-3 Annual mean instantaneous and maximum water depths (Max) associated with vegetation community types, 1988 through 1995.

Hydrologic Regimes By Habitat Type

The range of average conditions calculated for instantaneous and maximum water depths found during the study period in the PEM, PFO, PAB, and PSS communities are displayed in Figure 3-3. The solid bars in the figure show 50% of observations and, with the tails, represent 80% of observations. The ranges shown for each habitat type represent conditions for all seasons together and, therefore, extend considerably. Forested communities were, as expected, the driest of the community types with a median elevation of 62 cm above typical water levels, and ranged from about 12 cm inundation to approximately 210 cm above the water surface. Aquatic bed habitats were the wettest, being inundated or saturated all year. The biggest variation from wet to dry conditions throughout the year was observed in scrub-shrub communities, many of which exhibited higher average maximum water depths than typical (instantaneous) water depths. This corresponds to field observations of two different types of scrub-shrub communities. Willow-dominated communities (*Salix lucida* var. *lasiandra*, *S. sitchensis*) including red stem dogwood (*Cornus sericea*) were observed in wetter shrub habitats, while Scouler willow (*Salix scouleriana*), salmonberry (*Rubus spectabilis*), and black twinberry (*Lonicera involucrata*) were observed in drier areas.

Annual water level fluctuation averaged 21 cm among all scrub-shrub habitats, as compared with about 12 cm in the aquatic bed communities and 14 cm in the emergent habitats. Forested habitats were usually at an elevation above surface inundation, so water level fluctuation was not a significant factor. Aquatic bed communities were observed to have very high water level fluctuations, averaging 60 cm as compared with 11 and 18 cm, respectively, for emergent and scrub-shrub habitats. Figure 3-4 shows the median and range of water level fluctuation calculated in each zone for all four seasons. Open water and scrub-shrub habitats showed the greatest variation in water level fluctuation between seasons while emergent habitats were fairly consistent. The median WLF for aquatic bed habitats was less than 20 cm among all seasons, but WLF ranged significantly higher during peak events in aquatic bed habitats than in emergent, scrub-shrub, or forested habitats.

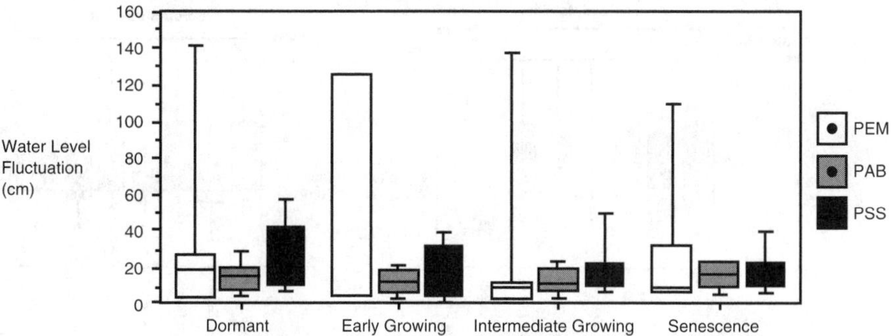

FIGURE 3-4 Water level fluctuation associated with vegetation community types in each season.

The ranges of instantaneous and maximum water levels for all seasons were plotted for each habitat and are shown in Figure 3-5. The seasonal changes in each community are demonstrated in the box plots. Negative numbers represent the number of centimeters of inundation. Positive numbers are above the water surface (equal to zero in the plots). The period of senescence, September through November, is definitely drier in all habitats, including the aquatic bed communities. Most aquatic bed habitats had no surface water during this period except during storm events although many were observed to have saturated soils. Most emergent habitats were inundated until the late growing season and during senescence. In general, forested wetland habitats were driest, followed by scrub-scrub and with emergent and aquatic bed habitats being the wettest.

Hydrologic Regimes of Some Species

The hydrologic conditions observed for some common wetland species was calculated in the same manner as for wetland habitats. The range of hydrology observed for different plants is reported in Table 3-5. The table shows that many species occur in a wide range of conditions while others are observed within a very narrow range.

Douglas spirea *(Spirea douglasii)* was found in wetland conditions that ranged from mostly dry through the year, to frequent temporary inundation, to complete inundation through both growing seasons, which may account for why this species was among the most widely distributed in our study. In addition Douglas spirea was found in wetlands with some of the highest water level fluctuations measured, averaging as high as 57 cm in the dormant season and 35 cm in the early growing season.

Analysis of the water level conditions where species were observed often showed seasonal differences that may account for differences in distribution. For example, two common wetland trees, red alder (*Alnus rubra)* and Oregon ash (*Fraxinus latifolia)* were both more prevalent on drier sites. Red alder differed, however, in that it was often found in areas subjected to mean WLFs of greater than 20 cm during the early growing season, whereas Oregon ash was observed mostly in areas with stable water levels in the early growing season. Oregon ash was often observed in organic soils and remained saturated for most of the growing season

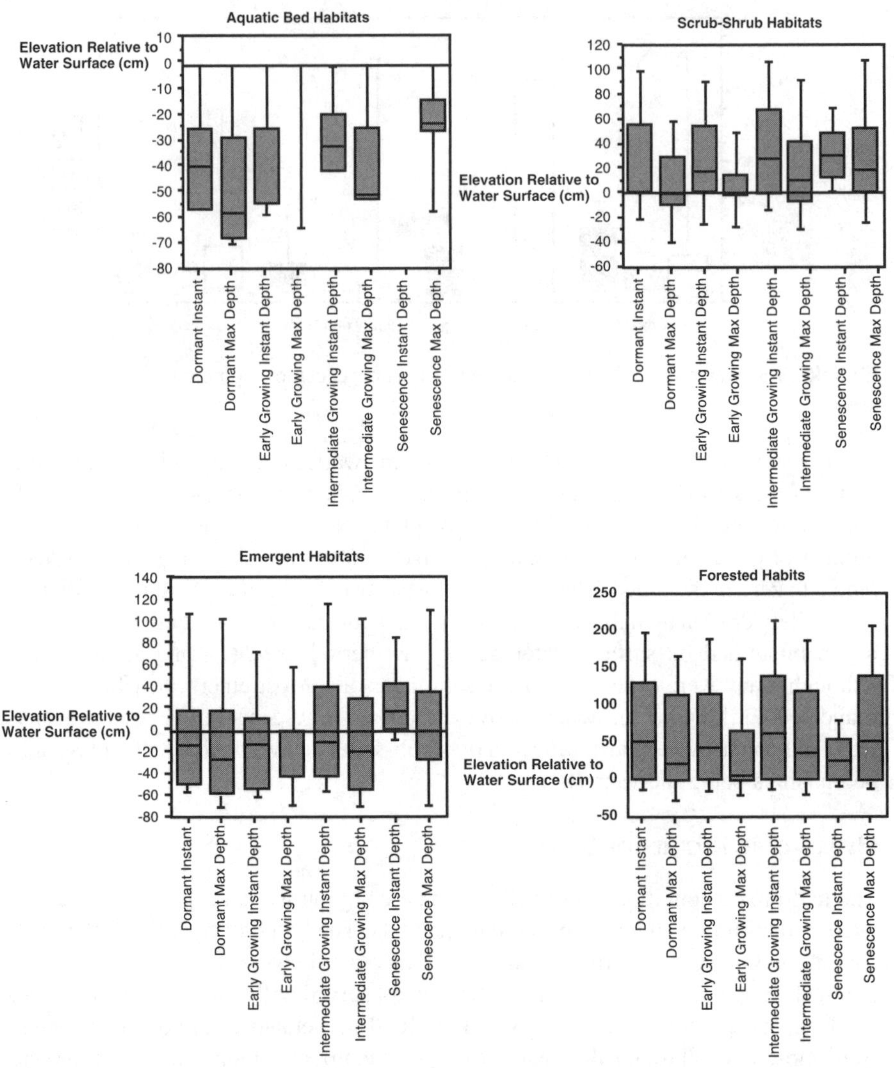

FIGURE 3-5 Seasonal hydrology associated with different habitat types.

while red alder was mostly found growing in mineral soils that typically became dry in the summer.

Both tapertip rush (*Juncus acuminatus*) and soft rush (*Juncus effusus*) were found growing in a similar range of conditions, but tapertip rush was found more often in areas that were slightly drier. Soft rush was usually found in areas shallowly inundated during the early spring. Slough sedge (*Carex obnupta)*, which is very common in wetlands around the region, was found in drier areas above water during the early and intermediate growing seasons, as compared with inflated sedge (*Carex exsiccata*)

TABLE 3-5
Maximum, Median, and Low Water Depths Found Where Plant Species Were Observed

Plant Species	Sample Size	Hydrologic Conditions Where Species Was Observed								
		Early Growing Season			Intermediate Growing Season			Senescence & Dormant Seasons		
		Max Water Depth (cm)	Median Water Depth (cm)	Low Water Depth (cm)	Max Water Depth (cm)	Median Water Depth (cm)	Low Water Depth (cm)	Max Water Depth (cm)	Median Water Depth (cm)	Low Water Depth (cm)
Carex lenticularis	28	63	39	14	68	41	14	61	39	18
Lemna minor	60	31	7	–17	23	1	–22	30	6	–19
Potamogeton natans	13	56	33	9	51	27	3	58	38	18
Utricularia vulgaris	152	80	51	22	90	53	16	84	49	14
Bidens cernua	65	29	0	–29	29	–4	–38	32	0	–32
Carex exsiccata (vesicaria)	48	12	0	–12	9	–19	–47	35	11	–12
Carex utriculata (rostrata)	16	68	–6	–79	40	–34	–108	45	–31	–108
Galium trifidum	62	38	0	–38	37	–17	–72	42	–20	–82
Gaultheria shallon	159	40	0	–40	33	–11	–55	33	–12	–58
Geranium robertianum	265	29	3	–36	8	–9	–27	26	–3	–32
Juncus acuminatus	28	20	–3	–26	–4	–23	–43	11	–11	–32
Juncus effusus	56	20	0	–20	9	–13	–35	22	0	–23
Lycopus uniflorus	41	44	–5	–55	25	–25	–74	41	–11	–63
Oenanthe sarmentosa	69	23	–5	–34	29	–19	–66	26	–19	–64
Polygonum hydropiper	44	27	0	–27	17	–11	–40	29	0	–29
Pseudotsuga menziesii	52	42	0	–42	30	–17	–63	50	–18	–86
Rubus laciniatus	68	31	–6	–43	23	–15	–54	38	–9	–56
Scirpus acutus	46	17	0	–17	17	22	–62	33	–7	–46
Sparganium emersum	79	32	2	–28	29	–11	–51	42	2	–38

TABLE 3-5 *(continued)*
Maximum, Median, and Low Water Depths Found Where Plant Species Were Observed

		Hydrologic Conditions Where Species Was Observed								
		Early Growing Season			Intermediate Growing Season			Senescence & Dormant Seasons		
Plant Species	Sample Size	Max Water Depth (cm)	Median Water Depth (cm)	Low Water Depth (cm)	Max Water Depth (cm)	Median Water Depth (cm)	Low Water Depth (cm)	Max Water Depth (cm)	Median Water Depth (cm)	Low Water Depth (cm)
Spirea douglasii	283	31	–6	–42	20	–21	–62	31	–15	–61
Torreyochloa pauciflora (Puccinellia p.)	29	36	0	–36	35	–23	–82	45	–13	–70
Utricularia minor	138	39	10	–19	28	–2	–31	36	7	–22
Anaphalis margaritacea	11	27	9	–9	19	0	–19	37	19	1
Epilobium angustifolium	179	9	0	–9	23	0	–23	56	0	–56
Hypericum anagalloides	25	47	16	–14	52	14	–25	46	14	–18
Hypericum formosum	25	17	0	–17	80	9	–63	81	13	–55
Juncus bufonius	46	64	19	–27	64	13	–38	66	19	–29
Juncus ensifolius	14	23	0	–23	32	5	–21	39	11	–18
Ludwigia palustris	22	38	2	–33	40	1	–37	51	14	–22
Lycopus americanus	49	69	13	–43	70	13	–44	64	13	–37
Myosotis laxa	23	41	0	–40	50	6	–37	46	13	–20
Nuphar polysepalum	14	155	14	–128	87	2	–83	182	13	–155
Potentilla palustris	23	18	0	–17	21	0	–21	23	6	–12
Typha latifolia	83	52	21	–10	45	13	–19	49	20	–9
Ledum groenlandicum	46	47	20	–8	46	14	–18	49	19	–12
Acer circinatum	123	7	–34	–76	3	–53	–110	3	–50	–103
Agrostis capillaris (tenuis)	15	19	–32	–83	–1	–61	–120	23	–28	–78

Alnus rubra	298	21	–25	–71	16	–41	–97	17	–35	–88
Athyrium filix–femina	272	22	–20	–62	22	–30	–82	23	–26	–75
Blechnum spicant	70	38	–13	–64	33	–25	–82	31	–25	–81
Carex arcta	50	22	–16	–54	23	–29	–82	35	–18	–71
Carex deweyana	40	42	–26	–93	53	–23	–99	32	–35	–102
Carex obnupta	42	17	–20	–57	12	–31	–73	15	–26	–67
Cornus sericea (stolonifera)	95	9	–25	–59	9	–31	–71	6	–29	–65
Dicentra formosa	20	37	–25	–86	19	–49	–117	–43	–111	–178
Dryopteris expansa (austriaca)	39	28	–13	–54	28	–25	–78	23	–33	–89
Epilobium ciliatum (watsonii)	93	14	–11	–37	12	–19	–49	10	–16	–43
Equisetum arvense	63	12	–33	–78	5	–46	–97	12	–35	–83
Fraxinus latifolia	19	2	–23	–49	1	–34	–69	–25	–42	–59
Galium aparine	21	19	–28	–75	19	–41	–100	20	–28	–76
Galium cymosum	18	15	–15	–46	7	–29	–65	16	–15	–45
Geum macrophyllum	32	23	–23	–70	21	–25	–71	23	–19	–62
Glyceria borealis	11	69	–20	–108	49	–33	–115	67	–23	–115
Glyceria grandis	82	3	–42	–87	–6	–59	–113	–2	–48	–94
Ilex aquifolia	28	50	–42	–134	28	–66	–160	34	–55	–143
Lonicera involucrata	132	13	–22	–57	12	–30	–71	9	–29	–66
Luzula parviflora	27	5	–20	–46	7	–31	–68	23	–41	–106
Lysichitum americanum	158	15	–23	–64	19	–32	–83	13	–30	–72
Maianthemum dilatatum	61	27	–23	–74	12	–51	–115	4	–55	–113
Malus fusca (Pyrus f.)	74	9	–37	–82	2	–53	–108	5	–44	–93
Menziesia ferruginea	36	16	–20	–56	16	–35	–87	7	–48	–103
Oplopanax horridus	25	46	–13	–72	35	–25	–84	9	–48	–105
Phalaris arundinacea	186	5	–35	–74	–5	–46	–87	12	–32	–77

TABLE 3-5 *(continued)*
Maximum, Median, and Low Water Depths Found Where Plant Species Were Observed

		Hydrologic Conditions Where Species Was Observed								
		Early Growing Season			Intermediate Growing Season			Senescence & Dormant Seasons		
Plant Species	Sample Size	Max Water Depth (cm)	Median Water Depth (cm)	Low Water Depth (cm)	Max Water Depth (cm)	Median Water Depth (cm)	Low Water Depth (cm)	Max Water Depth (cm)	Median Water Depth (cm)	Low Water Depth (cm)
Picea sitchensis	42	29	–7	–42	25	–17	–58	18	–40	–98
Populus tremuloides	340	78	–18	–114	36	0	–36	78	–22	–121
Prunus emarginata	14	44	–38	–120	45	–41	–128	22	–61	–143
Pteridium aquilinum	95	36	–15	–66	32	–25	–83	31	–26	–83
Ranunculus repens	67	1	–38	–77	1	–52	–105	13	–33	–79
Rhamnus purshiana	178	23	–16	–56	17	–30	–78	16	–33	–83
Ribes lacustre	15	33	–35	–102	32	–48	–129	36	–34	–104
Rubus procerus	87	11	–39	–89	7	–44	–95	19	–32	–82
Rubus parviflorus	19	30	–43	–115	18	–61	–141	–9	–87	–165
Rubus spectabilis	305	20	–28	–75	15	–42	–99	17	–38	–93
Rubus ursinus	165	26	–25	–75	22	–38	–97	16	–43	–102
Salix lucida var. lasiandra	96	30	–19	–68	20	–30	–81	23	–21	–67
Salix pedicellaris	14	9	–37	–83	–33	–88	–143	–22	–74	–126
Salix scoulerleriana	64	5	–22	–49	6	–28	–62	9	–26	–61
Salix sitchensis	91	23	–8	–39	17	–15	–48	28	0	–28
Sambucus racemosa	144	23	–31	–86	17	–46	–109	14	–43	–99
Scutellaria lateriflora	27	29	–12	–53	4	–45	–93	17	–30	–78
Solanum dulcamara	147	2	–31	–65	–1	–42	–82	7	–29	–65
Sorbus americana	20	29	–59	–147	28	–67	–163	40	–48	–136
Stellaria media	30	12	–11	–34	6	–25	–55	8	–17	–43

Thuja plicata	118	26	–12	–49	18	–22	–62	29	–17	–63
Tiarella trifoliata	15	5	–16	–38	–4	–32	–60	5	–31	–67
Tsuga heterophylla	140	41	–11	–64	34	–23	–83	36	–25	–85
Urtica dioica	84	9	–41	–91	–4	–56	–108	10	–40	–90
Vaccinium parvifolium	158	31	–11	–52	30	–23	–78	32	–25	–82
Veronica americana	74	27	–16	–59	22	–30	–82	31	–17	–64
Veronica scutellata	40	26	–11	–49	20	–25	–69	25	–17	–58
Callitriche heterophylla	23	–14	–50	–86	–20	–60	–100	–14	–51	–89
Equisetum telmateia	37	–1	–46	–91	–17	–62	–107	–6	–50	–95
Glyceria elata	21	–1	–11	–21	–17	–25	–33	–4	–16	–28
Holodiscus discolor	130	–21	–40	–59	–40	–66	–93	–35	–55	–75
Linnaea borealis	88	–5	–37	–69	–20	–52	–84	–25	–46	–68
Populus balsamifera	60	–3	–55	–106	–21	–75	–129	–6	–57	–108
Ribes divaricatum	19	–11	–56	–101	–14	–70	–126	–9	–55	–101
Rosa gymnocarpa	23	–15	–47	–78	5	–42	–89	–25	–56	–88
Salix alba	27	–15	–46	–76	–1	–36	–71	9	–25	–59
Scirpus microcarpus	27	–1	–31	–62	5	–34	–74	2	–29	–60
Stachys cooleyae	26	0	–58	–115	–5	–73	–141	4	–65	–134
Acer macrophyllum	42	–4	–61	–119	–7	–73	–138	–30	–85	–139
Carex hendersonii	16	–33	–76	–118	–54	–94	–135	–25	–79	–133
Corylus cornuta	44	–33	–90	147	–28	–93	–158	–34	–96	–157
Crataegus monogyna	65	–64	–111	–159	–75	–124	–173	–63	–110	–158
Equisetum hyemale	31	–25	–80	–135	–41	–102	–162	–22	–75	–128
Oemleria cerasiformis	59	–17	–80	–143	–37	–99	–161	–27	–86	–145
Petasites frigidus	41	–25	–118	–211	–18	–123	–229	–18	–118	–218
Rosa pisocarpa	12	–21	–71	–121	–34	–83	–132	–27	–74	–120
Smilacena racemosa	11	–11	–60	–109	–23	–73	–122	–12	–62	–112
Tolmiea menziesii	68	–7	–65	–122	–13	–75	–137	0	–71	–143
Trillium ovatum	17	–3	–63	–124	–57	–119	–181	–35	–112	–190

(old name *C. vesicaria*), which was observed almost exclusively in relatively undisturbed wetlands inhabiting saturated soils and areas of shallow inundation. Both species were found inundated during the dormant season and both were found in conditions of high water level fluctuation.

Small-fruited bulrush, *Scirpus microcarpus*, observed in disturbed wetlands, and wooly sedge, *Scirpus atrocinctus* (old name *S. cyperinus*), found in relatively undisturbed wetlands, were observed growing in similar hydrologic conditions. Wooly sedge, however, was found in slightly wetter conditions during the early growing season. In addition, small-fruited bulrush was found in wetlands with high WLF throughout the growing season whereas wooly sedge was not.

Several dominant species including soft rush, reed canarygrass (*Phalaris arundinacea)*, and cattail (*Typha latifolia)* were evaluated to see if there were hydrologic conditions common to dominant species. Of the three, reed canarygrass grew in the driest areas, and cattail in the wettest. Reed canarygrass was found in many wetlands with very high seasonal WLF, whereas cattail and soft rush were found in areas with low WLF except during the dormant period. All were consistently found within a broad range of hydrologic conditions distributed mostly within the emergent habitats of wetlands.

DISCUSSION

Wetland management regulations, for the most part, classify wetlands on the basis of area, the number and type of vegetation communities, and the presence of threatened or endangered species.[6,15] Although larger wetlands are often observed to be more diverse systems, the rarity or richness of the plant community was found to be unrelated to wetland area and more related to the number of habitats present. Regardless of size, wetlands were often found to have unique assemblages of plants and high species richness.

Factors other than wetlands size and endangered species, such as hydrologic regime and the kinds and frequency of disturbance, appeared more critical in determining the diversity, uniqueness, and character of the wetland plant communities we studied. For example, less common and less dominant species were observed within narrower hydrologic ranges than more common species, suggesting uncommon species could be extirpated if a change in wetland hydroperiod occurred that exceeded that range. In contrast, many of the more common and pervasive wetland species were found to grow in a wide range of drought to peak inundation conditions. Wetlands with the richest and most diverse plant communities were typically characterized by more complex hydrology and more variable morphology, providing many surfaces at different gradients for plant species to inhabit. Wetlands with simpler vegetation communities were more often topographically uniform, resulting in simpler hydrologic regimes. These differences may be traced, to some extent, to types and frequency of historical disturbance, including direct and indirect disturbances to the hydrologic regime. For example, the wetlands with the greatest diversity of habitats and wetland plants were wetlands with beaver dams or, otherwise, leaky outlet structures and a preponderance of large woody debris. The leaky dam outlet provided a fairly stable hydroperiod in these wetlands and woody debris added structure improving habitat functions.

Wetlands are frequently defined in terms of the frequency, duration, and timing of flooding or soil saturation, yet few species of vascular plants identified as occurring in wetlands of the U.S. have been evaluated to determine numeric hydrologic criteria for their presence or absence, particularly over more than a year or two.[5,7,8,16,17] Studies have shown that plant species composition is related to water level, which is dependent on wetland position in groundwater flow systems.[18-21] Human-induced changes in the relative proportion of precipitation, groundwater, and surface water inputs to wetlands can change species composition and reduce species diversity.[22-24] Therefore it is critical that we begin to better understand the relationships between wetland plant communities and hydroperiod.

While our data are a useful guideline, it is important to point out that we did not measure plant responses or vigor with respect to hydrologic conditions, but simply quantified the range of conditions in which a species was observed. In a related study, Ewing measured and analyzed actual tree responses, and observed that *Alnus rubra* was stressed by repeated cycles of inundation while *Fraxinus latifolia* showed no significant response to repeated inundation, provided the duration of the flooding was less than two to three days.[25] Our methods did not measure such detailed impacts. We also did not accurately account for soil saturation and soil type, which clearly also affect species distributions.

Herbs were a significant component of community structure within most habitat types. The variety of herbs and shrubs comprises much of total plant richness and may be important links to habitat value in wetlands. The effectiveness of wetland protection measures could be at least partially evaluated by observing the effects of stormwater management on the herb and shrub communities. Wetland restoration and construction efforts may be benefit by specifying a variety of herbs and shrubs in all wetland habitats and by encouraging herb establishment early on.

Eleven distinct wetland plant communities were identified that are typical of the region. These communities were mostly found in assemblages interspersed throughout individual wetlands. Several wetland plant habitat types with transition habitats between them characterized all of the wetlands we studied. In general, when multiple habitats were present, plant richness was higher within individual habitats, as many species were observed to transition between habitat types.

Generalized classifications of vegetation structure, such as forested, scrub-shrub, and emergent, do not reveal the presence or absence of unusual plant species, plant associations, or the contribution of a wetland to the regional landscape. It may be that regulating protection based on wetland size and the presence of certain types of wetland plant communities will not fully protect regional wetland values and functions. These results suggest wetland management should focus toward protecting or creating the conditions required to sustain wetland habitats. In addition to preserving large wetlands with diverse hydrologic regimes, we should consider addressing land use and development constraints to limit the extent of alterations in hydroperiod occurring in wetlands due to changing watershed conditions. Controlling the frequency of disturbances and monitoring invasive species presence also provide reasonable tools for maintaining species richness and regional biodiversity.

ACKNOWLEDGMENTS

Special thanks to all those who persevered through blackberries and spirea to help gather botanical data for 10 years including Catherine Conolly, Sarah Cassat, Mike Emers, Meredith Savage, Richard Pratt, Colleen Rasmussen, Phil Dinsmore, Greg Mazer, Elizabeth Zemke, Catherine Houck, and Ola Edwards. We would also like to thank Dr. Fred Weinmann for his comments and suggestions for this chapter, Catherine Houck for her informative work on the distribution of invasive species, and finally, Dr. Kern Ewing, who provided insight and contributed greatly to our research.

REFERENCES

1. Barbour, M. G., J. H. Burke, and W. P. Pitts, *Terrestrial Plant Ecology*, Benjamin/Cummings, Menlo Park, CA, 1987.
2. Gauch, H. G., *Multivariate Analysis in Community Ecology*, Cambridge University Press, Cambridge, U.K., 1982.
3. Cooke, S. S., R. R. Horner, C. Conolly, O. Edwards, M. Wilkinson, and M. Emers, Effects of Urban Stormwater Runoff on Palustrine Wetland Vegetation Communities — Baseline Investigation (1988). Report to U.S. Environmental Protection Agency, Region 10, by King County Resource Planning Section, Seattle, WA, 1989.
4. Hitchcock, C. L., A. Cronquist, M. Ownbey, and J. W. Thompson, *Vascular Plants of the Pacific Northwest*, University of Washington Press, Seattle, 1969.
5. Cowardin, L. M., V. Carter, F. C. Goulet, and E. T. LaRoe, Classification of Wetlands and Deepwater Habitat of the United States, U. S. Fish and Wildlife Service, Washington, D.C., 1979.
6. Toshach, S., Washington State Wetlands Rating System for Western Washington, Washington State Department of Ecology, Publication # 91-58, Olympia, WA, 1991.
7. Reed, P., National List of Plant Species that Occur in Wetlands: Northwest Region 9, U.S. Fish and Wildlife Service, Biol. Rep. 88, Washington, D.C., 1988.
8. Reed, P., Addendum to the National List of Plant Species that Occur in Wetlands: Northwest Region 9, U.S. Fish and Wildlife Service, Washington, D.C., 1993.
9. Hill, M. O., *TWINSPAN: A Fortran Program for Arranging Multi-Variate Data in an Ordered Two-Way Table of Classification of the Individuals and Attributes*, Cornell University Press, Ithaca, NY, 1979.
10. Hill, M. O., *DECORANA: A Fortran Program for Detrending Correspondence Analysis and Reciprocal Averaging*, Cornell University Press, Ithaca, NY, 1979.
11. Clifford, H. T. and W. T. Williams, Classification dendograms and their interpretation, *Aust. J. Bot.*, 21, 151, 1973.
12. Causton, D. R., *An Introduction To Vegetation Analysis*, Unwin Hyman, London, 1988.
13. Reinelt, L. E. and R. R. Horner, Characterization of the Hydrology and Water Quality of Palustrine Wetlands Affected by Urban Stormwater, King County Resource Planning, King County, WA, 1990.

14. Houck, C. A., The Distribution and Abundance of Invasive Plant Species in Freshwater Wetlands of the Puget Sound Lowlands, King County, Washington, M.S. thesis, University of Washington, Seattle, 1996.
15. King County, The King County Wetlands Inventory Notebook, Vol. I-III, King County Environmental Section, Parks and Recreation Division, Bellevue, WA, 1990.
16. Mitsch, W. J. and J. G. Gosselink, *Wetlands*, 2nd ed., Van Nostrand Reinhold, New York, 1993.
17. Bedford, B. L., The need to define hydrologic equivalence at the landscape scale for freshwater wetland mitigation, *Ecol. Appl.,* 6(1), 57, 1996.
18. Ingram, H. A. P., Problems of Hydrology and Plant Distribution in Mires, *J. Ecol.,* 55, 711, 1967.
19. Green, B. H. and M. C. Pearson, The ecology of Wybunbury Moss, Cheshire I. The present vegetation and some physical, chemical and historical factors controlling its nature and distribution, *J. Ecol.,* 56, 245, 1968.
20. Daniels, R. E. and M. C. Pearson, Ecological studies at Roydon Common, Norfolk, *J. Ecol.,* 62, 127, 1974.
21. Giller, K. E. and B. D. Wheeler, Acidification and succession in a flood-plain mire in the Norfolk Broadland, U.K., *J. Ecol. Res.*, 5, 849, 1988.
22. Wassen, M. J and A. Barendregt, Topographic transition and water chemistry of fens in a Dutch river, *J. Veg. Sci.,* 3, 447, 1992.
23. Wassen, M. J., R. Van Diggelen, L. Wolejko, and J. T. A. Verhoeven, A comparison of fens in natural and artificial landscapes, *Vegetatio,* 126(1–4), 5, 1996.
24. Glaser, P. H., J. A. Janssens, and D. I Siegle, The response of vegetation to chemical and hydrologic gradients in the Lost River peatland, Northern Minnesota, *J. Ecol.,* 78, 1021, 1990.
25. Ewing, K., Tolerance of four wetland plant species to flooding and sediment deposition, *J. Exp. Bot.,* Elsevier Science, 1996.

APPENDIX A
List of Plant Species and Frequency Found Among 19 Central Puget Sound Basin Palustrine Wetlands

Plant Species	Number of Wetlands	Percent of All Wetlands
Polystichum munitum	26	100
Rubus spectabilis	26	100
Rubus ursinus	26	100
Alnus rubra	25	0.96
Athyrium filix-femina	25	0.96
Salix scoulerleriana	25	0.96
Spirea douglasii	25	0.96
Lonicera involucrata	23	0.92
Pteridium aquilinum	23	0.92
Rhamnus purshiana	23	0.92
Rubus laciniatus	23	0.92
Vaccinium parvifolium	23	0.92
Dryopteris expansa (austriaca)	23	0.88
Epilobium ciliatum (watsonii)	23	0.88
Gaultheria shallon	22	0.85
Sambucus racemosa	22	0.85
Oemleria cerasiformis	21	0.81
Acer circinatum	20	0.77
Lysichitum americanum	20	0.77
Polypodium glycyrrhiza	20	0.77
Pseudotsuga menziesii	20	0.77
Salix lucida var. lasiandra	20	0.77
Solanum dulcamara	20	0.77
Thuja plicata	20	0.77
Tsuga heterophylla	20	0.77
Veronica americana	20	0.77
Blechnum spicant	19	0.73
Carex deweyana	19	0.73
Malus fusca (Pyrus f.)	19	0.73
Salix sitchensis	19	0.73
Equisetum arvense	18	0.69
Glyceria grandis	18	0.69
Luzula parviflora	18	0.69
Phalaris arundinacea	18	0.69
Populus balsamifera	18	0.69
Acer macrophyllum	17	0.65
Carex obnupta	17	0.65
Ranunculus repens	17	0.65
Rubus procerus (discolor)	17	0.65
Urtica dioica	17	0.65
Cornus sericea (stolonifera)	16	0.62
Geum macrophyllum	16	0.62

APPENDIX A *(continued)*
List of Plant Species and Frequency Found Among 19 Central Puget Sound Basin Palustrine Wetlands

Plant Species	Number of Wetlands	Percent of All Wetlands
Oenanthe sarmentosa	16	0.62
Corylus cornuta	15	0.58
Juncus effusus	15	0.58
Lemna minor	15	0.58
Prunus emarginata	15	0.58
Stellaria media	15	0.58
Trillium ovatum	15	0.58
Berberis nervosa	14	0.54
Carex utriculata =(rostrata)	14	0.54
Maianthemum dilatatum	14	0.54
Salix alba	14	0.54
Tolmiea menziesii	14	0.54
Bidens cernua	13	0.5
Carex arcta	13	0.5
Galium trifidum	13	0.5
Ilex aquifolia	13	0.5
Picea sitchensis	13	0.5
Sorbus americana	13	0.5
Equisetum hyemale	12	0.46
Equisetum telmateia	12	0.46
Holcus lanatus	12	0.46
Menziesia ferruginea	12	0.46
Polygonum hydropiper	12	0.46
Typha latifolia	12	0.46
Agrostis capillaris (tenuis)	11	0.42
Callitriche heterophylla	11	0.42
Epilobium angustifolium	11	0.42
Lycopus uniflorus	11	0.42
Rubus parviflorus	11	0.42
Scutellaria lateriflora	11	0.42
Sparganium emersum	11	0.42
Torreyochloa pauciflora (Puccinellia p.)	11	0.42
Veronica scutellata	11	0.42
Salix pedicellaris	10	0.38
Scirpus microcarpus	10	0.38
Carex hendersonii	9	0.35
Circium arvense	9	0.35
Dicentra formosa	9	0.35
Rosa gymnocarpa	9	0.35
Scirpus atrocinctus	9	0.35
Sorbus scopulina	9	0.35
Sphagnum spp.	9	0.35

APPENDIX A *(continued)*
List of Plant Species and Frequency Found Among 19 Central Puget Sound Basin Palustrine Wetlands

Plant Species	Number of Wetlands	Percent of All Wetlands
Carex lenticularis	8	0.31
Dactylis glomerata	8	0.31
Geranium robertianum	8	0.31
Glyceria elata	8	0.31
Juncus acuminatus	8	0.31
Juncus ensifolius	8	0.31
Rhododendron groenlandicum (Ledum g.)	8	0.31
Nuphar polysepalum	8	0.31
Oplopanax horridus	8	0.31
Potentilla palustris	8	0.31
Ribes lacustre	8	0.31
Stachys cooleyae	8	0.31
Symphoricarpos albus	8	0.31
Tiarella trifoliata	8	0.31
Anaphalis margaritacea	7	0.27
Festuca rubra	7	0.27
Galium cymosum	7	0.27
Glyceria borealis	7	0.27
Holodiscus discolor	7	0.27
Juncus bufonius	7	0.27
Lycopus americanus	7	0.27
Myosotis laxa	7	0.27
Potamogeton natans	7	0.27
Rumex crispus	7	0.27
Agrostis gigantea (alba)	6	0.23
Crataegus monogyna	6	0.23
Fraxinus latifolia	6	0.23
Galium aparine	6	0.23
Hypericum formosum	6	0.23
Ludwigia palustris	6	0.23
Rubus leucodermis	6	0.23
Smilacena racemosa	6	0.23
blue-green algae	6	0.23
Agrostis oregonensis	5	0.19
Amelanchier alnifolia	5	0.19
Betula papyrifera	5	0.19
Circaea alpina	5	0.19
Convolvulus arvensis	5	0.19
Digitalis purpurea	5	0.19
Drosera rotundifolia	5	0.19
Hedera helix	5	0.19
Lonicera ciliosa	5	0.19

APPENDIX A *(continued)*
List of Plant Species and Frequency Found Among 19 Central Puget Sound Basin Palustrine Wetlands

Plant Species	Number of Wetlands	Percent of All Wetlands
Ribes divaricatum	5	0.19
Rorippa calycina	5	0.19
Rosa pisocarpa	5	0.19
Sium suave	5	0.19
Utricularia minor	5	0.19
Azolla filiculoides	4	0.15
Eleocharis ovata	4	0.15
Eleocharis palustris	4	0.15
Hieracium pratense	4	0.15
Hypericum anagalloides	4	0.15
Iris pseudacorus	4	0.15
Kalmia microphylla	4	0.15
Linnaea borealis	4	0.15
Mentha arvensis	4	0.15
Myosotis scorpioides	4	0.15
Petasites frigidus	4	0.15
Physocarpus capitatus	4	0.15
Plantago lanceolata	4	0.15
Plantago major	4	0.15
Populus tremuloides	4	0.15
Solidago canadensis	4	0.15
Spirodela polyrhiza	4	0.15
Streptopus amplexifolius	4	0.15
Vaccinium oxycoccos	4	0.15
Actea rubra	3	0.12
Alisma plantago-aquatica	3	0.12
Carex exsiccata (vesicaria)	3	0.12
Circium vulgare	3	0.12
Cytisus scoparius	3	0.12
Dulichium arundinaceum	3	0.12
Elodea canadensis	3	0.12
Lolium perenne	3	0.12
Mimulus guttatus	3	0.12
Montia siberica	3	0.12
Nasturtium officinale	3	0.12
Phleum pratense	3	0.12
Ribes sanguineum	3	0.12
Nasturtium officinale	3	0.12
Taxus brevifolia	3	0.12
Utricularia vulgaris	3	0.12
Viola glabella	3	0.12
Adenocaulon bicolor	2	0.08

APPENDIX A *(continued)*
List of Plant Species and Frequency Found Among 19 Central Puget Sound Basin Palustrine Wetlands

Plant Species	Number of Wetlands	Percent of All Wetlands
Alnus sinuata	2	0.08
Alopecurus pratensis	2	0.08
Azolla mexicana	2	0.08
Bromus ciliatus	2	0.08
Claytonia lanceolata	2	0.08
Cornus nuttallii	2	0.08
Elytrigia repens (Agropyron repens)	2	0.08
Eriophorum chamissonis	2	0.08
Glecoma hederacea	2	0.08
Gnaphalium uliginosum	2	0.08
Gymnocarpium dryopteris	2	0.08
Hercaleum lanatum	2	0.08
Hypochaeris radicata	2	0.08
Impatiens noli-tangere	2	0.08
Lolium mulitflorum	2	0.08
Lotus corniculatus	2	0.08
Menyanthes trifoliata	2	0.08
Pinus monticola	2	0.08
Polygonum amphibium	2	0.08
Rhynchospora alba	2	0.08
Rumex obtusifolius	2	0.08
Salix hookeriana	2	0.08
Smilacena stellata	2	0.08
Sparganium eurycarpum	2	0.08
Streptopus roseus	2	0.08
Taraxacum officinale	2	0.08
Trfolium repens	2	0.08
Vaccinium uliginosum	2	0.08
Vallisneria americana	2	0.08
Adiantum pedatum	1	0.04
Agrostis scabra	1	0.04
Aira caryophyllea	1	0.04
Anthoxanthum odoratum	1	0.04
Arbutus menziesii	1	0.04
Asaurum caudatum	1	0.04
Berberis aquifolium	1	0.04
Brasenia schribneri	1	0.04
Carex athrostachya	1	0.04
Carex stipata	1	0.04
Chenopodium alba	1	0.04
Cladina rangiferina	1	0.04
Convolvulus sepium	1	0.04

APPENDIX A *(continued)*
List of Plant Species and Frequency Found Among 19 Central Puget Sound Basin Palustrine Wetlands

Plant Species	Number of Wetlands	Percent of All Wetlands
Cornus canadensis	1	0.04
Echinochloa crusgalii	1	0.04
Festuca pratensis	1	0.04
Fragaria virginiana	1	0.04
Goodyera oblongifolia	1	0.04
Hippurus vulgaris	1	0.04
Hydrocotyl ranunculoides	1	0.04
Hydrophyllum tenuipes	1	0.04
Juncus supiniformis	1	0.04
Lamium purpurea	1	0.04
Lythrum salicaria	1	0.04
Melilotus alba	1	0.04
Nymphaea odorata	1	0.04
Poa palustris	1	0.04
Poa pratensis	1	0.04
Potamogeton diversifolius	1	0.04
Potamogeton gramineus	1	0.04
Ranunculus acris	1	0.04
Rhinanthus crista-galli	1	0.04
Ribes bracteosum	1	0.04
Rorippa curvisiliqua	1	0.04
Rosa nutkana	1	0.04
Rosa rugosa	1	0.04
Rumex acetosella	1	0.04
Sagittaria latifolia	1	0.04
Scirpus acutus	1	0.04
Solanum nigrum	1	0.04
Stellaria longifolia	1	0.04
Tanacetum vulgare	1	0.04
Trifolium pratense	1	0.04
Vaccinium ovatum	1	0.04
Vicia sativa	1	0.04

4 Macroinvertebrate Distribution, Abundance, and Habitat Use

Klaus O. Richter

CONTENTS

Introduction 97
Methods 99
Field Methods 99
Statistical Methods 100
Results 102
Richness and Abundance of Aquatic and Semiaquatic Insects 102
Nematocera 114
Chironomids 114
Nonchironomids 126
Brachycera 129
Total Insect and Chironomid Species Richness and Wetland Chracteristics 129
Terrestrial Arthropod Richness and Abundance 133
Discussion 134
Acknowledgments 137
References 137

INTRODUCTION

Invertebrates are diverse, abundant, and ecologically significant components of freshwater ecosystems. Insects, especially, are pivotal in aquatic food webs, multiplying trophic connections with their richness, abundance, and diverse feeding strategies. As nymphs and larvae, invertebrates filter, shred, or scrape for food in sediment, on vegetation, and within the water column. They convert and assimilate microorganisms and plant tissue into biomass available to other aquatic fauna. As terrestrial and flying adults, they provide nutrition and energy to secondary and tertiary consumers feeding over wet and dry land. At all life stages, they furnish food for predatory arthropods including other insects, arachnids, and crustaceans. Additionally, invertebrates furnish much of the nutritional requirements of vertebrates, namely amphibians, water and shore birds, and small aquatic mammals.[1,2] Aquatic inverte-

1-56670-386-7/00/$0.00+$.50

brates are especially important to rearing fish (e.g., Salmonidae, game fishes) and breeding birds (e.g., waterfowl) contributing to commercial and sport activities.

The distribution, abundance, and health of aquatic macroinvertebrates have long been considered in assessing the condition of aquatic ecosystems. As early as the 1900s, the distribution, abundance, and health of aquatic macroinvertebrates were used to evaluate sewage pollution in rivers.[3] The evaluation of lotic environments continues to clarify the relationship between aquatic invertebrates and their habitats.[4-6] For many years now, developing biological criteria for designated uses of streams has received considerable attention and success.[7-10,63,64] Within the Pacific Northwest ecoregion, studies have focused on habitat use by stream invertebrates of the Coast Range ecoregion and on stream invertebrates in bioassessment.[19,20]

The invertebrates of select lotic environments have also been described and widely studied. One of the main goals of research has been to determine the relationships between taxa and oxygen concentrations, organic pollutants, and other water-quality conditions associated with lake eutrophication.[11-14] More recent studies have focused on the relationship between invertebrates, water permanence, and water-level manipulations, as well as invertebrate distributions among macrophytes within littoral zones of lakes.[6,15-18]

Aquatic invertebrates have traditionally been studied in streams and lakes, but sometimes also in some wetlands. Delta Marsh in south-central Manitoba, for example, has been studied to clarify the interactions between invertebrates and wetland hydrology, water quality, vegetation, and water birds under natural and manipulated conditions.[6,21,22] In Iowa, invertebrates were studied in relationship to changing marsh vegetation.[6] Other works have described the invertebrates of temporary ponds and vernal pools.[23] Studies relating aquatic invertebrates to wetland toxicity have been performed and reviewed by various authors.[24,25]

Overall, however, invertebrates of wetlands have received little attention. Only with recent concern for the dramatic loss of wetlands and their functions, combined with an understanding of both the ecological importance of invertebrates in streams and lakes and their success as bioindicators within these two aquatic systems, has invertebrate research in natural wetlands increased.[19,20,24,26,27] Most studies, however, are of ponds and marshes in sparsely populated areas, primarily because of the importance of invertebrates to game fish and waterfowl production.[28] An exception is invertebrates in wetlands receiving wastewater.[29-31] Few studies have investigated the distribution, abundance, and ecological nature of invertebrates within natural wetlands, especially those in urbanizing environments. Notable are studies that have documented invertebrate distributions along gradients of urban development.[32-35] These studies are especially important because urban wetlands, and wetlands in urbanizing watersheds, are rapidly disappearing or degrading. Moreover, the role of urban wetlands in flood control and water-quality enhancement may not be compatible with a high production of aquatic invertebrates and concomitant food-chain support essential to diverse and abundant fish, amphibians, birds and mammal populations. Such fish and wildlife populations are increasingly important to urban residents, including a growing group of outdoor recreationists and nature enthusiasts.

Clearly, a description of wetland invertebrate communities, starting with accurate characterization of species and abundance, is the first step to understanding their distribution, productivity, and habitat requirements. Consequently, in this chapter we characterize the macroinvertebrates of natural palustrine wetlands in King County, Washington describing the distribution and abundance of taxa collected in emergence traps at 19 wetlands in 1989, 1993, and 1995. We also investigate wetland flooding and watershed urbanization as an initial framework to describe and account for our captures.

METHODS

FIELD METHODS

Wetlands in this study are 2 to 20 ha palustrine systems with hydrology, water quality, and plant communities as characterized in Chapters 1 to 3 and by Cowardin et al.[36] We used emergence traps to collected adult macroinvertebrates (Figure 4-1). These function by funneling emerging invertebrates upward into a glass jar at the top of a trap containing pure ethylene glycol as the preserving liquid. Emergence traps rather than dip nets (e.g., sweep nets) or benthic core sampling were used

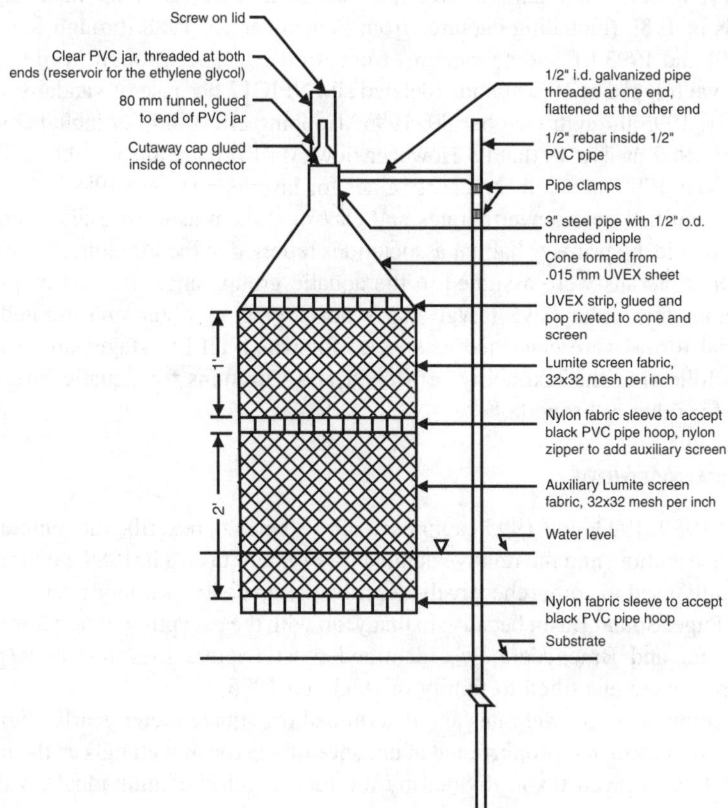

FIGURE 4-1 Cross section of aquatic macroinvertebrate emergence trap constructed by Robert Wisseman and modeled after Cook and Horn and Davies.[60,61]

because captures in emergent traps represent the final stage of insect production, readily allowing quantification of cumulative production over variable time periods. Emergence traps also sort species according to their ability to climb or fly into the collecting chamber, thereby facilitating identification. The traps provide better quantification of insect taxa, especially more precise estimates of the usually abundant chironomids, than sediment sampling or dip nets.[37]

Traps were constructed from cellulose acetate and Lumite 32 mesh-per-inch netting. A weighted PVC ring sewn into the base held traps flush to the substrate. Traps were supported by an internal PVC pipe crossbar fastened to two external rebars, which were hammered into the substrate. Substrate and vegetation at the trapping locations are characterized in Table 4-1.

In September 1988 we installed three circular 0.25 m^2 replicate traps within approximately 1 m of each other in the deepest (maximum 1 m) location at 14 wetlands. In May, 1989 we added traps at an additional five wetlands. Traps were emptied semimonthly except during mid-November through March because of low or nonexistent winter emergence. We attempted to collect invertebrates from all traps at each wetland within a three-day period. Due to unforeseen problems we took up to 19 days to collect all samples. We summarized macroinvertebrate data for all 19 wetlands in 1989 (including captures from September 15, 1988 through September 31, 1989) and 1993 (including captures from April 10, 1993 through April 9, 1994). In 1995 we trapped at 18 wetlands (deleted site NFIC12 because of vandalism) from January 1, 1995 through October 30, 1995. Trapping effort did not include captures for April 1989 at five wetlands. However, low April production in 1993 and 1995 suggests our 1989 data still represent values for invertebrate year 1989.

We identified macroinvertebrates and classified them as terrestrial, and aquatic or semiaquatic, by primary habitat associations reported in the literature.[38-40] Accordingly, all Dipterans were assigned to the aquatic group, since the vast majority of taxa within this order have larval stages developing in water or saturated soils. Terrestrial forms were classified as species for which all life stages are terrestrial habitats following the taxonomy of Merritt and Cummins for aquatic insects and Pennak for other arthropoda.[38,39]

Statistical Methods

We used 1989, 1993, and 1995 captures at all wetlands to describe the temporal and spatial distribution, and the relative abundance of major taxa. The1989 captures were additionally used to comprehensively describe the diversity of minor taxa within the Central Puget Sound Basin because in that year, with the exception of nonchironomid Nematocera and Brachycera, we identified most Diptera to genus and species. Brachycera were classified to family in 1993 and 1995.

We counted all invertebrates and determined per square meter yearly abundance (m^{-2}/yr). We calculated proportional abundance of a taxon at wetlands as the number of animals in a given taxon divided by the total number of individuals within all taxa of the grouping. Wetlands were categorized as being permanently or seasonally flooded, and within watersheds of low, medium, and high urbanization as a framework for our findings. These classifications were used in our initial attempt to explore

TABLE 4-1
Substrate and Vegetation Found at Emergence Traps

	Flow		Substrate				Vegetation																					
Wetland	Still Water	Discernible Flow	Gravel	Sand	Silt/mud	Moss	Periphyton	*Alnus*	*Athyrium*	*Carex*	*Epilobium*	*Equisetum*	*Glyceria*	*Juncus*	*Lemna*	*Lysichitum*	*Nuphar*	*Oenanthe*	*Phalaris*	*Potentilla*	*Salix*	*Scirpus*	*Solanum*	*Sparganium*	*Spirea*	*Typha*	*Veronica*	Unidentified Plants
AL3	X					X																						
B31		X	X	X	X			X		X		X							X			X						
BBC24	X																					X				X	X	
ELS39	X				X		X			X				X											X			
ELS61	X						X								X											X		
ELW1		X			X	X			X				X						X								X	
FC1		X			X		X						X		X				X									
HC13	X				X			X					X						X			X						
JC28	X				X			X								X												
LCR93		X															X	X							X	X		X
LPS9		X																X	X									
MGR36	X												X		X					X	X					X	X	
NFIC12	X				X	X	X							X														
PC12	X									X			X		X				X			X			X	X		
RR5	X									X				X								X						
SC4	X				X	X	X	X			X																X	
SC84	X					X	X											X					X				X	
SR24	X									X														X				
TC13	X				X	X	X											X				X				X	X	

Source: Ludwa, K.A., M.S. thesis, University of Washington, Seattle, 1994.

relationships between invertebrate distributions and wetland conditions hypothesized to account for our findings. PC-ORD was used to run summary community statistics and several ordination routines.[41,42] For summary statistics S = Richness and represents the total identified taxa, E = Evenness and is calculated as H/ln (Richness), and H′ = Diversity expressed as $-S(p_i \times \ln(P_i))$.[62] We chose Non-Metric Multidimensional Scaling (NMDS) using Sorensen's distance measure as our ordination method. Finally, we carried out Multi-Response Permutation Procedures (MRPP) to test differences in the aquatic and semiaquatic insect community between wetlands of differing flooding regimes and between wetlands within watersheds exhibiting different levels of urbanization.

RESULTS

We captured five orders of mostly aquatic and semiaquatic insects including Ephemeroptera (mayflies), Odonata (dragonflies/damselflies), Plecoptera (stoneflies), Trichoptera (mayflies), and Diptera (true flies) (Table 4-2). We also captured terrestrial representatives of eight additional insect orders (Coleoptera (beetles), Collembola (springtails), Hemiptera (suborder Homoptera, hoppers), Hymenoptera (wasps/bees), Lepidoptera (butterflies/moths), Neuroptera (spongilla flies/lacewings), Psocoptera (barklice), and Thysanoptera (thrips), as well as members of the arthropod class Arachnida (spiders/mites) (Table 4-3).

RICHNESS AND ABUNDANCE OF AQUATIC AND SEMIAQUATIC INSECTS

We identified a total of 115 aquatic and semiaquatic insect taxa among 19 wetlands during 1989 alone. Richness ranged from a high of 51 taxa at BBC24 to a low of 9 at both NFIC12 and ELS39 (Table 4-4). Among the wetlands with lower richness (<15 species), ELW1 exhibited the highest evenness, meaning that species identified tended to be equally abundant (E = 0.871). Conversely, of the wetlands with relatively high insect diversity, MGR36 with 28 taxa exhibited the second highest evenness value of 0.803. Lowest evenness among all wetlands, of 0.174, was found at ELS39 with only 9 taxa. Highest insect diversity was at BBC24 with an H′ of 2.788 and ELS 39 with the lowest value of 0.382.

Ephemeroptera were captured at a total of 10 wetlands (53%) during our study. They were found at 9 wetlands in 1989, 0 wetlands in 1993, and 4 wetlands in 1995 (Table 4-2). Strikingly, *Callibaetis* and *Paraleptophlebia*, the only two genera identified, were not captured at the same wetland except at PC12 (Table 4-5a). Overall Ephemeroptera abundance was low (<25) in both permanently and seasonally flooded wetlands except for Paraleptophlebia at LCR93 and JC28 where maximum numbers in 1989 of 232 m^{-2}/yr and 105 m^{-2}/yr were identified, respectively (Table 4-5b). At their highest abundance in 1989, Ephemeroptera represented only 2.4% of total insect captures. They exhibited an average density of 11.21 m^{-2}/yr during our 3-year census across all wetlands.

Ephemeroptera richness is greatest at permanently flooded wetlands but their abundance is highest at seasonally flooded wetlands. Moreover, Ephemeroptera are

TABLE 4-2
Total and Proportional Abundance (m^{-2}/yr) of Major Aquatic and Semiaquatic Insect Taxa[a]

		Insects Only										
		Ephemeroptera		Odonata		Plecoptera		Trichoptera		Diptera		
Wetland	Year	No.	%	No.	%	No.	%	No.	%	No.	%	Total
AL3	1989	0		0		0		9	0.20	4399	99.80	4408
	1993	0		0		0		0		2134	100.00	2134
	1995	1	0.08	0		0		36	2.72	1286	97.20	1323
B3I	1989	0		0		0		1	0.03	3035	99.97	3036
	1993	0		0		0		0		2360	100.00	2360
	1995	0		0		0		0		735	100.00	735
BBC24	1989	3	0.03	24	0.27	1	0.01	132	1.49	8698	98.19	8858
	1993	0		0		0		0		7515	100.00	7515
	1995	1	0.03	0		0		15	0.50	3004	99.47	3020
ELS39	1989	0		0		0		9	0.12	7328	99.88	7337
	1993	0		0		0		0		4229	100.00	4229
	1995	0		0		0		5	0.16	3199	99.84	3204
ELS61	1989	19	0.09	1	0.00	0		27	0.13	20781	99.77	20828
	1993	0		0		0		0		10844	100.00	10844
	1995	0		0		7	0.43	5	0.31	1600	99.26	1612
ELW1	1989	0		0		0		0		1238	100.00	1238
	1993	0		0		0		0		339	100.00	339
	1995	0		0		0		0		256	100.00	256
FC1	1989	0		0		0		1	0.02	4734	99.98	4735
	1993	0		0		0		0		6767	100.00	6767
	1995	0		0		0		5	0.17	2894	99.83	2899
HC13	1989	0		0		5	0.06	40	0.46	8708	99.49	8753
	1993	0		0		0		0		2272	100.00	2272
	1995	0		0		1	0.07	21	1.38	1500	98.55	1522
JC28	1989	105	9.26	0		32	2.82	3	0.26	994	87.65	1134
	1993	0		0		0		0		2900	100.00	2900
	1995	206	29.34	0		7	1.00	24	3.42	465	66.24	702
LCR93	1989	232	2.39	0		1575	16.26	61	0.63	7821	80.72	9689
	1993	0		0		0		0		6234	100.00	6234
	1995	0		0		101	4.49	9	0.40	2141	95.11	2251
LPS9	1989	0		0		0		3	0.06	5124	99.94	5127
	1993	0		0		0		1	0.09	1075	99.91	1076
	1995	0		0		0		0		21501	100.00	21501
MGR36	1989	7	0.10	0		0		35	0.48	7324	99.43	7366
	1993	0		0		0		1	0.01	6698	99.99	6699
	1995	8	0.27	2	0.07	0		231	7.79	2723	91.87	2964
NFIC12	1989	0		0		0		7	0.08	8863	99.92	8870
	1993	0		0		0		0		13047	100.00	13047
	1995	no data		no data		no data		no data		no data		no data
PC12	1989	11	0.19	0		0		36	0.61	5845	99.20	5892
	1993	0		0		0		0		5683	100.00	5683
	1995	0		0		3	0.19	23	1.43	1580	98`.38	1606
RR5	1989	3	0.03	0		0		28	0.32	8591	99.64	8622
	1993	0		0		0		1	0.04	2412	99.96	2413
	1995	0		0		0		13	0.41	3146	99.59	3159

TABLE 4-2 *(continued)*
Total and Proportional Abundance (m^{-2}/yr) of Major Aquatic and Semiaquatic Insect Taxa[a]

		Insects Only										
		Ephemeroptera		Odonata		Plecoptera		Trichoptera		Diptera		
Wetland	Year	No.	%	No.	%	No.	%	No.	%	No.	%	Total
SC4	1989	0		0		5	0.17	12	0.41	2935	99.42	2952
	1993	0		0		0		0		2186	100.00	2186
	1995	0		0		42	2.03	5	0.24	2020	97.73	2067
SC84	1989	3	0.08	0		0		0		3689	99.92	3692
	1993	0		0		0		0		1106	100.00	1106
	1995	0		0		0		5	0.30	1684	99.70	1689
SR24	1989	24	0.43	0		0		27	0.48	5547	99.09	5598
	1993	0		0		0		0		2506	100.00	2506
	1995	0		0		0		1	0.15	661	99.85	662
TC13	1989	0		0		0		1	0.02	4656	99.98	4657
	1993	0		0		0		0		2116	100.00	2116
	1995	0		0		0		1	0.15	661	99.85	662

[a] Data collected in 1989, 1993, and 1995 from three emergence traps in each of 19 wetlands in King County, Washington.

patchily distributed in wetlands along a gradient in watersheds ranging from low to high urbanization.

Odonata (Zygoptera) were captured at three wetlands in very low numbers during our surveys (Table 4-2). In 1989 they represented 0.3% of total insect production. Three-year average captures are 0.43 animals m^{-2}/yr. We identified three species from three genera representing one family (i.e., Coenagrionidae: *Ischnura cervula*, *Enallagma boreale*, and *Coenagrion* sp.). Collectively, these were captured at only two wetlands, and in very low numbers (1 at ELS61 and 24 at BBC24) (Table 4-5a). Interestingly, all three species were found at BBC24, whereas only *Ischnura cervula* was captured at ELS61. As expected, these species were captured within permanently flooded wetlands as they require year-round standing water for development.

Plecoptera, generally considered a lotic water insect order, were encountered at seven wetlands (Table 4-2). The 1989 captures represented eight taxa including a new species of *Capnia* (Table 4-5a,b). Taxa richness was highest at LCR93, where we identified all eight species. *Podmosta delicatula* was numerically dominant (1325) representing 84% of all Plecoptera at this wetland. Overall we found Plecoptera in relatively large numbers of 1575 m^{-2}/yr at LCR93 in 1989, moderate numbers of 101 m^{-2}/yr at LCR93 in 1995, and low numbers of 42 and 32 m^{-2}/yr at SC4 and JC28 in 1995 and 1989, respectively (Table 4-2). At other wetlands and years, they were collected in very low number (<10 m^{-2}/yr). Their greatest abundance at LCR93 in 1989 never exceeded 16% of yearly total insect production. Plecopterans were found within both permanently and seasonally flooded wetlands in watersheds with all levels of urbanization.

TABLE 4-3
Total Abundance (m^{-2}/yr) of Terrestrial Arthropod Taxa from 19 Wetlands in King County, Washington

			Insecta														Total Terrestrial
										Hymenoptera				Homoptera			
Wetland	Year		Collembola	Thysanoptera	Psocoptera	Hemiptera	Neuroptera	Coleoptera	Lepidoptera	Total	Other	Parasitoid	Formicidae	Total	Other	Aphididae	Arthropods
AL3	1989	7	3	0	11	0	0	11	2	5	0	5	0	2	1	1	64
	1993	3	0	0	0	0	0	0	1	5	0	5	0	0	0	0	20
	1995	1	4	1	8	0	0	1	1	13	0	13	0	0	0	0	57
B3I	1989	35	124	13	41	9	9	11	7	216	0	212	4	123	10	113	1047
	1993	4	1	4	20	1	3	12	7	61	7	51	3	110	5	105	367
	1995	3	37	1	20	0	0	0	0	51	0	51	0	165	1	163	379
BBC24	1989	33	179	17	144	0	7	56	7	173	1	171	1	55822	25	55797	56854
	1993	5	20	0	3	0	1	88	3	49	0	49	0	3	1	1	362
	1995	14	68	1	17	0	0	119	8	59	0	59	0	3	2	1	534
ELS39	1989	67	1043	33	4	1	0	40	8	514	4	505	5	289	156	133	3075
	1993	5	13	1	7	1	0	1	1	56	0	55	1	36	9	27	235
	1995	11	144	0	77	0	0	40	3	94	2	92	0	140	8	132	740
ELS61	1989	19	223	7	68	3	4	143	5	337	12	325	0	25377	164	25213	27012
	1993	12	72	4	0	3	0	37	24	235	1	234	0	210	13	197	1128
	1995	13	337	4	4	0	0	11	9	44	1	43	0	31	23	8	561
ELW1	1989	19	131	0	47	12	0	1	1	81	1	79	1	44	27	17	500
	1993	13	4	0	24	0	0	8	0	21	0	21	0	3	1	1	123
	1995	8	36	3	15	1	0	11	1	14	1	13	0	1	1	0	130

TABLE 4-3 *(continued)*
Total Abundance (m^{-2}/yr) of Terrestrial Arthropod Taxa from 19 Wetlands in King County, Washington

			Insecta														Total Terrestrial
										Hymenoptera				Homoptera			
Wetland	Year		Collembola	Thysanoptera	Psocoptera	Hemiptera	Neuroptera	Coleoptera	Lepidoptera	Total	Other	Parasitoid	Formicidae	Total	Other	Aphididae	Arthropods
FC1	1989	73	768	41	0	0	1	15	0	175	0	172	3	458	142	316	1897
	1993	20	7	5	9	0	0	8	0	17	0	17	0	48	25	23	156
	1995	9	11	5	70	0	0	3	1	12	0	12	0	1	1	0	140
HC13	1989	31	23	1	27	11	1	13	1	165	1	161	3	35	22	13	653
	1993	9	3	4	4	0	0	16	0	9	0	9	0	68	0	68	147
	1995	5	3	1	39	1	0	3	0	16	0	16	0	1	0	1	104
JC28	1989	11	32	1	20	4	0	1	0	21	0	21	0	6	1	5	139
	1993	1	1	0	0	0	3	0	0	3	0	3	0	0	0	0	17
	1995	3	7	1	11	1	0	3	1	0	0	0	0	1	1	0	32
LCR93	1989	88	21	57	129	15	3	101	1	442	0	439	3	4219	164	4055	6065
	1993	11	0	5	1	0	0	9	0	11	0	11	0	8	3	5	76
	1995	12	5	1	68	1	0	17	1	47	0	47	0	16	11	5	280
LPS9	1989	57	197	15	4	5	5	55	1	388	8	380	0	2140	56	2084	3704
	1993	20	15	0	0	1	0	5	0	94	5	89	0	18	5	13	346
	1995	16	19	0	0	0	2	9	1	69	1	68	0	100	0	100	366

MGR36	1989	49	85	36	41	5	5	51	1	147	0	145	1	7646	39	7607	8416
	1993	43	5	4	5	1	0	7	0	174	0	174	0	4	1	3	598
	1995	4	13	2	3	6	0	6	0	28	0	28	0	0	0	0	124
NFIC12	1989	7	8	5	11	0	0	3	3	27	3	24	0	10	1	9	134
	1993	4	11	4	16	5	1	3	0	231	4	227	0	6	1	5	747
	1995	nd	nd	nd	nd	nd	nd	nd	nd	nd	nd	nd	nd	nd	nd	nd	nd
PC12	1989	25	213	13	7	0	0	39	6	103	0	134	1	184	159	25	873
	1993	32	13	8	0	0	0	16	1	37	0	36	1	11	4	7	209
	1995	23	132	11	0	0	0	15	8	97	0	94	3	9	4	5	512
RR5	1989	21	137	15	39	3	0	261	23	228	1	213	13	219	144	75	1685
	1993	1	3	3	1	0	0	19	3	27	0	27	0	4	1	3	137
	1995	3	11	1	3	0	0	7	24	47	3	44	0	55	55	0	276
SC4	1989	64	53	7	1	13	0	40	3	108	4	97	13	19	16	3	573
	1993	1	28	3	3	1	1	7	15	30	1	29	0	15	14	1	187
	1995	21	41	0	1	0	0	17	3	43	0	43	0	0	0	0	232
SC84	1989	12	11	5	3	0	0	1	0	59	7	43	9	24	23	1	234
	1993	4	0	0	10	2	0	0	0	7	0	7	0	2	2	0	39
	1995	8	12	0	35	0	0	0	0	7	0	7	0	23	23	0	99
SR24	1989	3	31	15	3	0	0	9	0	155	1	152	1	2375	168	2207	2909
	1993	5	0	1	3	0	0	12	0	7	0	7	0	8	5	3	62
	1995	5	15	0	4	0	0	5	0	5	0	5	0	10	10	0	59
TC13	1989	11	17	8	3	0	0	7	0	39	0	39	0	1	0	1	171
	1993	9	3	7	1	0	0	4	0	9	0	9	0	2	1	1	57
	1995	3	8	1	7	0	0	5	0	5	0	5	0	0	0	0	44
Total		996	4401	375	1092	106	46	1382	185	5117	69	5017	66	100105	1549	98553	125687

Note: nd = no data.

TABLE 4-4
Richness (S), Evenness (E), and Diversity (H′) of All Aquatic and Semiaquatic Insect Taxa Identified in 1989 (n = 115 taxa)

	Permanently Flooded				Seasonally Flooded			
Urbanization	**Wetland**	**S**	**E**	**H′**	**Wetland**	**S**	**E**	**H′**
Low	BBC24	51	0.709	2.788	NFIC12	9	0.395	0.867
	HC13	24	0.709	2.253	TC13	13	0.59	1.514
	MGR36	28	0.803	2.677				
	RR5	41	0.739	2.744				
	SR24	33	0.751	2.626				
Moderate	ELW1	15	0.871	2.359	AL3	10	0.613	1.411
	PC12	33	0.603	2.109	LCR93	47	0.569	2.192
	SC84	17	0.482	1.364	SC4	17	0.46	1.303
High	ELS61	32	0.519	1.799	B3I	16	0.663	1.837
	FC1	20	0.635	1.902	ELS39	9	0.174	0.382
					JC28	26	0.642	2.092
					LPS9	10	0.734	1.691

Trichoptera richness is high and distribution widespread; 25 taxa, including one unknown group, were identified during 1989 (Table 4-5a,b). They were captured at 18 of the wetlands (95%). Additionally, they were found in 2 years at 13 of the wetlands (68%) (Table 4-2). Trichoptera were absent from ELW1. Regardless of wetland, the most abundant trichopterans belonged to the family Hydroptilidae (i.e., *Oxyethira* spp. and Limnephilidae i.e., *Lenarchus vastus* and *Limnephilus* spp.) *Oxyethira* is almost exclusively found at BBC24. Members of the Limnephilidae are identified at numerous wetlands. Overall, Trichoptera numbers are low (≤ 132 m^{-2}/yr) with the exception of 1995 at MGR36. Here, 231 animals m^{-2}/yr, representing roughly 8% of the wetland's total yearly insect production, were identified.

Diptera consistently dominated insect emergence in distribution, diversity, and abundance (Table 4-2). Numerically, this insect order most often represented more than 99% of total captures. Unexpectedly, in 1995 they represented an uncommonly low proportional abundance of 66% of all species at JC28 (465 m^{-2}/yr). We identified 14 families of nonchironomid Nematocera, 76 taxa of chironomid Nematocera, and 10 families of Brachycera.

On average, 4531 Diptera m^{-2}/yr were captured across all 19 wetlands during the 3-year monitoring period. Diptera abundance was highest at LPS9 with 21,501 m^{-2}/yr counted in 1995, followed by 20,781 m^{-2}/yr at ELS61 in 1989, and 13,047 m^{-2}/yr at NFIC12 in 1993. Lowest values are recorded at ELW1 with only 256 m^{-2}/yr tallied in 1995 and 339 m^{-2}/yr in 1993 (Table 4-2). Nevertheless, the 1995 low production of 256 m^{-2} /yr at ELW1 is less than five times the 1989 high production of 1238 m^{-2}/yr. Relatively low dipteran numbers of fewer than 1000 m^{-2}/yr in 2 out of 3 years were identified only at JC28. As expected, a significantly greater number of aquatic Nematocera than the mostly terrestrial suborder Brachycera were captured (Table 4-6).

TABLE 4-5a
Total and Proportional Abundance (/m^{-2}) of Plecoptera, Odonata, Ephemeroptera, and Trichoptera in Permanently Flooded Wetlands of Non-, Moderately, and Highly Urbanized Watersheds in 1989

		Wetland																				
		Non Urban										Moderately Urban						Highly Urban				
		BBC24		HC13		MGR36		RR5		SR24		ELW1		PC12		SC84		ELS61		FC1		
	Taxa	No.	%	No.	%	No.	%	No.	%	No.	%	No.	%	No.	%	No.	%	No.	%	No.	%	Total
Plecoptera	Total	1		5		0		0		0		0		0		0		0		0		6
	Capnia nr. oregona																					0
	Malenka																					0
	Ostracerca dimicki																					0
	Paraleuctra? vershina			4	80																	4
	Podmosta delicatula	1	100	1	20																	2
	Soyedina interrupta																					0
	Taenionema																					0
	Zapada cinctipes																					0
Odonata	Total	24		0		0		0		0		0		0		0		1		0		25
	Coenagrion	1	4																			1
	Enallagma boreale	3	13																			3
	Ischnura cervula	20	83															1	100			21
Ephemeroptera	Total	3		0		7		3		24		0		14		3		19		0		73
	Callibaetis	3	100							24	100			13	93	3	100	19	100			62
	Paraleptophlebia					7	100	3	100					1	7							11
Trichoptera	Total	131		40		34		27		26		0		0		0		27		1		286
	Banksiola crotchi							1	4													1
	Clisteronia					16	47															16
	Clostoeca disjuncta																					0
	Glyphopsyche irrorata	1	1																			1

TABLE 4-5a *(continued)*
Total and Proportional Abundance (/m^{-2}) of Plecoptera, Odonata, Ephemeroptera, and Trichoptera in Permanently Flooded Wetlands of Non-, Moderately, and Highly Urbanized Watersheds in 1989

	Wetland																				
	Non Urban										Moderately Urban						Highly Urban				
	BBC24		HC13		MGR36		RR5		SR24		ELW1		PC12		SC84		ELS61		FC1		
Taxa	**No.**	**%**	**No.**	**%**	**No.**	**%**	**No.**	**%**	**No.**	**%**	**No.**	**%**	**No.**	**%**	**No.**	**%**	**No.**	**%**	**No.**	**%**	**Total**
Halesochila taylori							1	4													1
Hydroptila					1	3															1
Limnephilus externus	3	2					8	30	24	92							5	19			40
L. fagus			3	8																	3
L. Fenestratus Gr.																					0
L. harrimani																					0
L. nogus																					0
L. sp. (female)																					0
L. spinatus																					0
L.occidentalis	8	6			17	50											15	56			40
Lenarchus rho							4	15	1	4							3	11			8
L. vastus			36	90			12	44													48
Lepidostoma																			1	100	1
Lepidostoma cinereum																					0
Limnephilus cerus																	4	15			4
Nemotaulius hostilis	1	1																			1
Neophylax			1	3																	1
Oxyethira	112	85					1	4													113
Polycentropus flavus	5	4																			5
Psychoglypha																					0
Ptilostomis ocellifera	1	1							1	4											2

TABLE 4-5b
Total and Proportional Abundance (/m^{-2}) of Plecoptera, Odonata, Ephemeroptera, and Trichoptera in Seasonally Flooded Wetlands of Non-, Moderately, and Highly Urbanized Watersheds in 1989

		Wetland																		
		Non Urban						Moderately Urban						Highly Urban						
		NFIC12		TC13		AL3		LCR93		SC4		B31		ELS39		JC28		LPS9		
	Taxa	No.	%	No.	%	No.	%	No.	%	No.	%	No.	%	No.	%	No.	%	No.	%	Total
Plecoptera	Total	0		0		0		1575		5		0		0		32		0		1612
	Capnia nr. oregona							1	0											1
	Malenka							83	5											83
	Ostracerca dimicki							1	0	4	80									5
	Paraleuctra? vershina							0												0
	Podmosta delicatula							1325	84	1	20					1	3			1327
	Soyedina interrupta							137	9							31	97			168
	Taenionema							27	2											27
	Zapada cinctipes							1	0											1
Odonata	Total	0		0		0		0		0		0		0		0		0		0
	Coenagrion																			0
	Enallagma boreale																			0
	Ischnura cervula																			0
Ephemeroptera	Total	0		0		0		232		0		0		0		105		0		337
	Callibaetis																			0
	Paraleptophlebia							232	100							105	100			337
Trichoptera	Total	7		1		9		61		12		1		10		3		3		107
	Banksiola crotchi																			0
	Clisteronia																			0
	Clostoeca disjuncta							1	2	1	8									2
	Glyphopsyche irrorata																			0

TABLE 4-5b *(continued)*
Total and Proportional Abundance ($/m^{-2}$) of Plecoptera, Odonata, Ephemeroptera, and Trichoptera in Seasonally Flooded Wetlands of Non-, Moderately, and Highly Urbanized Watersheds in 1989

	Wetland																		
	Non Urban						Moderately Urban						Highly Urban						
	NFIC12		TC13		AL3		LCR93		SC4		B31		ELS39		JC28		LPS9		
Taxa	No.	%	No.	%	No.	%	No.	%	No.	%	No.	%	No.	%	No.	%	No.	%	Total
Halesochila taylori																			0
Hydroptila																			0
Limnephilus externus																			0
L. fagus							19	31											19
L. Fenestratus Gr.							7	11											7
L. harrimani							9	15	7	58			3	30					19
L. nogus							4	7	4	33	1	100	3	30			3	100	15
L. sp. (female)							1	2											1
L. spinatus	1	14					8	13											9
L. occidentalis							4	7											4
Lenarchus rho	3	43																	3
L. vastus	3	43			9	100	4	7					4	40					20
Lepidostoma															3	100			3
Lepidostoma cinereum							3	5											3
Limnephilus cerus																			0
Nemotaulius hostilis																			0
Neophylax			1	100															1
Oxyethira																			0
Polycentropus flavus																			0
Psychoglypha							1	2											1
Ptilostomis ocellifera																			0

TABLE 4-6
Total and Proportional Abundance (/m^{-2}) of Major Aquatic and Semiaquatic Diptera Taxa Collected from Three Emergence Traps in Each of 19 Wetlands in King County, Washington

			Nematocera				
		Brachycera	Non-Chironomidae		Chironomidae		
Wetland	Year	Total	No.	%	No.	%	Total
AL3	1989	2428	724	36.73	1247	63.27	1971
	1993	184	584	29.95	1366	70.05	1950
	1995	225	354	33.36	708	66.73	1061
B3I	1989	888	634	29.53	1513	70.47	2147
	1993	1107	616	49.16	637	50.84	1253
	1995	168	460	81.13	108	19.05	567
BBC24	1989	267	203	2.41	8228	97.59	8431
	1993	136	145	1.97	7234	98.03	7379
	1995	94	93	3.20	2816	96.77	2910
ELS39	1989	1016	3552	56.27	2760	43.73	6312
	1993	404	1970	51.50	1855	48.50	3825
	1995	304	2299	79.41	596	20.59	2895
ELS61	1989	5488	3368	22.02	11925	77.98	15293
	1993	3779	2902	41.08	4163	58.92	7065
	1995	133	484	32.99	983	67.01	1467
ELW1	1989	583	485	73.93	171	26.07	656
	1993	55	269	94.72	15	5.28	284
	1995	140	105	90.52	11	9.48	116
FC1	1989	1575	461	14.59	2699	85.41	3160
	1993	73	301	4.50	6393	95.50	6694
	1995	15	35	1.22	2844	98.78	2879
HC13	1989	2169	3525	53.92	3013	46.08	6538
	1993	44	270	12.12	1958	87.88	2228
	1995	33	435	29.65	1032	70.35	1467
JC28	1989	69	169	18.27	756	81.73	925
	1993	7	53	1.83	2840	98.17	2893
	1995	28	134	30.66	303	69.34	437
LCR93	1989	2925	2580	52.70	2316	47.30	4896
	1993	128	193	3.16	5913	96.84	6106
	1995	952	877	73.76	313	26.32	1189
LPS9	1989	1160	3313	83.58	651	16.42	3964
	1993	432	504	78.38	139	21.62	643
	1995	880	1605	7.78	19016	92.22	20621
MGR36	1989	2884	1072	24.14	3368	75.86	4440
	1993	39	226	3.39	6433	96.61	6659
	1995	86	382	14.49	2255	85.51	2637
NFIC12	1989	2984	1127	19.17	4752	80.83	5879
	1993	368	1340	10.57	11340	89.44	12679
	1995	no data	no data	no data	no data	no data	no data

TABLE 4-6 *(continued)*
Total and Proportional Abundance (/m^{-2}) of Major Aquatic and Semiaquatic Diptera Taxa Collected from Three Emergence Traps in Each of 19 Wetlands in King County, Washington

			Nematocera				
		Brachycera	Non-Chironomidae		Chironomidae		
Wetland	Year	Total	No.	%	No.	%	Total
PC12	1989	484	440	8.21	4921	91.79	5361
	1993	288	437	8.10	4958	91.90	5395
	1995	234	944	70.13	402	29.87	1346
RR5	1989	884	500	6.49	7207	93.51	7707
	1993	35	132	5.55	2245	94.45	2377
	1995	902	313	13.95	1931	86.05	2244
SC4	1989	887	1232	60.16	816	39.84	2048
	1993	626	678	43.46	882	56.54	1560
	1995	307	1676	97.84	38	2.22	1713
SC84	1989	1377	449	19.42	1863	80.58	2312
	1993	20	181	16.67	904	83.24	1086
	1995	189	367	24.55	1128	75.45	1495
SR24	1989	467	815	16.04	4265	83.96	5080
	1993	12	82	3.29	2412	96.71	2494
	1995	6	77	12.13	558	87.87	635
TC13	1989	1043	407	11.26	3207	88.74	3614
	1993	9	104	4.94	2003	95.06	2107
	1995	76	96	16.41	489	83.59	585

Nematocera

We identified a high of 62 Nematocera taxa at BBC24 representing a yield of 8431 animals per square meter in 1989 (Tables 4-7a,b and 4-8). The lowest richness of roughly one-third this value was observed at NFIC12, ELS39, and AL3, where 15, 15, and 18 taxa were recorded, respectively. Densities were lowest at 116 m^{-2}/yr nematocerans in 1995 at ELW1 (Table 4-6).

Chironomids

Of the nematoceran Diptera, members of the family Chironomidae (midges) most likely represented the greatest number of taxa and highest densities (Table 4-7a,b) even though we did not identify the nonchironomids to species level. Four subfamilies — Podonominae, Prodiamesinae, Orthocladiinae, and Tanypodinae, and two tribes — Chironomini and Tanytarsini, characterized the chironomid fauna at wetlands. We identified a total of 76 Chironomidae taxa within these subfamilies. The greatest richness of 39 taxa was identified at BBC24 (Table 4-9) and the highest abundance of 19,016 individuals m^{-2}/yr was counted at LPS9 in 1995 (Table 4-6).

TABLE 4-7a
1989 Diptera: Chironomid Nematocera Distribution and Total Abundance (/m^{-2}) in Permanently Flooded Wetlands of Non-, Moderately and Highly Urbanized Watersheds

	Wetland																				
	Non Urban										Moderately Urban						Highly Urban				
	BBC24		HC13		MGR36		RR5		SR24		ELW1		PC12		SC84		ELS61		FC1		
Species	No.	%	No.	%	No.	%	No.	%	No.	%	No.	%	No.	%	No.	%	No.	%	No.	%	Total
Total Podonominae	0		8		145		0		0		0		0		0		0		0		153
Genus *Boreochlus*		0.00	8	100.00	145	100.00		0.00		0.00		0.00		0.00		0.00		0.00		0.00	153
Total Prodiamesinae	0		0		0		0		0		0		0		0		0		0		0
Genus *Odontomesa*		0.00		0.00		0.00		0.00		0.00		0.00		0.00		0.00		0.00		0.00	0
Prodiamesa		0.00		0.00		0.00		0.00		0.00		0.00		0.00		0.00		0.00		0.00	0
Total Orthocladiinae	301		2345		996		5884		1586		104		741		1220		10796		1564		25537
Unidentified Orthocladiinae	119	39.53	1637	69.81	660	66.27	4685	79.62	882	55.61	78	75.00	529	71.39	721	59.10	8433	78.11	1395	89.19	19139
Genus *Brillia*	5	1.66		0.00	15	1.51		0.00		0.00	9	8.65	1	0.13		0.00	5	0.05	1	0.07	36
Chaetocladius		0.00		0.00		0.00		0.00		0.00		0.00		0.00		0.00		0.00		0.00	0
Corynoneura	40	13.29	35	1.49	161	16.16	40	0.68	523	32.98		0.00	83	11.20		0.00	11	0.10	24	1.53	917
Cricotopus	25	8.31		0.00		0.00		0.00	65	4.10	1	0.96	24	3.24		0.00		0.00		0.00	115
Doithrix		0.00		0.00		0.00		0.00		0.00		0.00		0.00		0.00		0.00	4	0.26	4
Limnophyes	12	3.99	261	11.13	103	10.34	531	9.02	36	2.27	7	6.73	51	6.88	63	5.16	799	7.40	133	8.50	1996
Mesosmittia		0.00		0.00		0.00		0.00		0.00		0.00		0.00	1	0.08		0.00		0.00	1
Metriocnemus	1	0.33		0.00		0.00		0.00	17	1.07	4	3.85		0.00		0.00	13	0.12	3	0.19	38
Nanocladius	8	2.66		0.00		0.00	43	0.73		0.00		0.00		0.00		0.00		0.00		0.00	51
Orthocladius	21	6.98	1	0.04		0.00	83	1.41	36	2.27		0.00		0.00	3	0.25		0.00		0.00	144
Parakiefferiella		0.00		0.00		0.00		0.00		0.00	1	0.96		0.00		0.00		0.00		0.00	1
Parametriocnemus		0.00	24	1.02		0.00	188	3.20	16	1.01	1	0.96	3	0.40		0.00	1439	13.33		0.00	1671
Paraphaenocladius		0.00		0.00	17	1.71		0.00		0.00		0.00		0.00		0.00		0.00		0.00	17
Poryophaenocladius	5	1.66		0.00		0.00		0.00		0.00		0.00		0.00		0.00		0.00		0.00	5
Psectrocladius	21	6.98		0.00	1	0.10	285	4.84	11	0.69		0.00	32	4.32		0.00	15	0.14		0.00	365

TABLE 4-7a *(continued)*
1989 Diptera: Chironomid Nematocera Distribution and Total Abundance (/m^{-2}) in Permanently Flooded Wetlands of Non-, Moderately and Highly Urbanized Watersheds

		Wetland																				
		Non Urban										Moderately Urban						Highly Urban				
		BBC24		HC13		MGR36		RR5		SR24		ELW1		PC12		SC84		ELS61		FC1		
	Species	No.	%	No.	%	No.	%	No.	%	No.	%	No.	%	No.	%	No.	%	No.	%	No.	%	Total
	Pseudosmittia		0.00	273	11.64		0.00	29	0.49		0.00		0.00		0.00	425	34.84	71	0.66	4	0.26	802
	Rheocricotopus		0.00	3	0.13		0.00		0.00		0.00		0.00	1	0.13	1	0.08	1	0.01		0.00	6
	Smittia		0.00	111	4.73		0.00		0.00		0.00		0.00		0.00	5	0.41	9	0.08		0.00	125
	Thienemanniella	44	14.62		0.00	39	3.92		0.00		0.00		0.00	17	2.29	1	0.08		0.00		0.00	101
Species	*Cricotopus bifurcatus*		0.00		0.00		0.00		0.00		0.00	3	2.88		0.00		0.00		0.00		0.00	3
Total Chironomini		3481		463		487		856		927		39		1667		527		718		857		10022
	Unidentified Chironomini	1420	40.79	232	50.11	253	51.95	400	46.73	541	58.36	23	58.97	680	40.79	220	41.75	381	53.06	528	61.61	4678
Genus	Unk. Chironomini genus		0.00		0.00		0.00		0.00		0.00		0.00		0.00		0.00		0.00		0.00	0
	Dicrotendipes	47	1.35		0.00		0.00	4	0.47	8	0.86		0.00	11	0.66	3	0.57	27	3.76		0.00	100
	Glyptotendipes	104	2.99		0.00		0.00	1	0.12	79	8.52		0.00	5	0.30		0.00		0.00		0.00	189
	Paratendipes		0.00		0.00		0.00		0.00		0.00		0.00		0.00		0.00		0.00		0.00	0
	Polypedilum gr.1	864	24.82	95	20.52	105	21.56	109	12.73	128	13.81	4	10.26	851	51.05	19	3.61	128	17.83	55	6.42	2358
	Polypedilum gr. 2	419	12.04		0.00		0.00	71	8.29	65	7.01		0.00	61	3.66		0.00	72	10.03		0.00	688
	Stictochironomus		0.00		0.00		0.00	48	5.61		0.00		0.00		0.00		0.00		0.00	3	0.35	51
	Xestochironomus	5	0.14		0.00		0.00		0.00		0.00		0.00		0.00		0.00		0.00		0.00	5
Species	*Chironomus decorus gr.*	29	0.83	87	18.79	111	22.79	64	7.48	19	2.05		0.00	13	0.78	281	53.32	16	2.23	267	31.16	887
	Chironomus riparius		0.00		0.00		0.00		0.00		0.00		0.00		0.00		0.00		0.00		0.00	0
	Cladopelma viridula	68	1.95		0.00		0.00	20	2.34	7	0.76		0.00	1	0.06		0.00		0.00		0.00	96
	Demicryptochironomus nr. fas		0.00		0.00		0.00		0.00	5	0.54		0.00		0.00		0.00		0.00		0.00	5
	Endochironomus nigricans	19	0.55		0.00		0.00	5	0.58	56	6.04		0.00	24	1.44	1	0.19	71	9.89		0.00	176

	Endochironomus subtendens		0.00		0.00		0.00		0.00		0.00		0.00	4	0.24		0.00		0.00		0.00	4
	Kiefferulus dux		0.00	48	10.37		0.00		0.00		0.00		0.00		0.00		0.00		0.00		0.00	48
	Microtendipes pedellus var *p*	113	3.25		0.00	12	2.46		0.00		0.00		0.00		0.00		0.00		0.00		0.00	125
	Microtendipes pedellus var *s*	343	9.85		0.00		0.00	28	3.27		0.00		0.00		0.00		0.00		0.00		0.00	371
	Parachironomus cf. forceps		0.00		0.00		0.00	4	0.47		0.00		0.00		0.00		0.00		0.00		0.00	4
	Parachironomus monochromus		0.00		0.00		0.00		0.00		0.00		0.00		0.00		0.00		0.00	1	0.12	1
	Parachironomus sp. 1		0.00		0.00		0.00	44	5.14	4	0.43		0.00		0.00		0.00		0.00		0.00	48
	Parachironomus sp.2	7	0.20		0.00		0.00	47	5.49	15	1.62		0.00		0.00		0.00	1	0.14		0.00	70
	Paratendipes albimanus	8	0.23		0.00	3	0.62		0.00		0.00		0.00		0.00		0.00		0.00		0.00	11
	Phaenopsectra flavipes	5	0.14		0.00	3	0.62	7	0.82		0.00		0.00	17	1.02	3	0.57		0.00	3	0.35	38
	Phaenopsectra punctimes	4	0.11	1	0.22		0.00		0.00		0.00		0.00		0.00		0.00	3	0.42		0.00	8
	Polypedilum cf. simulans		0.00		0.00		0.00		0.00		0.00	1	2.56		0.00		0.00		0.00		0.00	1
	Polypedilum illinoense	13	0.37		0.00		0.00	4	0.47		0.00	11	28.21		0.00		0.00	19	2.65		0.00	47
	Polypedilum ophioides	13	0.37		0.00		0.00		0.00		0.00		0.00		0.00		0.00		0.00		0.00	13
Total Tanypodinae		859		31		874		254		703		13		1449		13		275		156		4627
Genus	*Ablabesmyia*	275	32.01		0.00	57	6.52	33	12.99	335	47.65		0.00	35	2.42		0.00	68	24.73	11	7.05	814
	Djalmabatista		0.00		0.00		0.00	3	1.18		0.00		0.00		0.00		0.00		0.00		0.00	3
	Labrundinia	75	8.73		0.00		0.00		0.00		0.00		0.00		0.00		0.00		0.00		0.00	75
	Larsia		0.00	1	3.23		0.00	51	20.08	36	5.12		0.00	5	0.35		0.00		0.00	3	1.92	96
	Macropelopiini		0.00		0.00	3	0.34		0.00		0.00		0.00		0.00		0.00		0.00		0.00	3
	Pentaneurini	19	2.21	13	41.94	283	32.38	64	25.20	75	10.67	1	7.69	33	2.28		0.00	7	2.55	9	5.77	504
Species	*Apsectrotanypus algens*	80	9.31		0.00	129	14.76	3	1.18		0.00		0.00		0.00		0.00		0.00		0.00	212
	Conchapelopia cf. currani	51	5.94		0.00	15	1.72	1	0.39		0.00		0.00	1	0.07		0.00		0.00		0.00	68
	Conchapelopia dusena		0.00	1	3.23		0.00		0.00		0.00		0.00		0.00		0.00		0.00	8	5.13	9
	Hayesomyia senata		0.00		0.00		0.00		0.00		0.00	1	7.69		0.00		0.00		0.00		0.00	1
	Meropelopia nr. americana		0.00		0.00		0.00		0.00		0.00	7	53.85		0.00		0.00		0.00		0.00	7

TABLE 4-7a *(continued)*
1989 Diptera: Chironomid Nematocera Distribution and Total Abundance (/m^{-2}) in Permanently Flooded Wetlands of Non-, Moderately and Highly Urbanized Watersheds

	Wetland																				
	Non Urban										Moderately Urban						Highly Urban				
	BBC24		HC13		MGR36		RR5		SR24		ELW1		PC12		SC84		ELS61		FC1		
Species	No.	%	No.	%	No.	%	No.	%	No.	%	No.	%	No.	%	No.	%	No.	%	No.	%	Total
Natarsia miripes		0.00		0.00		0.00		0.00		0.00		0.00		0.00		0.00		0.00		0.00	0
Paramerina smithae	21	2.44		0.00	161	18.42	7	2.76	12	1.71		0.00	3	0.21		0.00	4	1.45		0.00	208
Procladius bellus		0.00		0.00		0.00	59	23.23	8	1.14		0.00	5	0.35		0.00		0.00	1	0.64	73
Procladius nr. sp.?		0.00		0.00		0.00		0.00	1	0.14		0.00		0.00		0.00		0.00		0.00	1
Procladius nr. freemani	96	11.18		0.00		0.00	31	12.20	88	12.52		0.00	661	45.62		0.00	3	1.09		0.00	879
Procladius nr. sublettei	141	16.41		0.00	16	1.83	1	0.39	11	1.56		0.00	11	0.76		0.00		0.00		0.00	180
Psectrotanypus dyari	96	11.18	11	35.48	77	8.81	1	0.39	137	19.49	4	30.77	656	45.27	12	92.31	28	10.18	111	71.15	1133
Tanypus cf. parastellatus		0.00		0.00		0.00		0.00		0.00		0.00	4	0.28		0.00		0.00		0.00	4
Zavrelimyia fastuosa		0.00		0.00		0.00		0.00		0.00		0.00		0.00		0.00		0.00		0.00	0
Zavrelimyia sinuosa		0.00		0.00		0.00		0.00		0.00		0.00		0.00		0.00	61	22.18		0.00	61
Zavrelimyia thryptica	5	0.58	5	16.13	133	15.22		0.00		0.00		0.00	35	2.42	1	7.69	104	37.82	13	8.33	296
Total Tanytarsini	3568		168		824		212		1008		13		1062		96		136		121		7208
Unidentified Tanytarsini	2552	71.52	76	45.24	407	49.39	141	66.51	827	82.04	12	92.31	701	66.01	51	53.13	108	79.41	77	63.64	4952
Genus *Micropsectra gr.1*	67	1.88	75	44.64	384	46.60		0.00	19	1.88	1	7.69	257	24.20	4	4.17	12	8.82	44	36.36	863
Micropsectra gr.2	149	4.18	17	10.12	16	1.94	19	8.96	1	0.10		0.00	84	7.91	41	42.71	16	11.76		0.00	343
Rheotanytarsus		0.00		0.00		0.00		0.00		0.00		0.00		0.00		0.00		0.00		0.00	0
Tanytarsus	800	22.42		0.00	17	2.06	52	24.53	161	15.97		0.00	20	1.88		0.00		0.00		0.00	1050

Note: gr. — In a group of taxa where it is difficult to separate based on current known characteristics. nr. — Organism is very near or could be the name identified, but some uncertainty exists. cf. — Organism looks similar to the name identified, but greater uncertainty exists with the name. ? — Very uncertain of stated name.

TABLE 4-7b
1989 Diptera: Chironomid Nematocera Distribution and Total Abundance (/m^{-2}) in Seasonally Flooded Wetlands of Non-, Moderately, and Highly Urbanized Watersheds

	Wetland																		
	Non Urban				Moderately Urban						Highly Urban								
	NFIC12		TC13		AL3		LCR93		SC4		B31		ELS39		JC28		LPS9		
Species	No.	%	No.	%	No.	%	No.	%	No.	%	No.	%	No.	%	No.	%	No.	%	Total
Total Podonominae	0		0		0		0		3		0		0		0		0		3
Genus *Boreochlus*		0.00		0.00		0.00		0.00	3	100.00		0.00		0.00		0.00		0.00	3
Total Prodiamesinae		0.00		0.00		0.00		0.00		0.00	176	0.00		0.00	4	0.00		0.00	180
Genus *Odontomesa*		0.00		0.00		0.00		0.00		0.00	7	3.98		0.00	4	100.00		0.00	11
Prodiamesa		0.00		0.00		0.00		0.00		0.00	169	96.02		0.00		0.00		0.00	169
Total Orthocladiinae	4212		2271		852		1643		778		1207		2740		531		644		14878
Unidentified Orthocladiinae	3240	76.92	1688	74.33	716	84.04	1136	69.14	588	75.58	980	81.19	2150	78.47	329	61.96	442	68.63	11269
Genus *Brillia*		0.00		0.00		0.00	9	0.55		0.00	73	6.05		0.00	7	1.32		0.00	89
Chaetocladius		0.00		0.00		0.00		0.00		0.00	1	0.08		0.00		0.00		0.00	1
Corynoneura		0.00	1	0.04		0.00	37	2.25		0.00		0.00		0.00	17	3.20		0.00	55
Cricotopus		0.00		0.00		0.00		0.00		0.00	63	5.22		0.00		0.00		0.00	63
Doithrix		0.00		0.00		0.00		0.00	1	0.13		0.00		0.00		0.00		0.00	1
Limnophyes	849	20.16	167	7.35	124	14.55	289	17.59	160	20.57	12	0.99	574	20.95	152	28.63	67	10.40	2394
Mesosmittia		0.00		0.00		0.00		0.00		0.00		0.00		0.00		0.00		0.00	0
Metriocnemus		0.00		0.00		0.00	13	0.79	1	0.13	75	6.21	4	0.15		0.00	21	3.26	114
Nanocladius		0.00		0.00		0.00	39	2.37		0.00		0.00		0.00	1	0.19		0.00	40
Orthocladius		0.00		0.00		0.00	29	1.77		0.00		0.00		0.00	8	1.51		0.00	37
Parakiefferiella		0.00		0.00		0.00	3	0.18	3	0.39		0.00		0.00		0.00		0.00	6
Parametriocnemus		0.00	7	0.31		0.00	29	1.77	1	0.13	1	0.08		0.00	11	2.07	7	1.09	56
Paraphaenocladius	1	0.02		0.00		0.00		0.00		0.00	1	0.08		0.00		0.00		0.00	2
Poryophaenocladius		0.00		0.00		0.00		0.00		0.00		0.00		0.00		0.00		0.00	0
Psectrocladius		0.00	15	0.66		0.00		0.00		0.00		0.00		0.00		0.00		0.00	15

TABLE 4-7b *(continued)*
1989 Diptera: Chironomid Nematocera Distribution and Total Abundance (/m^{-2}) in Seasonally Flooded Wetlands of Non-, Moderately, and Highly Urbanized Watersheds

		Wetland																		
		Non Urban				Moderately Urban						Highly Urban								
		NFIC12		TC13		AL3		LCR93		SC4		B31		ELS39		JC28		LPS9		
	Species	No.	%	No.	%	No.	%	No.	%	No.	%	No.	%	No.	%	No.	%	No.	%	Total
	Pseudosmittia	1	0.02	385	16.95		0.00	4	0.24	7	0.90	1	0.08	8	0.29	5	0.94	9	1.40	420
	Rheocricotopus		0.00		0.00		0.00	29	1.77		0.00		0.00		0.00	1	0.19	43	6.68	73
	Smittia	121	2.87		0.00	12	1.41	19	1.16	17	2.19		0.00	4	0.15		0.00	55	8.54	228
	Thienemanniella		0.00	8	0.35		0.00	7	0.43		0.00		0.00		0.00		0.00		0.00	15
Species	*Cricotopus bifurcatus*		0.00		0.00		0.00		0.00		0.00		0.00		0.00		0.00		0.00	0
Total Chironomini		539		584		253		165		29		5		8		80		8		1671
	Unidentified Chironomini	336	62.34	347	59.42	184	72.73	116	70.30	17	58.62	4	80.00	4	50.00	43	53.75	3	37.50	1054
Genus	Unk. Chironomini genus		0.00		0.00		0.00	4	2.42		0.00		0.00		0.00		0.00	1	12.50	5
	Dicrotendipes		0.00		0.00		0.00		0.00		0.00		0.00		0.00		0.00		0.00	0
	Glyptotendipes		0.00		0.00		0.00		0.00		0.00		0.00		0.00		0.00		0.00	0
	Paratendipes		0.00		0.00		0.00	8	4.85		0.00		0.00		0.00	21	26.25		0.00	29
	Polypedilum gr. 1	15	2.78	7	1.20	16	6.32	28	16.97	4	13.79		0.00		0.00	13	16.25	1	12.50	84
	Polypedilum gr. 2		0.00	1	0.17		0.00		0.00		0.00		0.00		0.00		0.00	3	37.50	4
	Stictochironomus		0.00		0.00		0.00		0.00		0.00		0.00		0.00	1	1.25		0.00	1
	Xestochironomus		0.00		0.00		0.00		0.00		0.00		0.00		0.00		0.00		0.00	0
Species	*Chironomus decorus gr.*	188	34.88	229	39.21	52	20.55	1	0.61	8	27.59		0.00	4	50.00	1	1.25		0.00	483
	Chironomus riparius		0.00		0.00	1	0.40		0.00		0.00		0.00		0.00		0.00		0.00	1
	Cladopelma viridula		0.00		0.00		0.00		0.00		0.00	1	20.00		0.00		0.00		0.00	1
	Demicryptochironomus nr. fas		0.00		0.00		0.00		0.00		0.00		0.00		0.00		0.00		0.00	0
	Endochironomus nigricans		0.00		0.00		0.00	3	1.82		0.00		0.00		0.00		0.00		0.00	3
	Endochironomus subtendens		0.00		0.00		0.00		0.00		0.00		0.00		0.00		0.00		0.00	0
	Kiefferulus dux		0.00		0.00		0.00		0.00		0.00		0.00		0.00		0.00		0.00	0

	Microtendipes pedellus var. p		0.00		0.00		0.00		0.00		0.00		0.00		0.00		0.00		0.00	0
	Microtendipes pedellus var. s		0.00		0.00		0.00		0.00		0.00		0.00		0.00		0.00		0.00	0
	Parachironomus cf. forceps		0.00		0.00		0.00		0.00		0.00		0.00		0.00		0.00		0.00	0
	Parachironomus monochromus		0.00		0.00		0.00		0.00		0.00		0.00		0.00		0.00		0.00	0
	Parachironomus sp. 1		0.00		0.00		0.00		0.00		0.00		0.00		0.00		0.00		0.00	0
	Parachironomus sp. 2		0.00		0.00		0.00	5	3.03		0.00		0.00		0.00		0.00		0.00	5
	Paratendipes albimanus		0.00		0.00		0.00		0.00		0.00		0.00		0.00		0.00		0.00	0
	Phaenopsectra flavipes		0.00		0.00		0.00		0.00		0.00		0.00		0.00	1	1.25		0.00	1
	Phaenopsectra punctimes		0.00		0.00		0.00		0.00		0.00		0.00		0.00		0.00		0.00	0
	Polypedilum cf. simulans		0.00		0.00		0.00		0.00		0.00		0.00		0.00		0.00		0.00	0
	Polypedilum illinoense		0.00		0.00		0.00		0.00		0.00		0.00		0.00		0.00		0.00	0
	Polypedilum ophioides		0.00		0.00		0.00		0.00		0.00		0.00		0.00		0.00		0.00	0
Total Tanypodinae		0		345		13		272		4		127		0		131		0		892
Genus	*Ablabesmyia*		0.00		0.00		0.00	1	0.37		0.00		0.00		0.00		0.00		0.00	1
	Djalmabatista		0.00		0.00		0.00		0.00		0.00		0.00		0.00		0.00		0.00	0
	Labrundinia		0.00		0.00		0.00		0.00		0.00		0.00		0.00		0.00		0.00	0
	Larsia		0.00		0.00		0.00		0.00		0.00		0.00		0.00		0.00		0.00	0
	Macropelopiini		0.00		0.00		0.00	113	41.54		0.00	5	3.94		0.00		0.00		0.00	118
	Pentaneurini		0.00		0.00	4	30.77		0.00	3	75.00	7	5.51		0.00	3	2.29		0.00	17
Species	*Apsectrotanypus algens*		0.00		0.00		0.00	15	5.51		0.00		0.00		0.00	117	89.31		0.00	132
	Conchapelopia cf. currani		0.00		0.00		0.00		0.00		0.00	3	2.36		0.00		0.00		0.00	3
	Conchapelopia dusena		0.00		0.00		0.00		0.00		0.00	112	88.19		0.00		0.00		0.00	112
	Hayesomyia senata		0.00		0.00		0.00		0.00		0.00		0.00		0.00		0.00		0.00	0
	Meropelopia nr. americana		0.00		0.00		0.00		0.00		0.00		0.00		0.00		0.00		0.00	0
	Natarsia miripes		0.00		0.00		0.00	111	40.81		0.00		0.00		0.00	3	2.29		0.00	114
	Paramerina smithae		0.00		0.00		0.00	19	6.99		0.00		0.00		0.00	3	2.29		0.00	22
	Procladius bellus		0.00		0.00		0.00		0.00		0.00		0.00		0.00		0.00		0.00	0
	Procladius nr. sp.?		0.00		0.00		0.00		0.00		0.00		0.00		0.00		0.00		0.00	0
	Procladius nr. freemani		0.00		0.00		0.00		0.00		0.00		0.00		0.00		0.00		0.00	0
	Procladius nr. sublettei		0.00		0.00		0.00		0.00		0.00		0.00		0.00		0.00		0.00	0
	Psectrotanypus dyari		0.00	345	100.00	5	38.46	4	1.47		0.00		0.00		0.00		0.00		0.00	354
	Tanypus cf. parastellatus		0.00		0.00		0.00		0.00		0.00		0.00		0.00		0.00		0.00	0

TABLE 4-7b *(continued)*
1989 Diptera: Chironomid Nematocera Distribution and Total Abundance (/m^{-2}) in Seasonally Flooded Wetlands of Non-, Moderately, and Highly Urbanized Watersheds

	Wetland																		
	Non Urban				Moderately Urban						Highly Urban								
	NFIC12		TC13		AL3		LCR93		SC4		B31		ELS39		JC28		LPS9		
Species	**No.**	**%**	**No.**	**%**	**No.**	**%**	**No.**	**%**	**No.**	**%**	**No.**	**%**	**No.**	**%**	**No.**	**%**	**No.**	**%**	**Total**
Zavrelimyia fastuosa		0.00		0.00		0.00		0.00		0.00		0.00		0.00	1	0.76		0.00	1
Zavrelimyia sinuosa		0.00		0.00		0.00		0.00		0.00		0.00		0.00		0.00		0.00	0
Zavrelimyia thryptica		0.00		0.00	4	30.77	9	3.31	1	25.00		0.00		0.00	4	3.05		0.00	18
Total Tanytarsini	0		7		1		234		1		0		12		8		0		263
Unidentified Tanytarsini		0.00	3	42.86		0.00	141	60.26	1	100.00		0.00		0.00	4	50.00		0.00	149
Genus *Micropsectra gr.1*		0.00	1	14.29		0.00	21	8.97		0.00		0.00	12	100.00		0.00		0.00	34
Micropsectra gr.2		0.00	3	42.86		0.00	43	18.38		0.00		0.00		0.00	1	12.50		0.00	47
Rheotanytarsus		0.00		0.00	1	100.00		0.00		0.00		0.00		0.00		0.00		0.00	1
Tanytarsus		0.00		0.00		0.00	29	12.39		0.00		0.00		0.00	3	37.50		0.00	32

Note: gr. — In a group of taxa where it is difficult to separate based on current known characteristics. nr. — Organism is very near or could be the name identified, but some uncertainty exists. cf. — Organism looks similar to the name identified, but greater uncertainty exists with the name. ? — Very uncertain of stated name.

TABLE 4-8
Nonchironomid Nematocera Abundance (/m^{-2}) at 19 Wetlands in King County, Washington in 1989, 1993, and 1995

Wetland	Year	Anisopodidae	Bibionidae	Cecidomyiidae	Ceratopogonidae	Chaoboridae	Culicidae	Dixidae	Mycetophilidae	Phoridae	Psychodidae	Scatopsidae	Sciaridae	Simuliidae	Tipulidae	Other Nonchironomids
AL3	1989	0	0	159	53	5	3	100	52	0	120	0	137	0	95	0
	1993	0	0	13	130	19	75	9	57	1	185	0	15	0	81	0
	1995	0	0	5	25	72	13	19	4	1	21	0	9	0	185	0
B3I	1989	1	0	407	71	0	1	1	8	0	32	4	13	0	84	1
	1993	1	0	246	1	0	0	5	76	61	16	0	21	0	249	0
	1995	0	0	301	3	0	0	0	9	35	5	0	19	0	122	0
BBC24	1989	0	0	11	121	11	1	33	0	0	1	0	9	0	0	15
	1993	0	0	3	25	35	39	28	0	7	8	0	5	0	3	0
	1995	0	0	5	15	33	8	7	4	3	9	0	11	0	1	0
ELS39	1989	0	2	226	16	0	0	2	106	0	14	0	3152	0	32	2
	1993	0	0	108	152	39	208	59	23	60	839	31	463	0	51	0
	1995	0	0	139	24	6	13	1	94	208	109	0	1906	0	7	0
ELS61	1989	0	0	24	819	12	25	7	504	0	1449	1	336	0	183	8
	1993	0	0	68	1886	0	37	20	218	315	315	0	153	0	205	0
	1995	0	0	4	243	0	23	1	68	17	27	0	21	0	97	0
ELW1	1989	0	0	35	223	0	1	0	16	0	16	0	77	0	115	3
	1993	0	0	177	0	0	0	0	27	20	1	0	23	0	41	0
	1995	0	0	25	0	0	0	0	5	65	4	0	61	0	9	0
FC1	1989	1	0	32	48	0	53	116	40	0	120	1	5	0	33	11
	1993	0	0	1	29	0	33	181	0	3	52	0	3	0	1	0
	1995	0	0	3	3	0	0	18	0	0	3	0	8	0	1	0

TABLE 4-8 *(continued)*
Nonchironomid Nematocera Abundance (/m^{-2}) at 19 Wetlands in King County, Washington in 1989, 1993, and 1995

Wetland	Year	Anisopodidae	Bibionidae	Cecidomyiidae	Ceratopogonidae	Chaoboridae	Culicidae	Dixidae	Mycetophilidae	Phoridae	Psychodidae	Scatopsidae	Sciaridae	Simuliidae	Tipulidae	Other Nonchironomids
HC13	1989	3	0	104	116	0	1	7	39	5	2421	0	509	4	308	8
	1993	0	0	1	7	80	83	72	8	8	11	0	1	0	8	0
	1995	0	0	4	20	106	39	55	188	0	13	0	4	0	7	0
JC28	1989	0	0	1	65	0	0	7	3	0	55	0	11	0	16	11
	1993	0	0	8	0	0	0	29	8	3	5	0	1	0	1	0
	1995	0	0	11	19	0	1	15	16	8	15	0	40	0	19	0
LCR93	1989	31	0	241	764	0	0	191	177	1	367	0	129	152	297	229
	1993	0	0	11	21	0	1	48	31	11	0	0	3	25	53	0
	1995	0	0	138	67	0	0	20	67	0	61	0	448	4	72	0
LPS9	1989	4	0	1096	68	0	0	7	49	0	75	0	1528	0	483	4
	1993	0	0	130	9	0	0	1	82	26	1	0	266	0	15	0
	1995	0	0	138	67	0	0	20	67	73	0	0	448	4	72	0
MGR36	1989	0	0	105	92	0	24	100	15	0	609	1	40	0	51	35
	1993	0	0	12	57	8	12	93	0	9	25	0	19	0	0	0
	1995	0	0	7	289	0	25	22	4	0	13	0	21	0	0	0
NFIC12	1989	0	0	352	72	4	15	1	21	0	145	0	489	0	27	0
	1993	0	0	100	596	43	25	4	80	57	96	0	323	0	73	0
	1995	nd	nd	nd	nd	nd	nd	nd	nd	nd	nd	nd	nd	nd	nd	nd

Chironomid abundance varied widely between wetlands and within wetlands by year (Table 4-6). A few wetlands such as BBC24 and MGR36 consistently ranked high in production with 3-year averages of 6093 and 4019 animals m^{-2}/yr, respectively. ELW1 consistently ranked low, with production values of only 171, 15, and 11 m^{-2}/yr during 1989, 1993, and 1995, respectively. Often, capture rates differed up to a factor of 10 between years, as observed at PC12, SC4, LCR93 B3I, ELS61, LPS9, PC12, and ELS61.

Orthocladiinae, followed in decreasing order by Chironomini and Tanytarsini, were the chironomid family and tribes of greatest numerical representation. Combined, they accounted for 62, 18, and 11% of total classified chironomids, respectively (Table 4-7a,b). *Limnophyes* was the predominant Orthocladiinae genus representing 31% of this family in permanently flooded wetlands and 66% in seasonally flooded wetlands. *Polypedilum* gr.1 accounted for the highest percentage of Chironomini, representing 82% and 14% of taxa in permanent and seasonal wetlands, respectively. *Psectrotanypus dyari* within Tanypodinae and *Microspectra* gr.1 within Tanytarsini proportionately represented the most abundant taxa within these two groupings.

The range of taxa richness at individual wetlands varies considerably within our basic wetland characterization schemes. Average taxa richness of semiaquatic and aquatic chironomid insects was higher (>20 taxa) in permanently rather than seasonally flooded wetlands (Table 4-9). Moreover, some chironomids, e.g., *Meropelopia americana* and *Zavrelimia sinuosa*, are exclusive to permanently flooded wetlands, whereas others e.g., *Odontomesa* and *Natarsia miripes*, are found exclusively in seasonally flooded wetlands. Several other taxa appear to be found in either permanently or seasonally flooded wetlands (Table 4-10). Overall abundance of chironomids was also higher at permanently flooded wetlands. Flooded wetlands exhibited roughly twice the chironomid numbers found at seasonally flooded wet-

TABLE 4-9
Total Richness (S), Evenness (E), Diversity (H′) and Average Richness ($\bar{x}$S) of Chironomid Taxa Identified in 1989 (n = 76)

	Permanently Flooded					Seasonally Flooded				
Urbanization	Wetland	E	H′	S	$\bar{x}$S	Wetland	E	H′	$\bar{x}$S	S
Low	BBC24	0.73	2.69	39	29	NFIC12	0.463	0.83	6	9
	HC13	0.73	2.14	19		TC13	0.607	1.508	12	
	MGR36	0.82	2.61	24						
	RR5	0.76	2.68	34						
	SR24	0.76	2.55	29						
Moderate	ELW1	0.87	2.36	15	21	AL3	0.59	1.295	9	16
	PC12	0.61	2.09	31		LCR93	0.756	2.518	28	
	SC84	0.49	1.35	16		SC4	0.406	1.008	12	
High	ELS61	0.53	1.72	36	23	B3I	0.675	1.827	15	15
	FC1	0.64	1.89	19		ELS39	0.159	0.286	16	
						JC28	0.589	1.821	22	
						LPS9	0.746	1.64	9	

TABLE 4-10
1989 Chironomid Taxa and Numbers of Individuals Found Almost Exclusively Either Under Permanently Flooded or Seasonally Flooded Conditions Using ≤ 5 Individuals Collectively Detected at All Wetlands as the Threshold Value

		Permanently Flooded Wetlands	Seasonal Flooded Wetlands
Subfamily	**Genera/Species**	**No.**	**No.**
Podonominae	*Boreochlus*	13	3
Phodiamesinae	*Odontomesa*	0	11
Orthocladiinae	*Paraphaenoceadius*	17	2
Chironominae	*Paratendipes sp. 1*	0	29
	Phaenopsectra flavipes	38	1
	Stictochironomus	51	1
Tanypodinae	*Alabesmyia*	814	1
	Conchapelopia cf.	68	3
	Macropelopiini	3	118
	Meropelopia nr. Americana	7	0
	Natarsia miripes	0	114
	Zavrelimia sinuosa	61	0

lands. Finally, MRPP testing confirmed a statistical difference in taxa characteristics between these two hydrologically distinct wetland types ($p = .018$) suggesting that permanently flooded wetland chironomid communities indeed differ from those at seasonally flooded wetlands. Chironomid taxon distribution and abundance, however, are not unique to wetlands within watersheds of different levels of urbanization (MRPP, $p = .381$).

Nonchironomids

Nonchironomid taxa also varied widely in distribution and abundance (Tables 4-8 and 4-11a,b). For example, in abundant years nonchironomid production was as much as 4.7 and 6.6 times the values in low years as recorded at MGR36 and LPS9, respectively. The range between low and high counts at most other wetlands differed less dramatically (1 to 3 times) between years (Table 4-8). Despite these yearly differences in total numbers within wetlands the standing of lowest and highest abundances of certain families remained the same over the years. B3I and ELS61, for example, ranked in the top three wetlands in Cecidomyiidae and Tipulidae abundance in two of three years.

Numerically important families include the Sciaridae, Cecidomyiidae, Ceratapagonidae, and Psychodidae. In these families we counted more than 1000 individuals m^{-2}/yr in wetlands during a single year. Other less abundant nonchironomid families with up to 500 captures m^{-2}/yr include the Dixidae, Mycetophilidae, and Tipulidae. Rarely captured families, with less than 50 individuals captured m^{-2}/yr,

TABLE 4-11a
Total and Proportional Abundance (/m^{-2}) of Nonchironomid Nematocera and Brachycera in Permanently Flooded Wetlands of Non-, Moderately, and Highly Urbanized Watersheds in 1989

	Wetland																				
	Non-Urban										Moderately Urban						Highly Urban				
	BBC24		HC13		MGR36		RR5		SR24		ELW1		PC12		SC84		ELS61		FC1		
Taxa	No.	%	No.	%	No.	%	No.	%	No.	%	No.	%	No.	%	No.	%	No.	%	No.	%	Total
Nonchiron. Nematocera	202		3525		1072		500		816		486		440		449		3368		460		11318
Anisopodidae		0	3	0		0		0		0		0	1	0		0		0	1	0	5
Bibionidae		0		0		0		0		0		0		0		0		0		0	0
Cecidomyiidae	11	5	104	3	105	10	11	2	107	13	35	7	15	3	65	14	24	1	32	7	509
Ceratopogonidae	121	60	116	3	92	9	241	48	189	23	223	46	84	19	33	7	819	24	48	10	1966
Chaboridae	11	5		0		0	7	1	7	1		0		0	7	2	12	0		0	44
Culicidae	1	0	1	0	24	2		0	19	2	1	0	23	5	1	0	25	1	53	12	148
Dixidae	33	16	7	0	100	9		0	48	6		0	1	0	12	3	7	0	116	25	324
Mycetophilidae		0	39	1	15	1	16	3	163	20	16	3	8	2	9	2	504	15	40	9	810
Phoridae		0	5	0		0		0		0		0	3	1		0		0		0	8
Psychodidae	1	0	2421	69	609	57	101	20	84	10	16	3	179	41	84	19	1449	43	120	26	5064
Other Nematocera	15	7	8	0	35	3	16	3	132	16	3	1	24	5	4	1	8	0	11	2	256
Scatopsidae		0		0	1	0		0		0		0	1	0		0	1	0	1	0	4
Sciaridae	9	4	509	14	40	4	75	15	27	3	77	16	64	15	79	18	336	10	5	1	1221
Simuliidae		0	4	0		0		0		0		0		0		0		0		0	4
Tipulidae		0	308	9	51	5	33	7	40	5	115	24	37	8	155	35	183	5	33	7	955
Brachycera	267		2169		2884		884		467		583		484		1377		5488		1575		16178

TABLE 4-11b
Total and Proportional Abundance (/m^{-2}) of Nonchironomid Nematocera and Brachycera in Seasonally Flooded Wetlands of Non-, Moderately, and Highly Urbanized Watersheds in 1989

	Wetland																		
	Non-Urban						Moderately Urban						Highly Urban						
	NFIC12		TC13		AL3		LCR93		SC4		B3I		ELS39		JC28		LPS9		
TAXA	No.	%	No.	%	No.	%	No.	%	No.	%	No.	%	No.	%	No.	%	No.	%	Total
Nonchiron. Nematocera	1126		406		724		2579		1233		623		3552		169		3314		13726
Anisopodidae		0		0		0	31	1.2		0	1	0.2		0		0	4	0.1	36
Bibionidae		0		0		0		0		0		0	2	0.1		0		0	2
Cecidomyiidae	352	31	88	22	159	22	241	9.3	299	24	407	65	226	6.4	1	0.6	1096	33	2869
Ceratopogonidae	72	6.4	84	21	53	7.3	764	30	3	0.2	71	11	16	0.5	65	38	68	2.1	1196
Chaboridae	4	0.4	1	0.2	5	0.7		0		0		0		0		0		0	10
Culicidae	15	1.3		0	3	0.4		0	16	1.3	1	0.2		0		0		0	35
Dixidae	1	0.1	5	1.2	100	14	191	7.4		0	1	0.2	2	0.1	7	4.1	7	0.2	314
Mycetophilidae	21	1.9	11	2.7	52	7.2	177	6.9	7	0.6	8	1.3	106	3	3	1.8	49	1.5	434
Phoridae		0		0		0	1	0		0		0		0		0		0	1
Psychodidae	145	13	73	18	120	17	367	14	117	9.5	32	5.1	14	0.4	55	33	75	2.3	998
Other Nematocera		0	3	0.7		0	229	8.9	3	0.2	1	0.2	2	0.1	11	6.5	4	0.1	253
Scatopsidae		0	1	0.2		0		0		0	4	0.6		0		0		0	5
Sciaridae	489	43	89	22	137	19	129	5	728	59	13	2.1	3152	89	11	6.5	1528	46	6276
Simuliidae		0		0		0	152	5.9		0		0		0		0		0	152
Tipulidae	27	2.4	51	13	95	13	297	12	60	4.9	84	13	32	0.9	16	9.5	483	15	1145
Brachycera	2984		1043		2428		2925		887		888		1016		69		1160		13400

include the Anisopodidae and Scatopsidae. Hardly any Bibionidae were captured among the wetlands.

Overall, nonchironomid numbers are much lower than chironomid numbers, although at certain wetlands and in certain years nonchironomid densities exceed chironomid densities (Table 4-6). At ELS39 and ELW1, however, nonchironomids were always more abundant than chironomids regardless of year. Also, in 1989, nonchironomid densities were greater than chironomid densities at HC13, LCR93, LPS9, and SC4.

Psychodidae exhibited the greatest proportional representation of all families in 5 of the 10 permanently flooded wetlands (50%) and was absent from seasonally flooded wetlands. In contrast, Sciaridae exhibited the highest proportional abundance of organisms of all families within seasonally flooded wetlands. It was proportionately the most abundant in 5 of 9 seasonally flooded wetlands (56%) but in none of the permanently flooded wetlands.

Brachycera

Ten Brachycera families were identified both in 1993 and 1995 (Table 4-12a,b). Phoridae was the most widely distributed family, being found in 18 of 19 wetlands (95%). However, the overall abundance was similar to densities observed for other families. Clearly, Dolochipodidae, Ephydridae, and Empididae were also widely distributed, being found at 13–16 wetlands over both years. Least widely distributed were individuals belonging to Stratiomiidae. They were captured only at B3I in 1993, and then only four organisms were caught.

The most abundant Brachycera captured in 1993 belonged to the family Dolochipodidae, of which 4352 individuals were captured, followed by Ephydridae of which 956 organisms were captured. In contrast, in 1995 the most frequently captured Brachycera belonged to the family Empididae of which a total of 1775 were counted. The second most abundant numbers (829) are found among the Dolochipodidae (Table 4-12a,b). Empididae numbers were consistently high at LPS9, representing 88% and 79% of all families in 1995 and 1993, respectively.

Ephydridae appeared in moderate numbers at all permanently flooded wetlands (Table 4-12a) and in numbers of more than 5 individuals at only TC13, or 11% of seasonally flooded wetlands (Table 4-12b). This finding suggests this family may have a preference for wetlands characterized by permanent water.

Total Insect and Chironomid Species Richness and Wetland Characteristics

Diversity measures calculated for the full complement of aquatic and semiaquatic insects (Table 4-4) for 1989 indicate taxa richness is generally higher in permanently (n = 10, $\bar{x}$ = 29.4) than seasonally flooded wetlands (n = 9, $\bar{x}$ = 17.4). Our data also suggest that richness may be highest within wetlands of watersheds with low (n = 7, $\bar{x}$ = 28.4) levels of urbanization. Wetlands in watersheds of moderate development (n = 6, $\bar{x}$ = 23.1) exhibit greater richness than wetlands in watersheds with high development (n = 6, $\bar{x}$ = 18.8). Exceptions to this generality exist. LCR93, a seasonally flooded wetland within a watershed of moderate urbanization, is character-

TABLE 4-12a
Total and Proportional Abundance of Brachycera in Permanently Flooded Wetlands of Non-, Moderately, and Highly Urbanized Watersheds in 1993 and 1995

		Wetland																				
		Non-Urban										Moderately Urban						Highly Urban				
		BBC24		HC13		MGR36		RR5		SR24		ELW1		PC12		SC84		ELS61		FC1		
Taxa		No.	%	No.	%	No.	%	No.	%	No.	%	No.	%	No.	%	No.	%	No.	%	No.	%	Total
Total Brachycera	1993	136		44		38		34		12		55		289		19		3498		73		4198
Chloropidae		4	3		0	7	18	1	3		0		0	4	1		0	93	3	4	5	113
Dolochipodidae		43	32	29	66		0	5	15		0	7	13	112	39	3	16	2433	70	1	1	2882
Empididae			0	4	9	5	13		0		0	15	27	3	1		0	4	0	5	7	87
Ephydridae		77	57		0	12	32	17	50	12	100	7	13	98	34	10	53	301	9	53	73	933
Phoridae		7	5	8	18	9	24	3	9		0	20	36	37	13	1	5	35	1	3	4	234
Rhagionidae			0		0		0		0		0	1	2		0		0		0		0	3
Sciomyziidae			0		0		0		0		0		0		0	2	11	1	0		0	14
Sphaeroceridae			0		0	1	3		0		0		0	3	1	1	5	1	0	4	5	19
Stratiomyiidae			0		0		0		0		0		0		0		0		0		0	0
Syrphidae		1	1		0	1	3		0		0		0	1	0		0	379	11		0	397
Other Brachycera		4	3	3	7	3	8	8	24		0	5	9	31	11	2	11	251	7	3	4	389
Total Brachycera	1995	94		33		85		904		6		130		235		190		132		14		1823
Chloropidae		5	5	1	3	2	2	3	0		0		0	7	3		0	8	6		0	26
Dolochipodidae		5	5	8	24		0	677	75		0	11	8	8	3	100	53	20	15		0	829
Empididae		8	9		0		0	12	1		0	48	37	20	9	23	12	3	2		0	114
Ephydridae		53	56	12	36	67	79	39	4		0	13	10	29	12	4	2	41	31	9	64	267
Phoridae		3	3		0	4	5	15	2	2	33	55	42	79	34	3	2	17	13		0	178
Rhagionidae			0		0		0		0		0		0	4	2		0		0		0	4
Sciomyziidae			0	7	21	3	4		0		0		0	1	0	19	10	1	1	1	7	32
Sphaeroceridae			0	1	3	6	7	3	0		0		0	4	2		0	1	1	1	7	16
Stratiomyiidae			0		0		0		0		0		0		0		0		0		0	0
Syrphidae			0		0		0	3	0	4	67		0		0		0	1	1		0	8
Other Brachycera		20	21	4	12	3	4	152	17		0	3	2	83	35	41	22	40	30	3	21	349

TABLE 4-12b
Total and Proportional Abundance of Brachycera in Seasonally Flooded Wetlands of Non-, Moderately, and Highly Urbanized Watersheds in 1993 and 1995

		Wetland																		
		Non-Urban				Moderately Urban						Highly Urban								
		NFIC12		TC13		AL3		LCR93		SC4		B31		ELS39		JC28		LPS9		
Species		No.	%	No.	%	No.	%	No.	%	No.	%	No.	%	No.	%	No.	%	No.	%	Total
Total Brachycera	1993	367		9		184		128		627		1108		405		7		432		3267
Chloropidae			0		0		0	12	9		0	3	0		0		0	1	0	16
Dolochipodidae		45	12	1	11	169	92	63	49	308	49	649	59	226	56		0	9	2	1470
Empididae		29	8		0	7	4	33	26	239	38	20	2	63	16	1	14	341	79	733
Ephydridae		5	1	1	11		0	1	1	3	0	12	1		0		0	1	0	23
Phoridae		57	16	3	33	1	1	11	9	35	6	61	6	60	15	3	43	26	6	257
Rhagionidae			0		0		0		0		0	3	0		0		0	1	0	4
Sciomyziidae			0		0		0		0		0		0		0		0		0	0
Sphaeroceridae		1	0		0	3	2		0	3	0		0	5	1		0	3	1	15
Stratiomyiidae			0		0		0		0		0	4	0		0		0		0	4
Syrphidae		3	1		0		0		0		0	19	2	3	1		0	1	0	26
Other Brachycera		227	62	4	44	4	2	8	6	39	6	337	0	48	12	3	43	49	0	719
Total Brachycera	1995	no data		76		224		952		308		168		310		28		879		2945
Chloropidae					0	1	0	5	1		0	1	1	8	3		0		0	15
Dolochipodidae				37	49	113	50	270	28	56	18	27	16	14	5	11	39		0	528
Empididae					0	94	42	548	58	180	58	9	5	55	18	3	11	772	88	1661
Ephydridae				16	21	4	2	5	1	2	1	4	2		0		0		0	31
Phoridae				7	9	1	0	73	8	23	7	35	21	208	67	8	29	70	8	425
Rhagionidae					0		0		0		0	13	8		0		0	1	0	14
Sciomyziidae				3	4		0	16	2	4	1		0	6	2	5	18		0	34
Sphaeroceridae				5	7		0	3	0		0	4	2	2	1		0	1	0	15
Stratiomyiidae					0		0		0		0	2	1		0		0		0	2
Syrphidae					0		0		0		0	2	1		0		0		0	2
Other Brachycera				8	11	11	5	32	3	43	14	71	0	17	5	1	4	35	0	218

ized by 47 taxa and is the second highest in insect richness among wetlands. Alternately ELW1, a permanently flooded wetland in a moderately developed watershed, exhibits a relatively low richness of 15 taxa. Moreover, both NFIC12 and TC13, wetlands with low species richness, are located in least-disturbed watersheds. Nevertheless, the overall average richness for all wetlands in watersheds of low urbanization remains high. NFIC12 and TC13 may rank low because of the seasonality of flooding rather than watershed urbanization.

The diversity index and evenness value mimic trends of taxa richness. Permanent wetlands in watersheds of low urbanization again exhibit the highest average E and H′ values. Lowest average evenness and diversity are found in seasonally flooded wetlands of low-urbanization watersheds. However, if it weren't for NFIC12 and TC13 disproportionately reducing the average E and H′ values in the low watershed urbanization group, the lowest group for all diversity indicators would be seasonally flooded wetlands in watersheds of high urbanization. Clearly, the trend in total insect richness, evenness, and diversity is a reflection of Chironomid distribution and abundances because this taxon accounts for the greatest richness and abundance of all insects within wetlands.

Non-Metric Multidimensional Scaling (NMDS) distinguished between the overall invertebrate and Chironomid communities in permanent and seasonal wetlands along Axis 1 (Figure 4-2). Stress values for the one-dimensional solution performed on the distribution and abundance data was 50.31, suggesting an acceptable degree of fit. For aquatic insects in general and chironomids in particular, Axis 1 represents wetlands with a hydrological gradient progressing from seasonally to permanently flooded wetlands. With the exception of ELW1, all permanently and seasonally flooded wetlands readily separate along this gradient. The insect communities found at TC13 and LCR93 require further evaluation to identify conditions that may account for their overlap with SC84, HC13, and FC1. Generally, however, wetlands along the left side of Axis 1 are characterized by summer drought and harsher and less predictable environments than those along the right side of the axis. Consequently, they may be expected to have different invertebrate communities than observed in permanently flooded wetlands. In general, seasonally flooded wetlands should exhibit simpler food chains, with fewer linkages per species than permanently flooded, more stable systems, thereby accounting for reduced richness.

The graphs showing the urbanization gradient in Figure 4-2 suggest wetland total insect and chironomid-only communities are unrelated to watershed urbanization. Wetlands with high, medium, and low urbanization within the watersheds are not consistently grouped along any axis. Even though low urbanization wetlands BBC24, HC13, MGR36, RR5, and SR24 suggest similar communities along Axis 1, each is also characterized by permanent flooding. When considering all wetlands, this alignment better explains aquatic insect communities within the watershed. Though moderately and highly urbanized wetlands could be distinguished from seasonal wetlands, they displayed more separation on both Axis 1 and Axis 2 than the nonurbanized wetlands. For example, the highly urbanized seasonal wetland B3I, and the moderately urbanized permanently flooded wetland ELW1, are close to each other but unusually distant to all other sites indicating somewhat similar insect communities to each other but unique insect communities from other wetlands.

FLOODING

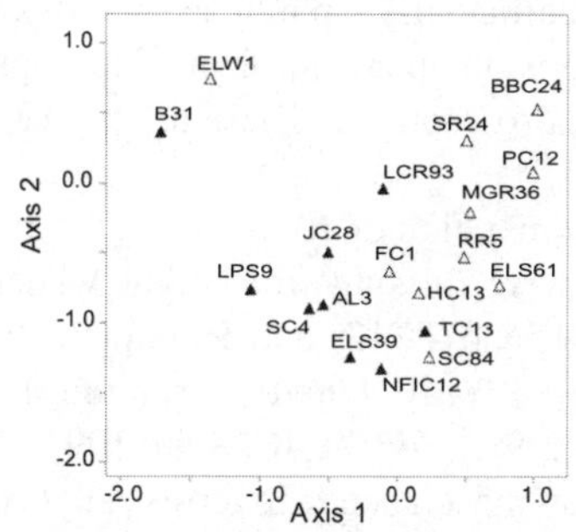

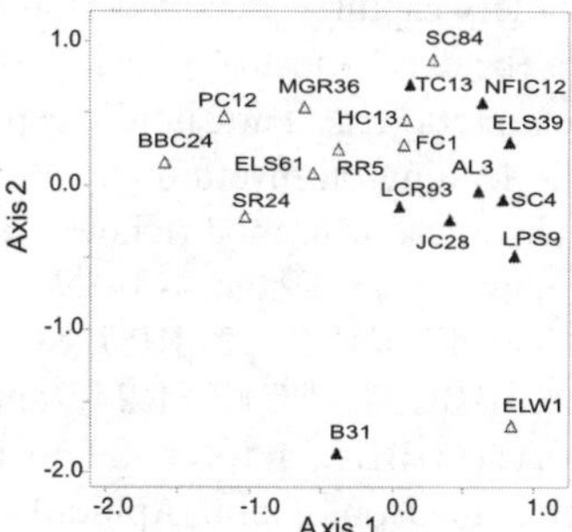

Nonmetric multidimensional scaling ordination of the distribution and abundance of **115 aquatic and semiaquatic invertebrate** taxa at 19 palustrine wetlands.

△ permanently flooded wetlands
▲ seasonally flooded wetlands

Nonmetric multidimensional scaling ordination of the distribution and abundance of **76 chironomid** taxa at 19 palustrine wetlands.

△ permanently flooded wetlands
▲ seasonally flooded wetlands

URBANIZATION

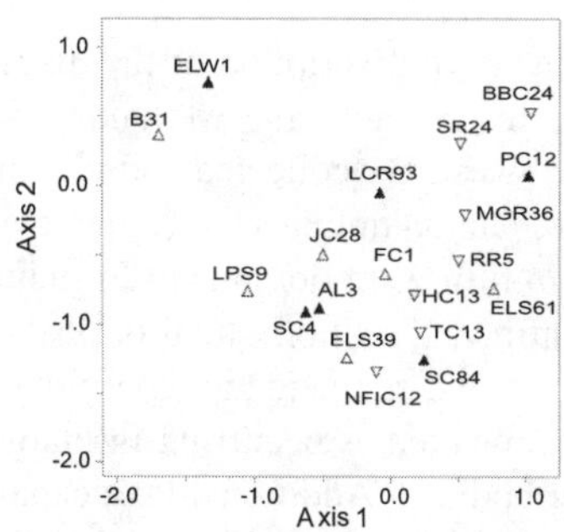

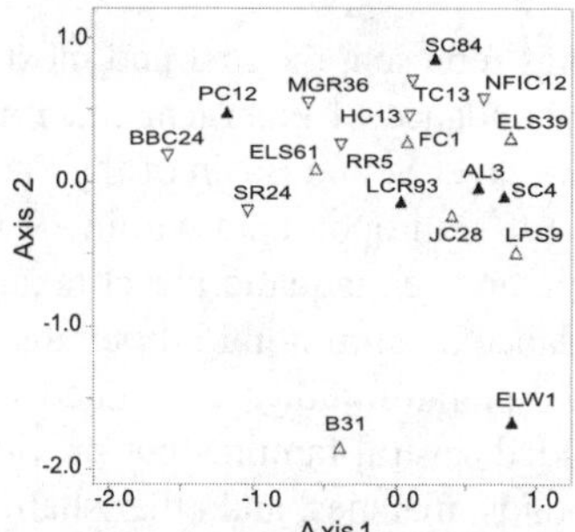

Nonmetric multidimensional scaling ordination of the distribution and abundance of **115 aquatic and semiaquatic invertebrate** taxa at 19 palustrine wetlands.

△ high watershed urbanization
▲ moderate watershed urbanization
▽ low watershed urbanization

Nonmetric multidimensional scaling ordination of the distribution and abundance of **76 chironomid** taxa at 19 palustrine wetlands.

△ high watershed urbanization
▲ moderate watershed urbanization
▽ low watershed urbanization

FIGURE 4-2 Comparison of the distribution and abundance of taxa identified at 19 palustrine wetlands in 1989 using nonmetric multidimensional scaling ordination.

TERRESTRIAL ARTHROPOD RICHNESS AND ABUNDANCE

Arachnids (spiders and mites) and eight primarily terrestrial insect orders were unexpectedly captured in emergence traps (Table 4-3). Spiders are common predators and mites are common parasites of arthropods and other invertebrates. Both are regularly found associated with aquatic insects. Generally, they are not captured in emergence traps, which are supposed to exclude such terrestrial taxa. Representatives of insect orders captured included Homoptera, particularly Aphididae (aphids) and Coleoptera (beetles), most of which are herbivores on emergent plants. Neuroptera and Hymenoptera (e.g., Parasitoid wasps) were also captured. These are predatory, and may be feeding on invertebrates in the traps.

Total terrestrial arthropod richness ranged from a high of 10 to a low of 7 major taxa in a single year (Table 4-3). Neuroptera were missing from eight wetlands (AL3, ELS39, ELW1, PC12, RR5, SC84, SR24, and TC13) and Hemiptera from five (AL3, BBC24, FC1, SR24, and TC13). Total densities ranged from 56,439 m^{-2}/yr in BBC24 for 1989 to a low of 9 m^{-2}/yr at JC28 in 1993. Not surprisingly, the Homoptera: Aphididae are the most abundant terrestrial taxon. Aphids are often found in extremely high densities on plants because of their reproductive strategies, communal feeding, and small size. Moreover, aphids frequently feed on aquatic vegetation (e.g. *Potamogeton* spp., personal observation) and therefore may be expected to be abundant on floating and emergent vegetation at wetlands such as BBC24. Alternately, Aphididae may be largely missing from forested and scrub-shrub wetlands such as JC28 and AL3 that are without floating, broad-leaved herbaceous species.

DISCUSSION

Our findings represent the first and most comprehensive description of the distribution and abundance of emergent macroinvertebrates at palustrine wetlands within the Central Puget Sound Basin of the Pacific Northwest. We collected and identified a total of 128 arthropod taxa within 19 wetlands. Our sampling was dominated by 115 aquatic and semiaquatic insect taxa, which in turn were dominated in richness and abundance by chironomid dipterans. We identified 17 out of a total of 35 North American dipteran families associated with aquatic and semiaquatic environments. This included several families not previously documented as occurring in marginal areas of ponds, marshes, and other shallow water bodies.[43] Additional taxa captured included eight mostly terrestrial insect orders dominated by Homoptera: Aphididae.

Total production for aquatic and semiaquatic insects ranged from a high of 21,000 organisms m^{-2}/yr to a low of 256 organisms m^{-2}/yr and averaging 4,590 organisms m^{-2}/yr across all wetlands for the three survey years. This biomass of invertebrates when extrapolated to similar habitats for entire wetlands represents a significant food source for predators including aquatic insects and other invertebrates, fish, amphibians, and birds.

Chironomids, as expected, were found to be the most diverse and abundant insect taxa because they are the most numerous in many regions of North America and represent more aquatic species than all other insect orders combined.[43,44] They are also readily captured in emergence traps, further accounting for their high representation within our wetlands. We identified 76 chironomid taxa in addition to several undescribed species in 1989 alone. We extended the range of several taxa not listed for the Pacific Northwest and nearly half of our taxa have not been previously reported in wetlands.[44,45] Nevertheless, our Chironomid richness may still be considered low compared to the number of species identified in the southeast and other regions of the U.S.[46] Our richness, however, will likely increase when additional wetland types and numbers across King County and the ecoregion are surveyed.

A comparison of invertebrate taxa diversity and abundance with other wetlands within the Central Puget Sound Basin or in North America is problematic, as this is the first intensive study of arthropods within numerous wetlands undertaken in the northwest. We did, however, find similar insect taxa and relative abundances obtained under almost identical trapping protocols at one palustrine wetland in the Coast Range Ecoregion.[37] In this study, Chironomidae, Tanypodinae, and Orthocladiinae were also the most abundant taxa, with Chironomidae representing more than 40% of total emergence. Low abundances of Ephemeroptera, Trichoptera, and Odonata were also common to both studies, reflecting sampling bias toward small invertebrates in emergence traps.

Interesting observations were made for several taxa. Unidentified Orthocladinae females were dominant at all wetlands in 1989. No positive male associations were found that would allow generic or specific identification. Therefore, these females may represent parthenogenetic forms of one or more taxa. Unassociated females were also common or abundant in most of the remaining subfamilies and tribes, suggesting parthenogenetic taxa in these groups as well.

Plecoptera nymphs typically inhabit cold perennial or temporary streams and with lentic sightings confined to cold oligotrophic bodies of water and wave-swept shores of lakes.[47] Captures of Plecoptera thus were unexpected at depressional-forested systems including JC28 and NFIC12. Captures at LCR93 were more likely because of its proximity to a stream. However, we did not expect to capture 1576 m^{-2}/yr, which represented 16% of total yearly insect production at this wetland in 1989. Findings such as these are not well documented and may be unique to wetlands of the Pacific Northwest. Plecoptera at these wetlands generally suggests good water quality because most species are particularly sensitive to pollution.

Other non-dipteran aquatic and semiaquatic insects at our wetlands, for the most part, are found elsewhere in similar habitats. The three odonate taxa we recorded (*Ischnura cervula, Enallagma boreal* and *Coenagrion*) are also commonly found in Canadian marshes.[48] These taxa are also relatively short damselflies (13–25 mm excluding caudal lamellae) that can crawl through the funnel opening and be captured. The few Odonata overall most certainly reflects sampling bias associated with emergence traps rather than depauperate populations, because we observed diverse and abundant adults of other species during our field work. Large adult darners (e.g., Aeschnidae [*Aeshna palmata, Anax junius*]); skimmers (Libbellulidea [*Libellula* spp., *Libellula forensis, L. pulchella, L. luctosa*]) and meadowhawks (Libbellulidea [*Sympetrum occidentale, S. internum*]) were commonly seen flying at wetlands and their nymphs were captured in dip nets among the sediment.

Trichoptera richness was greater at permanently than seasonally flooded wetlands. Trichoptera were also found at all wetlands of slightly urbanized watersheds, patchily represented at wetlands in moderately urbanized wetlands, and rare at our two highly urbanized wetlands. Their virtual absence from highly urbanized wetlands suggest that impacts from altered hydrology and pollutants may limit their presence since other habitat conditions capable of supporting Trichoptera were present. However, because our wetland population contained only two permanently flooded wetlands in heavily urbanized watersheds and two in seasonally flooded wetlands of lightly urbanized watersheds, this conclusion must be treated with caution.

Ephemeroptera, in general, inhabit both lentic and lotic waters with adequate supplies of dissolved oxygen and may be expected within some of our wetlands. However, the capture of *Paraleptophlebia* and *Callibaetis* were unexpected. *Paraleptophlebia*, common to northwest ephemeral streams, was found in both permanently as well as seasonally flooded wetlands. Adult *Callibaetis*, on the other hand, generally emerge from seasonally flooded wetlands.[47] Thus their absence from seasonally flooded wetlands but presence in permanent wetlands warrants further investigation.

Within our wetlands we found greater values for richness, Shannon diversity, and evenness at permanently flooded wetlands than at seasonally flooded wetlands. This was true for all insects combined and for just the chironomids. We expected this finding, as insect adaptations for survival are less difficult in permanently rather than seasonally flooded wetlands. Permanent wetlands also exhibit greater thermal constancy, physiochemical characteristics, and vegetation stability, all conditions fostering a rich and diverse invertebrate fauna. Another potential reason for higher richness in permanently flooded wetlands may be that sediments of deeper-water wetlands experience shorter or no periods of anaerobic conditions, enabling more diverse taxa, as fewer species are stressed by oxygen poor conditions.[21,49] Finally, permanently flooded wetlands exhibit more stable emergent vegetation, greater productivity, and recurring seasonal death in the persistently flooded zone, thereby providing greater quantities of organic litter than at seasonally flooded sites. Collectively, all these conditions may account for our greater chironomid taxa numbers among the collector-filterers, gatherers, and scrapers (e.g., Orthocladiinae [62%] Chironomini [18%], and Tanytarsini [11%]) found at permanently flooded wetlands. We also found greater numbers of predators (i.e., Tanypodinae) at permanently flooded wetlands. In general, taxa richness decreases with disturbance, although in some species (e.g., Chironomids) increasing density is found with increasing disturbance.[50]

We may not be able to readily identify unique insect assemblages associated with watershed urbanization because we had only two wetlands each within heavily and lightly urbanized watersheds and with differing hydrology. Consequently, our wetland numbers may be too small to adequately separate urbanization from flooding effects using our statistical methods. A second reason we were not able to differentiate by aquatic insect communities may be that landscape traits, such as urbanization within watersheds, may not be as significant in structuring the emergent insect community as water permanence or other unidentified factors. Numerous studies suggest that habitat architecture, vegetation, substrate, and water quality may determine invertebrate distribution. For example, data from rivers, streams, and lakes show that many invertebrates have preferences for specific substrates and macrophytes.[50,51] It is also well known that Chironomids are strongly affected by sediment characteristics.[52,53] Particle size, water temperature, vegetation to water ratio and vegetation interspersion, richness, and species all support a higher diversity of macroinvertebrates and play important roles in structuring aquatic communities.[1,5,17,53-59] Although we tried to control for some of these conditions it is clear from Table 4-1 that our trapping habitats still varied by flow, substrate, and dominant vegetation. Consequently, the extent to which microhabitat conditions, as opposed to landscape conditions such as flooding and urbanization, accounted for species needs significant further study. Nevertheless, urbanization clearly affects flooding and overall wetland ecology (this volume) and thus urbanization may indirectly account for macroinvertebrate communities by influencing these microhabitat attributes. Regardless, the aquatic insect-substratum relationship is extremely complex, depending on numerous interrelationships between physical and biological factors difficult to standardize between wetlands.[53]

Although we describe extensively the invertebrate communities at palustrine wetlands within King County our taxa-wetland associations should be used and interpreted with caution, as this is the first description from select areas of palustrine wetlands surveyed by emergence traps. We tried to minimize habitat variability by selecting lightly vegetated, similarly flooded sites. Nevertheless, the distribution and abundance of all taxa, particularly the numerically dominant Chironomidae, has been shown to be highly variable and will require monitoring under more rigorous control of substrate, vegetation, and flooding conditions. Considerable additional work needs to be undertaken to describe the benthic macroinvertebrate communities of palustrine wetlands in the Pacific Northwest, particularly in more wetlands in watersheds of high urbanization and varied hydrology. It is only then that we will understand their ecological relations to microhabitat and to wetland and watershed conditions that will help us determine their changing distribution and abundances and concomitant trophic interactions, including food-chain support functions for fish and wildlife.

ACKNOWLEDGMENTS

I would like to thank Robert Wisseman for his work in developing the trapping protocols identifying species and early contributions to this study, Dr. Leonard Ferrington (University of Kansas) for identifying Chironomidae, Dr. George Byers and Dr. Ernest May (University of Kansas) for identifying adult Tipulidae, and Dr. Greg Courtney (Smithsonian Institution) for advice on nematoceran families. Meredith Savage and Ken Ludwa collected invertebrates and provided other field assistance. Richard Robohm was kind enough to edit an early version of this chapter.

REFERENCES

1. Murkin, H. R. and B. D. J. Batt, The interactions of vertebrates and invertebrates in peatlands and marshes, *Mem. Entomol. Soc. Can.*, 140, 15, 1987.
2. Murkin, H. R. and D. A. Wrubleski, Aquatic invertebrates of freshwater wetlands: function and ecology, in *The Ecology and Management of Wetlands Volume 1: Ecology of Wetlands*, D. D. Hook, W. H. McKee, Jr., H. K. Smith, J. Gregory, V. Burrell, Jr., M. R. DeVoe, R. E. Sojka, S. Gilbert, R. Bani, L. H. Stolzy, C. Brooks, T. D. Matthews, and T. H. She, Eds., Timber Press, Portland, OR, 1988, p. 239.
3. Cairns, J. C., Jr. and J. R. Pratt, A history of biological monitoring using benthic macroinvertebrates, in *Freshwater Biomonitoring and Benthic Macroinvertebrates*, D. M. Rosenberg and V. H. Resh, Eds., Chapman and Hall, New York, NY, 1993, p. 10.

4. Hynes, H. B. N., *The Ecology of Running Waters*, University of Toronto Press, Toronto, 1970.
5. Voigts, D. K., Aquatic invertebrate abundance in relation to changing marsh vegetation, *Am. Mid. Nat.*, 95, 313, 1976.
6. Murkin H. R. and J. A. Kadlec, Responses by benthic macroinvertebrates to prolonged flooding of marsh habitat, *Can. J. Zool.*, 64, 65, 1986.
7. Hilsenhoff, W. L., An improved biotic index of organic stream pollution, *Great Lakes Entomol.*, 20, 31, 1987.
8. Ohio EPA, Biological Criteria for the Protection of Aquatic Life. Vol. I–III, Ohio Environmental Protection Agency, Division of Water Quality Monitoring and Assessment, Columbus, 1988.
9. Ohio EPA, The Use of Biocriteria in the Ohio EPA Surface Water Monitoring and Assessment Program, Ecological Assessment Section, Division of Water Quality, Planning and Assessment, Columbus, 1990.
10. Plafkin, J. L., M. T. Barbour, K. D. Porter, S. K. Gross, and R. M. Hughes, Rapid Bioassessment Protocols for Use in Streams and Rivers, Benthic Macroinvertebrates and Fish, Office of Water Regulations and Standards, EPA/444/4-89/001, U.S. Environmental Protection Agency, Washington, D.C., 1989.
11. Thienemann, A., Biologische Seetypen and die Gründung einer Hydrobiologischen Anstalt am Bodensee, *Arch. Hydrobiol.*, 13, 609, 1921.
12. Brundin, L., Chironomiden und andere Bodentiere der südschwedischen Urergebnissen, Institute of Freshwater Research Drottningholm, Rep. 30, 1949.
13. Brundin, L., Die bodenfaunistischen Seetypen und ihre Anwendbarkeit auf die Südhalbkugel, Zugleich eine Theorie der produktionsbiologischen Bedeutung der glazialen Erosion, Institute of Freshwater Research Drottningholm, Rep. 37, 1956, p. 186.
14. Saether, O. A., Chironomid communities as water quality indicators, *Holarctic Ecol.*, 2, 65, 1979.
15. Wiggins, G. B., R. J. Mackay, and I. M. Smith, Evolutionary and ecological strategies of animals in annual temporary pools, *Arch. Hydrobiol.*, 58, 99, 1980.
16. Hunt, P. C. and J. W. Jones, The effect of water level fluctuations on a littoral fauna, *J. Fish Biol.*, 4, 385, 1972.
17. Dvorak, J. and E. P. H. Best, Macro-invertebrate communities associated with macrophytes of Lake Vechten: structural and functional relationships, *Hydrobiologia*, 95, 115, 1982.
18. Wilcox, D. A. and J. E. Meeker, Implications for faunal habitat related to altered macrophyte structure in regulated lakes in northern Minnesota, *Wetlands*, 12, 192, 1992.
19. Adamus, P. R., Bibliography of aquatic, wetland, and riparian resources of the Pacific Northwest: unedited working draft, 11/1/98, http://biosys.bre.orst.edu/restore/bib_aquatic_resources.htm, 1998.
20. Fore, L. S., J. R. Karr, and R. W. Wisseman, Assessing invertebrate responses to human activities: evaluating alternative approaches, *J. N. Am. Benthol. Soc.*, 15, 212, 1996.
21. Kadlec, J. A., Nitrogen and phosphorus dynamics in inland freshwater wetlands, in *Waterfowl and Wetlands — An Integrated Review*, T. A. Bookhout, Ed., Proc. 1977 Symp. North-Central Section Wildlife Society, Wildlife Society, Madison, WI, 1979, p. 17.
22. Kaminski, R. M. and H. H. Prince, Dabbling duck and aquatic macroinvertebrate responses to manipulated wetland habitat, *J. Wildl. Manage.*, 45, 1, 1981.
23. Williams, D. D., *The Ecology of Temporary Waters*, Tomber Press, Portland, OR, 1987.

24. Adamus, P. R. and K. Brandt, Impacts on quality of inland wetlands of the United States: A survey of indicators, techniques and applications of community level biomonitoring data, EPA/600/3-90/073, USEPA Environmental Research Laboratory, Corvallis, OR, 1990.
25. Adamus, P. R., Bioindicators for assessing ecological integrity of Prairie wetlands, U.S. Environmental Protection Agency, National Health and Environmental Effects Research Laboratory, EPA/600/R-96/082, Corvallis, OR, 1996.
26. Turner, R. E., Secondary production in riparian wetlands, *Trans. 53rd Wildl. Natl. Res. Conf.*, 491, 1988.
27. Streever, W. J., K. M. Portier, and T. L. Crisman, A comparison of dipterans from 10 created and 10 natural wetlands, *Wetlands,* 16, 416, 1996.
28. Kusler, J. A, Urban wetlands and urban riparian habitat: Battleground or creative challenge for the 1990's, in *Urban Wetlands*, Proc. Natl. Wetland Symp., Oakland, CA, June 26–29, 1988, J. A. Kusler, S. Daly, and G. Brooks, Eds., Association of Wetland Managers, Berne, NY, 1988, p. 2.
29. McMahan, E. A. and L. Davis, Jr., Density and diversity of microarthropods in wastewater treated and untreated cypress domes, in *Cypress Wetlands for Water Management, Recycling and Conservation*, N. T. Odum and K. C. Ewel, Eds., University of Florida Press, Gainesville, FL, 1978, p. 429.
30. Karouna-Reinier, N., An Assessment of Contaminant Toxicity to Aquatic Macroinvertebrates in Urban Stormwater Treatment Ponds, M.S. thesis, University of Maryland, College Park, 1995.
31. Graves G. A., D. G. Strom, and B. E. Robson, Stormwater impact to the freshwater Savannas Preserve marsh, Florida, *Hydrobiologia,* 379, 111, 1998.
32. Pedersen, E. R. and M. A. Perkins, The use of benthic invertebrate data for evaluating impacts of urban runoff, *Hydrobiologia,* 139, 13, 1986.
33. Hicks, A. L., Impervious surface area and benthic macroinvertebrate response as an index of impact from urbanization on freshwater wetlands, M.S. thesis, University of Massachusetts, Amherst, 1995.
34. Hicks, A. L., *Aquatic Invertebrates and Wetlands: Ecology, Biomonitoring and Assessment of Impact from Urbanization*, Environmental Institute, University of Massachusetts Press, Amherst, 1996.
35. Richter, K. O., A. Azous, S. S. Cook, R. W. Wisseman, and R. R. Horner, Effects of Stormwater runoff on wetland zoology and wetland soils characteristics and analysis, DOE Rep., Fourth Year Comprehensive Research, Puget Sound Wetlands and Stormwater Management Program, King County Resources Planning, Bellevue, WA, 1991.
36. Cowardin, L. M., V. Carter, F. C. Goulet, and E. T. LaRoe, Classification of Wetlands and Deepwater Habitat of the United States, U.S. Fish and Wildlife Service, Washington, D.C., 1979.
37. Harenda, M. G., Evaluation of techniques to monitor wetland hydrology and macroinvertebrate community characteristics, M.S. thesis, Oregon State University, Corvallis, 1991.
38. Pennak, R. W., *Fresh-Water Invertebrates of the United States: Protozoa to Mollusca*, 3rd ed., John Wiley & Sons, New York, 1989.
39. Merritt, R. W. and K. W. Cummins, *An Introduction to the Aquatic Insects of North America*, 3rd ed., R. Kendall-Hunt, Dubuque, IA, 1996.
40. Courtney, G. W., R. W. Merritt, H. J. Teskey, and B. A. Foote, Aquatic Diptera. Part 1. Larvae of Aquatic Diptera, in *An Introduction to the Aquatic Insects of North America*, 3rd ed., R. W. Merritt and K. W. Cummins, Eds., Kendall-Hunt, Dubuque, IA, 1996, p. 484.

41. McCune, B. and M. J. Mefford, PC-ORD, Multivariate Analysis of Ecological Data, Ver. 2.0, MjM Software Design, Gleneden Beach, OR, 1995.
42. McCune, B. and M. J. Mefford, PC-ORD, Multivariate Analysis of Ecological Data, Ver. 3.0, MjM Software Design, Gleneden Beach, OR, 1997.
43. McCafferty, W. P., *Aquatic Entomology*, Jones and Bartlett, Boston, 1983.
44. Wrubleski, D. A., Chironomidae (Diptera) of peatlands and marshes in Canada, *Mem. Entomol. Soc. Can.* 140, 141, 1987.
45. Beck, W. M., Jr., Environmental Requirements and Pollution Tolerance of Common Freshwater Chironomidae, Environmental Monitoring and Support Laboratory, EPA-600/4-77-024, Office of Research and Development, Environmental Protection Agency, Cincinnati, OH, 1977.
46. Hudson, P. L., D. R. Lenat, B.A. Caldwell, and D. Smith, Chironomidae of the Southeastern United States: A Check List of Species and Notes on Biology, Distribution and Habitat, Res. Pap. 7, Fish and Wildlife Service, Fish and Wildlife, U.S. Department of Interior, Washington, D.C., 1990.
47. Ward, J. V., *Aquatic Insect Ecology*, John Wiley & Sons, New York, 1992.
48. Rosenberg, D. M. and H. V. Danks, Eds., *Aquatic Insects of Peatlands and Marshes in Canada*, Entomology Society of Canada, Ottawa, 1987.
49. Augenfeld, J. M., Effects of oxygen deprivation on aquatic midge larvae under natural and laboratory conditions, *Physiol. Zool.*, 40, 149, 1967.
50. Dougherty, J. E. and M. D. Morgan, Benthic community response (primarily Chironomidae) to nutrient enrichment and alkalinization in shallow, soft water humic lakes, *Hydrobiologia*, 215, 73, 1991.
51. Gorman, O. T. and J. R. Karr, Habitat structure and stream fish communities, *Ecology*, 59, 507, 1978.
52. McGarrigle, M. L., The distribution of Chironomid communities and controlling sediment parameters, in *Chironomidae: Ecology, Systematic, Cytology and Physiology*, D. A. Murray, Ed., Permagon Press, New York, 1980, p. 275.
53. Minshall, G. W., Aquatic insect-substratum relationships, in *The Ecology of Aquatic Insects*, V. H. Resh and D. M. Rosenberg, Eds., Praeger, New York, 1984, p. 358.
54. BiDanks, H. V., Some effects of photoperiod, temperature and food on emergence in three species of Chironomidae (Diptera), *Can. Entomol.* 110, 289, 1978.
55. Whitman, W. R., The Response of Macro-Invertebrates to Experimental Marsh Management, Ph.D. thesis, University of Maine, Orono, 1974.
56. Whitman, W. R., Impoundments for waterfowl, Canadian Wildlife Service, Pap. 22, 1976.
57. Cyr, H. and J. A. Downing, The abundance of phytophilous invertebrates on different species of submerged macrophytes, *Freshwater Biol.*, 20, 365, 1988.
58. Chapman, D. W., The relative contributions of aquatic and terrestrial primary producers to the trophic relations of stream organisms, *Pymantuing Lab. Ecol.*, Spec. Publ. No. 4, 116, 1966.
59. Lodge, D. M., Macrophyte-gastropod associations: observations and experiments on macrophyte choice by gastropods, *Freshwater Biol.*, 15, 695, 1985.
60. Cook, P. P., Jr. and H. S. Horn, A sturdy trap for sampling emergent Odonata, *Ann. Entomol. Soc. Am.*, 61, 1506, 1968.
61. Davies, I. J., Sampling aquatic insect emergence, in *A Manual On Methods For The Assessment Of Secondary Productivity in Freshwaters*, 2nd ed., J. A. Downing and F. H. Rigler, Eds., IBP Handbook No. 17, Blackwell Scientific, Oxford, U.K., 1984, p. 161.

62. Shannon, C. E. and W. Weaver, *The Mathematical Theory of Communication*, University of Illinois Press, Urbana, 1949.
63. Karr, J. R. and E. W. Chu, *Restoring Life in Running Waters: Better Biological Monitoring*, Island Press, Washington, D.C., 1999.
64. Ludwa, K. A., Urbanization Effects on Palustrine Wetlands: Empirical Water Quality Models and Development of Macroinvertebrate Community-Based Biological Index, M. S. thesis, University of Washington, Seattle, 1994.

5 Amphibian Distribution, Abundance, and Habitat Use

Klaus O. Richter and Amanda L. Azous

CONTENTS

Introduction 143
Methods 145
  Amphibian Surveys 145
  Wetland Characteristics 146
Results 147
  Distribution 147
  Abundance 151
  Watershed Urbanization 152
  Water Level Fluctuations 154
  Water Permanence 154
Discussion 155
Conclusions 158
Acknowledgments 158
References 159
Appendix 5-1: 1988 Capture Rates of Amphibians in Various Wetlands 162
Appendix 5-2: 1989 Capture Rates of Amphibians in Various Wetlands 163
Appendix 5-3: 1993 Capture Rates of Amphibians in Various Wetlands 164
Appendix 5-4: 1995 Capture Rates of Amphibians in Various Wetlands 165

INTRODUCTION

Amphibians are a diverse vertebrate class. Many breed and develop in wetlands and carry out the remainder of their life functions in nearby shrublands, forests, and other terrestrial habitats. Others may be entirely terrestrial, breeding, developing, and living entirely on land. Their potential abundance suggests a significant role in energy transfers, nutrient cycling, and food chain support functions. For example, in the Hubbard Brook Experimental Forest of New Hampshire, terrestrial-breeding salamander numbers (primarily red-backed salamander, *Plethodon cinereus*) regularly exceeded 2000 individuals per hectare with a concomitant biomass of 1.65 kg.[1] This

1-56670-386-7/00/$0.00+$.50

biomass equaled that of small mammals and was twice that of birds in the forest. Similar amphibian abundances have been estimated in southern Appalachian forests.[2]

At wetlands, large numbers of breeding amphibians and their larvae may reduce eutrophication by their net export of nitrogen. At some wetlands the nitrogen in tadpoles may be more than double that of residual pond nitrogen.[3] At others, it has been shown that amphibians (i.e., Leopard frog, *Rana pipiens*; bullfrog, *R. catesbeiana,* and mole salamanders, *Ambystoma* spp.) collectively export 6–12 times more nitrogen from the pond than imported by breeding adults. Equally important, tadpoles reduce the biomass of nitrogen-fixing blue-green algae and primary production by feeding on all forms of algae.[3,4]

In King County, wetlands are used by a wide array of amphibians.[5] Breeding Western toads, Pacific treefrogs, red-legged frogs, and bullfrogs may produce thousands of larvae and hundreds of post metamorphs and adults yearly. Anuran larvae (i.e., tadpoles) within these wetlands are opportunistic omnivores or detritivores thereby significantly influencing nutrient and energy dynamics.[6] Northwestern salamanders, long-toed salamanders, newts, and other caudates (i.e., salamanders and newts) generally produce fewer eggs and larvae than anurans, yet they similarly may be abundant at wetlands. Their larvae are carnivorous and recently hatched larvae feed extensively on zooplankton including water fleas [Cladocera], copepods [Copepoda] and other aquatic crustaceans [fairy, tadpole and clam shrimps: Eubranchiopoda].[7] Older larvae become voracious predators on larger aquatic invertebrates including midges [Chironomidae], mayflies [Ephemeroptera], and snails [Gastropoda]. Moreover, large caudate larvae feed on smaller frog and salamander larvae.[8-11]

Collectively, both anuran and caudate larvae provide food for larger predaceous aquatic invertebrates including insects and crustaceans.[10] Crayfish (Decapoda: Astacidae), dragonfly naiads [Odonata: Anisoptera], diving beetles [Coleoptera: Dytiscidae], and giant water bugs [Hemiptera: Belostomatidae] are well-known predators of amphibian larvae. Moreover, larvae provide food for salmon (*Oncorhynchus* spp.) and other fishes, reptiles (e.g., garter snakes (*Thamnophis* spp.), turtles (*Chrysemys* spp.), and wading birds (e.g., herons).[12] Upon metamorphosis to terrestrial forms, amphibians play pivotal roles in transferring biomass from wetland to adjacent terrestrial systems were they continue to be prey for reptiles, birds, and mammals. Finally, as adults all amphibians are carnivorous and, in turn, feed on a wide array of aquatic and terrestrial animals.

Along with our increasing recognition of the ecological importance of lentic-breeding amphibians in the food-chain dynamics of both aquatic and terrestrial ecosystems, studies have shown significant decreases of populations and the extinctions of others.[13-15] Some population declines have been difficult to confirm. For most we are uncertain of the proximate causes of amphibian decline because of inadequate information regarding their historical distribution, abundances, and habitat use.[16] This is clearly the situation for amphibians in the Central Puget Sound Basin and specifically for populations in urbanizing watersheds.

The occurrence of northwest amphibians depicted on range and spot maps indicates the historical distribution of at least 13 species (not counting the green frog) in King County;[17,18] 12 of these are associated with aquatic environments and 10 breed specifically in marshes, swamps, and bogs. We sighted seven lentic-breeding, one lotic-breeding, and two terrestrial-breeding species over a 2-year, 19-wetland survey in 1988 and 1989 in the Puget Lowland Ecoregion. [5] Furthermore, we previously reported that amphibian distribution was unrelated to wetland characteristics of size, number of vegetation classes, the presence of vertebrate predators, and water permanence. We found that lower species richness correlated with decreasing forest land within the watersheds of wetlands and hypothesized that altered wetland hydrology and, specifically, higher and more frequent water level fluctuations from greater impervious areas in urbanized watersheds, significantly contributed to this association.

In this chapter we continue to describe the geographic distribution of amphibians at the 19 wetlands in the Central Puget Sound Basin after an additional two years of surveys in 1993 and 1995. This work also provides information on the relative abundance of amphibians and we use these data to assess species distribution and population characteristics in the context of wetland and watershed condition.

METHODS

AMPHIBIAN SURVEYS

We determined the distribution and abundance of amphibians primarily by autumn pitfall trapping within wetlands buffers. We did not use drift fences because some wetlands were within heavily developed watersheds and we worried about drawing attention to traps, thereby encouraging vandalism. Trapping commenced with the first significant rains after our summer drought to maximize capture rates because amphibians are overall more active when the temperatures are cool and the soil surface and litter layer is wet. In autumn many northwest amphibians also move between wetlands and uplands and others migrate from summer feeding areas to winter hibernacula, increasing trapping rates. For all species excluding Pacific treefrogs, which were able to readily climb out of traps and bullfrogs which were no longer active in mid-October, we determined distribution and abundance from trapping surveys along two lines of 10 traps, each 10 m apart, and on opposite sides of wetlands. We ignored trap data when nightly temperatures dropped below 4°C and continued on warmer nights until 14 trap nights were logged to standardize trapping effort. Yearly and seasonal weather data for the region was obtained for Snoqualmie Falls and Landsburg stations, the closest sites to our wetlands.[19] Detailed site selection and trap installation procedures are described in a previous journal article.[5]

All trapped amphibians were identified to species and most were weighed and measured for length. Regulatory constraints prohibited us from removing or individually marking animals, hence we did not estimate abundances using mark-recapture methods. Rather, we use total captures as an index of abundances because we suspect little if any difference in the proportion of recaptured species either within a wetland or between wetlands. Moreover, our comparisons of the lengths and weights of trapped individuals suggests that very few, if any, animals were recaptures.

Spring egg survey counts were used to confirm red-legged frog and northwestern salamander breeding at wetlands. These included February through April searches of shoreline to 1 m deep in palustrine open water (POW), scrub-shrub (PSS), emergent (PEM), and forested (PFO) habitat types similar to those previously described.[20] Because of large egg masses and multiple surveys that individually identified egg masses we are confident that we detected most red-legged frog and northwestern salamander clutches and have a good index of their relative abundance.

Populations of red-legged frogs and northwestern salamanders were conservatively estimated from egg counts. We assumed each egg mass the yearly ovarian complement from one female and then extrapolated this value to total population associated with a wetland during a given year. In lieu of any empirical studies on sex ratios of breeding red-legged frogs we assumed a 50/50 male/female ratio. We also assume most red-legged frogs first successfully breed during the second year after metamorphosis. Adding an equal number of nonbreeding one-year-old frogs, the total yearly population of frogs is extrapolated to be four times the number of egg masses during a given year. For the northwestern salamander, a biased breeding sex ratio of three males to each female has been reported.[21] Also, northwestern salamanders take two years to metamorphose and we presume another year to reach sexual maturity, hence breeding may not occur until their fourth year. Therefore we estimate total yearly population at 12 times the number of egg masses. Although egg masses from breeding paedomorphs alter these extrapolations, some ad hoc aquatic funnel trapping and dip netting suggest paedomorphs are absent or in very low numbers. Consequently, for simplicity they are excluded from our extrapolations.

We also consider the presence of long-toed salamander clutches, larvae, and recently metamorphosed individuals of other species at wetlands as proof of breeding. However, because the eggs of these species are difficult to find (i.e., long-toed salamander, rough-skinned newt) or were not spawned till later after our spring census (western toad, bullfrogs), comprehensive surveys and counts for the eggs of these species were not undertaken.

Amphibian observations by knowledgeable biologists during small mammal, avifaunal, and other monitoring purposes augment our survey data.

Wetland Characteristics

Land use within the watershed directly affects hydrologic patterns (see Chapter 1) therefore we also monitored instantaneous and maximum water levels and calculated the average range of fluctuation to determine if these hydrologic descriptors affect the richness of amphibian communities. Methods for calculating these hydrologic descriptors are also presented in Chapter 1. Wetlands were classified as permanently or seasonally flooded based on whether water was present throughout the year. Locations and further descriptions of the wetlands are provided in Section I and in Chapters 1, 2, and 3 of Section II.

We determined wetland boundaries, wetland size, and habitat classes. We also obtained land cover within the watershed as well as within radii of 10, 100, 500, and 1000 m of each wetland from King County's Wetlands Inventory, King County Surface Water Management Division's GIS system, and from 1992 Landsat Thematic Map for the Puget Sound Region.[22,23] Using Landsat images we initially identified and characterized ten cover types,

1. Impervious surfaces,
2. Freeway/parking/gravel areas,
3. Cleared land,
4. Grasslands/golf courses,
5. Multifamily housing,
6. Single-family residential housing,
7. Single-family with forest,
8. Agriculture/pasture lands,
9. Forests, and
10. Open water.

These categories were collapsed to two cover types that we considered to be either favorable amphibian breeding, feeding, migration, and hibernation habitats (cover types 7–10) or unfavorable to these life functions (cover types 1–6).

We classified the ecoregion and identified wetland habitat structure using habitat classes mapped during historical wetland surveys superimposed on aerial photographs.[25,26] These were confirmed and updated by vegetation transect data and from site visits during our studies. Amphibian life history characteristics and habitat associations were taken from Nussbaum et al.[17] and our own observations. Nomenclature for caudates follows Petranka[27] and that for anurans is from Collins[28] with the exception of spotted frogs, for which we used Green et al.[29] We used *StatView SE+*® software to run statistical analyses.[59]

RESULTS

Distributions

We identified 10 of 14 amphibian species historically recorded in King County. This richness represents seven lentic-breeding, one lotic-breeding, and two terrestrial-breeding species (Table 5-1). Excluding the Cascades frog because of its higher elevational distribution beyond our wetland surveys, we identified all native lentic-breeding species of the regional fauna with the exception of the Oregon spotted frog. We sighted the nonnative-introduced bullfrog, however, we did not find a second introduced amphibian, the green frog.

Six native species representing 86% of possible native Central Puget Sound Basin lentic-breeding taxa were recorded across all wetlands. Most wetlands, however exhibited between three and four ($\bar{x}$ = 3.63) native lentic-breeding taxa or 48% of regional native lentic-breeding amphibians (Table 5-2). Five lentic-breeding species, 71% of collective richness, was the highest richness for any wetland and was recorded at both ELS61 and PC12. In contrast, no native species was identified at ELW1. B3I, with the second lowest richness, exhibited only two species, the Pacific treefrog and the long-toed salamander.

We sighted both terrestrial-breeding species native to the region. Six, or 32% of all wetlands, had both ensatina and the western red-backed salamander. Either one or the other was identified at 14 or 74% of total wetlands. Ensatina was found at 12 (63%) and western red-backed salamanders at 9 (47%) wetlands (Table 5-2).

TABLE 5-1
Amphibians Whose Range Includes King County and Their Habitat Associations (X)[17,18] (Includes Species Not Found During Our Studies)

Amphibians of King County — Taxa	Breeding Habitat			Non-Breeding Habitat		Native or Introduced
	Lentic	Lotic	Terrestrial	Aquatic	Terrestrial	
Frogs and Toads						
Red-legged frog (*Rana aurora*)	X			X	X	N
Oregon spotted frog (*Rana pretiosa*)	X			X	X	N
Cascade frog (*Rana cascadae*)	X			X	X	N
Pacific treefrog (*Hyla regilla*)	X				X	N
Bullfrog (*Rana catesbeiana*)	X			X		I
Green frog (*Rana clamitans*)	X			X		I
Tailed frog (*Ascaphus truei*)		X		X	X	N
Western toad (*Bufo boreas*)	X				X	N
Salamanders and Newts						
Northwestern salamander (*Ambystoma gracile*)	X				X	N
Long-toed salamander (*Ambystoma macrodactylum*)	X				X	N
Rough-skinned newt (*Taricha granulosa*)	X				X	N
Pacific giant salamander (*Dicamptodon tenebrosus*)		X			X	N
Western red-backed salamander (*Plethodon vehiculum*)			X		X	N
Ensatina (*Ensatina eschscholtzii*)			X		X	N

Although we identified identical regional amphibian faunas in surveys conducted in 1993 and 1995 to those species observed in 1988 and 1989, we unexpectedly found additional species at some wetlands where they were not identified in earlier years. Similarly, some species were missing from 1993 and 1995 that appeared in earlier surveys. At five wetlands, species richness (wetland and terrestrial breeders) in 1989 was higher than in 1995 whereas at four wetlands, fewer total species were captured. Consequently, captures during one year did not guarantee captures during other years. For example, red-legged frogs were captured at two wetlands in 1988, five in 1989, and only one in 1995. Only at two wetlands were red-legged frogs captured during all four years of trapping. Similarly, northwestern salamander were captured at three wetlands in 1988, three different wetlands in 1989, and only one in 1995. They were not captured at any wetland in 1993. The long-toed salamander exhibited similar capture patterns but was trapped during all four years at one wetland. Both the northwestern and long-toed salamanders were captured in only four wetlands in three of four years.

TABLE 5-2
Amphibian Distribution Within Wetland of the Puget Lowland Ecoregion[a]

Common Name	Scientific Name	AL3	B31	BBC24	ELS39	ELS61	ELW1	FC1	HC13	JC28	LCR93	LPS9	MGR36	NFIC12	PC12	RR5	SC4	SC84	SR24	TC13	Sum of Wetlands[a]	Percent of Total Wetlands
Bullfrog	*Rana catesbeiana*																				7	37%
Ensatina	*Ensatina eschscholtzii*																				12	63%
Long-toed Salamander	*Ambystoma macrodactylum*																				13	68%
Northwestern Salamander	*Ambystoma gracile*																				15	79%
Pacific Giant Salamander	*Dicamptodon tenebrosus*																				2	11%
Pacific Treefrog	*Hyla regilla*																				18	95%
Red-legged Frog	*Rana aurora*																				16	84%
Rough-skinned Newt	*Taricha granulosa*																				3	16%
Western Red-backed Salamander	*Plethodon vehiculum*																				9	47%
Western Toad	*Bufo boreas*																				4	21%
Total Number of Species in Wetland		5	2	7	5	6	1	4	8	6	6	5	5	5	7	4	5	5	8	5	mean:	5.21
Number of Native Species		5	2	6	5	5	0	3	7	6	6	5	5	4	7	4	5	5	7	5	mean:	4.84
% Total Richness of Native Species		56	22	67	56	56	0	33	78	67	67	56	56	44	78	44	56	56	78	56	mean:	54%

[a] FC1 and ELW1 were only sampled in 1988 with no captures in ELW1. AL3, NFIC12, SC84, and TC13 were not sampled in 1988.

TABLE 5-3
Presence of Breeding Amphibians Including Estimates of Minimum Relative Egg Mass Numbers. (The presence of western toads is based on tadpoles and recently metamorphosed toad sightings. Egg masses and larvae of other species were not censused.)

Wetland	Red-Legged Frog	Northwestern Salamander	Pacific Treefrog	Long-Toed Salamander	Western Toad
AL3	A	A	P	P	A
B3I	A	A	P	A	A
BBC24	1	4	P	P	M
ELS39	A	A	A	A	A
ELS61	2	3	P	P	A
ELW1	A	A	A	A	A
FC1	A	A	A	A	A
HC13	A	2	P	P	A
JC28	A	A	A	A	A
LCR93	2	A	P	P	A
LPS9	1	2	A	A	A
MGR36	1	2	P	A	A
NFIC12	A	A	A	A	A
PC12	2	3	P	P	L
RR5	2	4	P	P	M
SC4	A	A	A	A	A
SC84	A	1	A	A	A
SR24	2	4	P	P	A
TC13	3	1	1	P	A

Note: P = present; A = absent; 1 = 1–10; 2 = 11–50; 3 = 51–100; 4 = >100.

The Pacific treefrog is clearly the most widely distributed lentic-breeding amphibian among our wetlands, being sighted at 18 of 19 or 95% of our sites (Table 5-2). Red-legged frogs, long-toed salamanders, and northwestern salamanders were the next most broadly distributed lentic-breeding species, being sighted at 13 to 16 or 68 to 84% of wetlands. In contrast, the rough-skinned newt and the western toad exhibited the most restrictive distribution. Newts were only identified at three wetlands. Interestingly, western toads were last sighted in 1989, when they were trapped at three wetlands, although they were also identified in 1988 at one other wetland. Bullfrogs were identified at seven wetlands.

Neither the tailed frog nor Pacific giant salamander were expected because of the lotic-breeding biology of both species. Nonetheless, Pacific giant salamanders were captured at PC12 and SR24. Presumably, these animals dispersed from adjoining creeks and streams.

The northwestern salamander, long-toed salamander, red-legged frog, Pacific treefrog, and western toad were confirmed breeding at four wetlands (Table 5-3). At least 100 northwestern salamander egg masses were counted at BBC24, SR24, and RR5, suggesting minimum populations of 1200 salamanders associated with these wetlands. ELS61 and PC12 egg mass counts were between 50 and 100 indicating these wetlands had between 600 and 1200 salamanders nearby. Between 50 and 100 red-legged frog clutches at TC13 extrapolates to a minimum population of 200–400 frogs adjacent to these wetlands. Although both red-legged frogs and northwestern salamanders were found at both permanently and seasonally flooded wetlands, red-legged frogs bred in greater numbers at seasonally flooded wetlands and northwestern salamanders in permanently flooded wetlands. Moreover, when both species were found within wetlands, the number of egg masses of one species exceeded the other by more than 90%, again suggesting possible differences between red-legged frogs and northwestern salamander preferences for breeding wetlands.

Although the timing of oviposition varied by several weeks between years, little year-to-year variation in total egg masses existed at wetlands with the exception of declining numbers at PC12 and LPS9. At PC12 the beaver dam outlet was breached in 1990, draining most of the wetland. Consequently, over the next few years cattail, beggar's tick, and reed canarygrass invaded and colonized the site eliminating the red-legged frog breeding habitat, and resulting in a decrease from a high of 63 clutches in 1995 to two in 1997. Correspondingly, northwestern salamander eggs decreased from eight egg masses in 1995 to one in 1997. Western toads disappeared in 1989. At LPS9 egg masses declined from a high of 45 in 1993, nine in 1995, two in 1987, to zero in 1998 without any noticeable changes to wetland hydrology or land use.

Abundance

The red-legged frog, after the Pacific treefrog, is the second most abundant and widely distributed lentic-breeding amphibian as determined by pitfall captures. It was caught at 9 and 11 wetlands in 1988 and 1989, respectively, with capture rates exceeding one animal per 100 trap nights (TN) on six occasions during both years (Figure 5-1; Appendixes 1 to 4). LCR93 exhibited the highest red-legged frog populations with captures equal to or exceeding two frogs per 100 TN in 1988 and 1993. Long-toed salamanders, and northwestern salamanders exhibited a more restricted distribution and were captured in relatively modest numbers of 0.3 to 0.7 animals per 100 TN. Interestingly, adult red-legged frogs were captured in pitfalls at AL3, although neither spawn nor recently metamorphosed frogs were sighted during spring and summer surveys.

Capture rates of northwestern salamanders exceeded one per 100 TN on only two occasions, whereas long-toed salamanders exceed this abundance on only one occasion. The highest capture rate of any species was 9.7 northwestern salamanders per 100 TN and was biased by the capture of 29 salamanders during one night in 1989 at BBC24 (Figure 5-1; Appendix 2).

We found significant differences in the abundance of species captured within wetlands between 1988 and 1995; 1988 and 1989 were ranked similarly with average capture rates of 2.1 and 2.3 individuals per 100 TN, respectively, but differed significantly from 1993 and 1995 in which average capture rates were 0.46 and 0.83, respectively (Friedman test, $\chi^2 = 19.6$, $p = .0004$). Over the study period, the number of amphibian captures per 100 TN declined in 12 of the 19 wetlands. Six wetlands

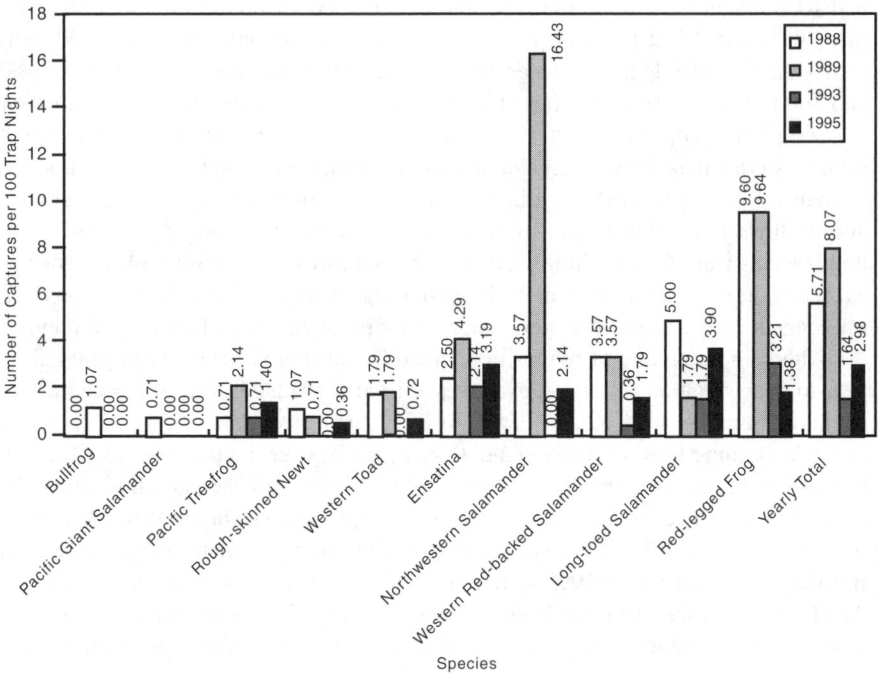

FIGURE 5-1 Species and abundance of amphibians captured in pitfall traps per 100 trap nights.

showed the highest capture rates in 1989 and then declined. Only one wetland, SC84, showed a slight (0.3) increase in capture rate between 1988 and 1995.

Among individual species, pitfall captures were too low to draw conclusions regarding population trends. The exceptions were significantly lower capture rates between 1988 and 1993 for red-legged frog (paired *t*-test, $p = .01$) and ensatina ($p = .005$).

Our review of temperature and precipitation during the survey years and the year prior to the surveys did not identify weather conditions to be unique enough in any given year to account for differences in amphibian abundances during pitfall trapping. Temperature and precipitation during autumn trapping, spring spawning, or summer metamorphosing were not different enough to account for declines (Table 5-4). Cold weather-deterring surface movement, either during breeding or autumn, was not identified during years of low captures. Correspondingly, hot dry summers, which may have dried out seasonally flooded wetlands prior to metamorphosis and possibly reducing recruitment, also were not recorded during the year or in the year prior to low abundances.

Watershed Urbanization

Urbanization in watersheds of wetlands was negatively correlated with combined totals of native lentic and terrestrial breeding amphibian richness at wetlands (Figure 5-2). Wetlands with increasing urbanization in contributing watersheds are signifi-

TABLE 5-4
Comparative Temperature and Precipitation Values for Survey Year and the Year Prior to Surveys Based on 30-Year Values. The Symbols >, <, and = Indicate Greater Than, Less Than, or Approximately Equal to the Mean Value, Respectively. The Decision Factor for Determining the Presence of a Difference Was 2.5°F (1.4°C) for Temperature and 1.75 in. (4.4 cm) for Precipitation

		1988		1989		1993		1995	
Survey Year		**L**	**S**	**L**	**S**	**L**	**S**	**L**	**S**
Trapping	Temp.	>	>	=	=	>	>	<	=
(Oct.)	Precip.	=	=	<	=	=	<	>	=
Metamorphosing	Temp.	=	=	=	=	=	=	=	=
(July–Sept.)	Precip.	=	>	<	=	=	=	=	=
Spawning	Temp.	=	=	<	<	=	=	=	=
(Feb.–Mar.)	Precip.	=	=	=	=	<	<	=	=

		1987		1988		1992		1994	
Previous Year		**L**	**S**	**L**	**S**	**L**	**S**	**L**	**S**
Trapping	Temp.	=	=	>	>	=	=	=	=
(Oct.)	Precip.	<	<	=	=	<	=	=	>
Metamorphosing	Temp.	=	=	=	=	=	=	=	>
(July–Sept.)	Precip.	<	<	=	>	=	=	=	<
Spawning	Temp.	=	>	=	=	>	>	=	=
(Feb.–Mar.)	Precip.	=	=	=	=	<	<	=	=

Note: L = Landsburg; S = Snoqualmie Falls.

cantly more likely to have lower amphibian richness of less than four native species than wetlands in less urbanized watersheds (Fisher's r to z, r = –0.55, p = .01). Two wetlands with high amphibian richness exceeding six native species had the lowest watershed urbanization. Not all urbanized wetlands had low native amphibian richness but the likelihood was much higher.

Urbanized land uses immediately adjacent to a wetland decreased native amphibian richness. We found that within concentric rings encompassed by radii of 10, 100, 500, and 1000 m from the wetland edge, amphibian richness was related to the percentage of forest land available. In general, wetlands adjacent to larger areas of forests were more likely to have richer populations of native amphibians. The significance of this relationship was weakest for the area circumscribed by a 10 m buffer around the wetland edge 10-m radius (Fisher's r to z, r = 0.38, p = .11) and strongest for the 1000-m buffer surrounding wetlands (r = .71, p = .0004). Almost all wetlands had high proportions of forest land within 10 m (most likely attributable to buffer requirements) and to a lesser extent at 100 m. But amphibian richness is

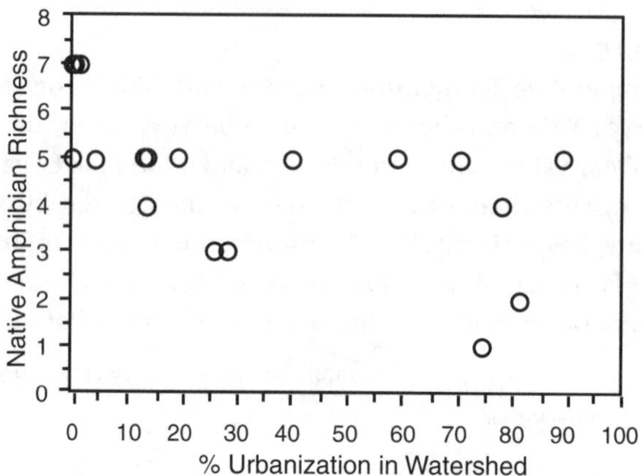

FIGURE 5-2 Relationship between native amphibians (including both lentic and terrestrial breeders) and percent of watershed urbanization.

highest in wetlands that retain at least 60% of adjacent area in forest land up to and exceeding 500 m from the wetland, and lowest in the wetlands that, while having a high proportion of forest land within 10 or 100 m, lost significant forest land at 500 m and further from the wetland.

Water Level Fluctuations

Of the four hydrological parameters investigated, minimum, maximum, average water level, and water level fluctuation (WLF), only average WLF showed a statistically significant relationship with lentic-breeding amphibian richness. When average WLF was 20 cm or more during the year, the number of native lentic-breeding amphibian species averaged three or fewer. Wetlands with lower WLFs (less than 20 cm) were significantly more likely to have a higher proportion of lentic-breeding amphibian richness, averaging four species (Figure 5-3, Mann Whitney, $p = .04$) as compared with an average of 2.9 species in wetlands with WLF exceeding 20 cm.

Water Permanence

Water permanence at wetlands was not correlated with native lentic-breeding amphibian richness (Mann-Whitney, $p = .13$). Bullfrog distribution, as expected, was related to water permanence, being found at seven of ten (70%) permanent wetlands and at zero out of nine (0%) semipermanent wetlands. We found bullfrogs at ELW1 and FC1 along Lake Washington, several wetlands on the Sammamish plateau (ELS61, NFIC12), and at wetlands of the Bear Creek (BBC24), Snoqualmie River (SR24), and Harris Creek drainages (HC13). They were not seen at PC12, presumably because of shallow water. Bullfrogs also were not identified at either MGR36 or RR5 most likely because these wetlands are the furthest from urbanization and exhibit little development within their watersheds. Interestingly, lentic-breeding amphibian richness at wetlands with bullfrogs ($N = 7$, $\bar{x} = 3.57$) did not

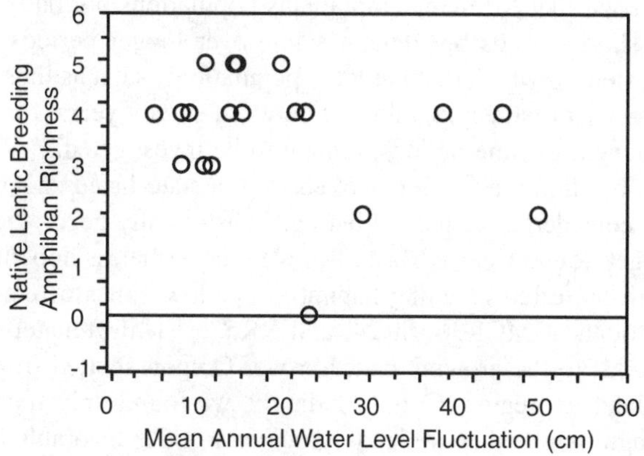

FIGURE 5-3 Relationship between native lentic breeding amphibian richness and mean annual water level fluctuation.

differ in richness from wetlands without bullfrogs (N = 12, $\bar{x} = 3.66$). Unexpectedly, the northwestern salamander, which requires at least one year of permanent flooding was found in seven of nine semipermanent wetlands (77%) and eight of ten (80%) permanent wetlands. Populations at LPS9 and possibly other seasonally flooded wetlands may be disjunct remnants of original populations.

DISCUSSION

During 1988 and 1989 we identified eight species of anurans and six species of caudates at 19 wetlands located in the Central Puget Sound Basin.[5] We identified the same collective richness after additional surveys in 1993 and 1995. At a few wetlands we identified additional species, and for the most part we captured fewer species across all wetlands in 1993 and 1995. We expected to get reduced richness with fewer captures.

Our study shows different amphibian richness and abundances between the survey years from pitfall trap captures, suggesting that sampling with multiple techniques and for several years is a prerequisite to the accurate identification and quantification of a wetlands' amphibian fauna. Explanations accounting for differences could include periodic fluctuations in abundance that may or may not be weather related.[30] However, if weather related, one would expect populations to respond similarly across all wetlands. Our data, however, did not demonstrate this, as richness increased at six wetlands while decreasing at five between successive years, suggesting that when looking at individual species population trends are also variable. For example, pitfall captures of *A. gracile*, increased at four wetlands, decreased at four, and remained unchanged at two wetlands between 1988 and 1989. Furthermore, capture values differ between our initial 1988, 1989, and later 1993 and 1995 surveys, implying that local factors, possibly physiognomy, microclimatology, or wetland condition rather than regional climate may account for species distribution, population sizes, and activity patterns. Other investigators, in analyzing data from long-term studies, have shown that for many amphibians populations normally fluctuate dramatically over short periods but remain stable over longer periods of five to ten years.[30,31] The extent to which distinct local populations, such as those found in our wetlands, vary asynchronously within and between survey years, for what reasons, and over what length of time periods, remain to be investigated.

We still did not find the Oregon spotted frog, a state-listed endangered species currently being considered for federal listing.[32] Historically never abundant in wetlands of the Puget Sound Central Basin, we were nevertheless hopeful to find them at wetlands with preferred breeding habitat of shallow zones of emergent grasses such as those found at MGR36, SR24, and RR5.[33,34] Unfortunately, our negative results further adds to the growing concern over Oregon spotted frog extinction in the Puget Lowland Ecoregion. Correspondingly, we found only four western toad localities although many other wetlands appeared to have favorable breeding habitats. This species has recently been listed as a state species of concern.[32] Our last sightings of western toads were in 1989, suggesting local toads may possibly be experiencing the same declines noted for other populations in the western U.S.[15]

Bullfrogs were identified at all permanently flooded wetlands near urbanized areas but not at wetlands at greater distances from established populations. However, we predict that bullfrogs will disperse to these uncolonized wetlands aided by urbanization. The decline of native amphibians associated with bullfrogs remains equivocal. Our surveys support previous studies which show no clear relationship between bullfrog presence and native species, primarily because of confounding influences and unexplained causal relationships.[35,36] Declines of endemic amphibians in the presence of bullfrogs, however, have been found in many surveys.[35,37-41] Furthermore, detrimental interactions within enclosure experiments suggest reduced survival of red-legged frog and Pacific treefrog in the presence of bullfrog larvae.[42] From our work reported here as well other surveys, we believe that in structurally simple wetlands bullfrogs extirpate many native species, either through outright predation or indirectly through competition and agonistic behavior, whereas in hydrologically, vegetatively, and overall structurally complex wetlands, native amphibians and bullfrogs may coexist.

Despite the low overall amphibian richness of the Puget Lowlands when compared to the diversity of the southeast and central states, the biomass of northwest species may be high.[43-45] The pitfall capture of 29 northwestern salamanders in one night and the presence of more than 100 egg masses at BBC24 clearly show the numerical and potential ecological importance of this species. A minimum of 26 (40 more recently counted) red-legged frog eggs at the 150-m^2 breeding pond at TC13 and 80 at PC12 underscores the abundance of red-legged frogs and again champions the ecological significance of amphibians at northwest wetlands. Although, the extrapolation of egg mass numbers to population for both red-legged frogs and northwestern salamanders remains speculative, these numbers suggest amphibians may still be present in high densities at some of our wetlands.

This research confirms our earlier survey findings that the number of amphibian species and wetland size are unrelated, a finding now supported by numerous other studies throughout the U.S.[46,47] Consequently, the importance of small wetlands as amphibian breeding sites and as population sources for larger wetlands is reconfirmed and strongly argues for protecting small wetlands to maintain biodiversity. The absence of a correlation between amphibian richness and the number of habitat classes and the correlation of richness with adjacent upland forest cover additionally indicates that vegetatively simple wetlands provide important breeding habitat if favorable nonbreeding upland habitat is available. Most important is the presence of a well-established thin-stemmed emergent vegetation zone composed of stems of herbaceous species, tiny branches and twigs of submerged vegetation on which breeding amphibians attach their eggs and in which larvae find food and shelter.[48]

We confirmed an inverse relationship between urbanization and native species richness. Additionally, our data suggest that the reduced richness may simultaneously be due to reduced breeding success attributable to differences in wetland hydrologic patterns from increased runoff and to a reduction of nonbreeding habitat from land-use conversion. Average WLF increases as the frequency of peak flood events increases from greater impervious surfaces resulting from development.[49,50] The eggs of lentic-breeding species die from freezing and desiccation as water levels recede after storms. In addition, we observed fewer egg masses of northwestern salamanders and red-legged frogs, and the disappearance of western toads at wetlands that have been drained, where water permanence decreased, and where previously flooded areas have become densely vegetated. These observations provide direct evidence of breeding habitat loss and corresponding population declines.

Terrestrial breeding species may also be affected at wetlands with high WLF from seasonal flooding of buffers and overall wet buffer conditions. Terrestrial breeders tend to avoid soaked or flooded sites, being more frequently found in cool, flat, well-drained soils.[51,52] Low numbers in wet riparian as opposed to dryer upland habitats have, for example, been documented for ensatina, in red alder, second-growth conifer, and unmanaged Douglas fir stands.[51-55] Flooded habitats also may have less litter and other organic material that provide habitat for invertebrate food sources and cover and breeding sites for these species.

Our surveys also supported the importance of forest lands adjacent to and up to 1000 m from wetlands. Such areas no doubt provide essential nonbreeding habitat for all of our native species and provide food, cover, migration corridors, and road-free environments. Others have similarly detected inverse relationships between urbanization and species richness and attribute these findings to habitat fragmentation, distances between neighboring wetlands, and to road density.[56,57] The distance isolating wetlands, in part, explains the distribution of bullfrogs at our sites and may also explain other findings including the low numbers of species at FC1, ELW1, and B3I, which are located in older urbanized areas.

Water permanence did not account for native lentic-breeding amphibian richness, implying that water in seasonally flooded wetlands remained long enough for successful reproduction of most lentic-breeding species. Unexpectedly, we found northwestern salamanders, which take two years to metamorphose, at seven seasonally flooded wetlands. At this time it remains undetermined whether breeders are older individuals returning to once permanent natal sites, individuals only successfully breeding during infrequent wet years when ponds remain permanently flooded, or whether the breeders are dispersers from nearby source populations.

At other sites amphibian observations are also not clearly explainable. The sighting of adults but no eggs at wetlands (e.g., red-legged frogs at AL3) suggests

1. Amphibians historically bred at a wetland but breeding conditions may have since become unfavorable;
2. Sighted individuals are dispersers from elsewhere and wetland conditions were never favorable for breeding;
3. The species may not breed every year; and
4. Breeding may have occurred but eggs remain undetected.

Only more comprehensive studies over longer intervals of time will be able to explain such observations.

We found differences in richness at wetlands depending on survey technique, suggesting that multiple methods should be employed to accurately assess a wetland's amphibian population. Northwestern salamanders were not captured at SR24 presumably indicating an absence of these species even though large numbers of egg masses indicated large numbers of breeding animals. Correspondingly, we captured rough-skinned newts in funnel traps at ELS61 yet never saw or captured them in pitfall traps originally suggesting this species absence.

CONCLUSIONS

In conclusion, our surveys suggest the distribution of amphibian populations are attributable to numerous interacting factors influencing amphibians throughout their varied life stages. We attribute some declines to egg and larval mortality from changes in depth, duration, and frequency of wetland flooding associated with an increase in impervious surfaces from urbanization. Simultaneously, we found the loss of forests from urbanization decreases nonbreeding habitat and increases wetland isolation, making it increasingly difficult for amphibians to interact with each other at the metapopulation level. We have yet to carry out landscape studies with individually identified animals to assess the roles of dispersal and source sink phenomenon. Such work is critical in developing a strategy for the conservation of amphibians before their breeding wetlands and adjacent nonbreeding landscapes are developed.

ACKNOWLEDGMENTS

Our thanks to Mike Emers, Mike Gratz, Timothy Quinn, Claus Svendsen, Heather Roughgarden, and the many wildlife biologists that helped in autumn pitfall trapping of both amphibians and small mammals.

REFERENCES

1. Burton, T. M. and G. E. Likens, Salamander populations and biomass in the Hubbard Brook Experimental Forest, New Hampshire, *Copeia,* 3, 541, 1975.
2. Cohn, J. P., Salamanders slip-sliding away or too surreptitious to count?, *BioScience,* 44, 219, 1994.
3. Seale, D. B., Influence of amphibian larvae on primary production, nutrient flux, and competition in a pond ecosystem, *Ecology,* 61, 1531, 1980.
4. Beebee, T. J. C., *Ecology and Conservation of Amphibians*, Chapman & Hall, London, 1996.
5. Richter, K. O and A. L. Azous, Amphibian occurrence and wetland characteristics in the Puget Sound Basin, *Wetlands,* 15, 305, 1995.
6. Hoff, K. vS, A. R. Blaustein, R. W. McDiarmid, and R. Altig, Behavior, interactions and their consequences, in *Tadpoles: the Biology of Anuran Larvae*, R. W. McDiarmid and R. Altig, Eds., University of Chicago Press, Chicago, 1999, p. 215.
7. Blaustein, L., Non-consumptive effects of larval *Salamandra* on crustacean prey: can eggs detect predators?, *Oecologia,* 110, 212, 1997.
8. Calef, G. W., Natural mortality of tadpoles in a population of *Rana aurora*, *Ecology,* 54, 742, 1973.
9. Licht, L. E., Survival of embryos, tadpoles and adults of the frogs *Rana aurora aurora* and *Rana pretiosa pretiosa* sympatric in southwestern British Columbia, *Can. J. Zool.*, 52, 613, 1974.
10. Woodward, B. D., Predator-prey interactions and breeding-pond use of temporary-pond species in a desert anuran community, *Ecology,* 64, 1549, 1983.
11. Hoffman, E. A. and D. W. Pfennig, Proximate causes of cannibalistic polyphenism in larval tiger salamanders, *Ecology,* 80, 1076, 1999.
12. Hecnar, S. J. and R. T. M'Closkey, The effect of predatory fish on amphibian species richness and distribution, *Biol. Conserv.,* 79, 123, 1997.
13. Wyman, R. L., What's happening to the amphibians?, *Biol. Conserv.,* 4, 350, 1990.
14. Wake, D. B., Declining amphibian populations, *Science,* 253, 860, 1991.
15. Corn, P. S., What we know and don't know about amphibian declines in the west, in *Sustainable Ecological Systems: Implementing an Ecological Approach to Land Management*, T. C. W. Covington and Leonard F. DeBano, Eds., U.S. Department of Agriculture, Rocky Mountain Forest and Range Experiment Station, Fort Collins, CO, 1994, p. 59.
16. Sarkar, S., Ecological theory and anuran declines, *BioScience,* 46, 199, 1996.
17. Nussbaum, R. A., D. B. Edmond, Jr., and R. M. Storm, *Amphibians & Reptiles of the Pacific Northwest*, University of Idaho Press, Moscow, ID, 1983.
18. Leonard, W. P., H. A. Brown, L. L. C. Jones, K. R. McAllister, and R. M. Storm, *Amphibians of Washington and Oregon,* Seattle Audubon Society, Seattle, WA, 1993.
19. NOAA, Washington Climate Summaries: Snoqualmie Falls (#82) and Landsburg (#83), Washington, D.C., http://www.wrcc.dri.edu/summary/mapwa.html, 1999.
20. Thoms, C., C. C. Corkran, and D. H. Olson, Basic amphibian survey for inventory and monitoring in lentic habitats, in *Sampling Amphibians in Lentic Habitats*, D. H. Olson, W. P. Leonard, and R. B. Bury, Eds., Northwest Fauna No. 4, Society for Northwestern Vertebrate Biology, Olympia, WA, 1997.
21. Stringer, A., Intensive study: northwestern salamander, Appendix B, in *Wildlife Use of Managed Forests: A Landscape Perspective,* Vol. 2, Aubrey, K. B., S. D. West, D. A. Manuwal, A. B. Stringer, J. L. Erickson, and S. Pearson, Timber, Fish and Wildlife Rep. No. TFW-WL4-98-002, 1997.

22. Wetlands Inventory — 3 Volumes, King County Environmental Division, Bellevue, WA, 1990.
23. Puget Sound Regional Council, Satellite remote sensing project-land cover and change detection, Seattle, WA, 1994.
24. Omernik, J. M., Ecoregions of the conterminous United States, *Ann. Assoc. Am. Geograph.,* 77, 118, 1987.
25. Cowardin, L. M., V. Carter, F. C. Goulet, and E. T. LaRoe, Classification of Wetlands and Deepwater Habitat of the United States, U.S. Fish and Wildlife Service, Washington, D.C., 1979.
26. Wetland Characterizations, Resource Planning Section, King County, WA, 1987.
27. Petranka, J. W., *Salamanders of the United States and Canada*, Smithsonian Institution Press, Washington, D.C., 1998.
28. Collins, J. T., Standard common and current scientific names of North American amphibians and reptiles, Herpetological Circular 19, Society for the Study of Amphibians and Reptiles, Department of Biology, St. Louis University, MO, 1990.
29. Green, M., H. Kaiser, T. F. Sharbel, J. Kearsley, and K. R. McAllister, Cryptic species of spotted frogs, *Rana pretiosa* complex, in western North America, *Copeia,* 1, 1997.
30. Pechmann, J. H. K., D. E. Scott, R. D. Semlitsch, J. P. Caldwell, L. J. Vitt, and J. W. Gibbons, Declining amphibian populations: The problem of separating human impacts from natural fluctuations, *Science,* 253, 892, 1991.
31. Hairston, N. G., *Community Ecology and Salamander Guilds*, Cambridge University Press, Cambridge, MA, 1987.
32. Washington Department of Fish and Wildlife, Species of Concern, Effective 16 March, URL: *http://www.wa.gov/wdfw/wlm/diversity/soc/soc.htm,* 1999.
33. McAllister, K. R. and B. P. Leonard, Past Distribution and Current Status of the Spotted Frog (*Rana pretiosa*) in Western Washington, 1989 Prog. Rep., Washington State Department of Wildlife, Olympia, 1990.
34. McAllister, K. R. and B. P. Leonard, Past Distribution and Current Status of the Spotted Frog in Western Washington, 1990 Prog. Rep., Washington Department of Wildlife, Olympia, 1991.
35. Hayes, M. P. and M. R. Jennings, Decline of ranid frog species in western North America: are bullfrogs (*Rana catesbeiana*) responsible?, *J. Herpetol.,* 20, 490, 1986.
36. Hayes, M. P. and M. R. Jennings, Habitat correlates of distribution of the California red-legged frog (*Rana boylii*): implications for management, in *Management of Amphibians, Reptiles and Small Mammals in North America*, R. Szaro, K. Severson, and D. Patton, Eds., U.S. Forest Service Tech. Rep. RM-166, Washington, D.C., 1988, p. 144.
37. Moyle, P. B., Effects of introduced bullfrog, *Rana catesbeiana*, on the native frogs of the San Joaquin Valley, California, *Copeia,* 1973, 18, 1973.
38. Schwalbe, C. R. and P. C. Rosen, Preliminary report on effects of bullfrogs on wetland herpetofaunas in southern Arizona, in *Management Of Amphibians, Reptiles And Small Mammals In North America*, R. Szaro, K. Severson, and D. Patton, Eds., U.S. Forest Service Tech. Rep., RM-166, Washington, D.C., 1988, p. 166.
39. Fisher, R. N. and H. B. Shaffer, The decline of amphibians in California's Great Central Valley, *Biol. Conserv.,* 10, 1387, 1996.
40. Jameson, D. L., Growth, dispersal and survival of the Pacific treefrog, *Copeia,* 1956, 25, 1956.
41. Green, D. M., Northern leopard frogs and bullfrogs on Vancouver Island, *The Canadian Field-Naturalist,* 92:78–79, 1978.

42. Adams, M. J., Correlated factors in amphibian decline: exotic species and habitat change in western Washington, *J. Wildl. Manage.*, 63, 1162, 1999.
43. Gibbons, J. M. and R. D. Semlitsch, *Guide to the Reptiles and Amphibians of the Savannah River Site,* University of Georgia Press, Athens, GA, 1991.
44. Clarke, R. F., An ecological study of reptiles and amphibians in Osage County, Kansas, *Emporia State Res. Studie,* 7, 1, 1958.
45. Clawson, M. E. and T. S. Baskett, Herpetofauna of the Ashland Wildlife Refuge area, Boone County, Missouri, *Trans. Missouri Acad. Sci.,* 16, 5, 1982.
46. Dodd, C. K., Jr. and B. S. Cade, Movement patterns and the conservation of breeding amphibians in small temporary wetlands, *Biol. Conserv.,* 12, 331, 1998.
47. Semlitsch, R. D. and J. R. Bodie, Are small, isolated wetlands expendable?, *Biol. Conserv.,* 12, 1129, 1998.
48. Richter, K. O. and Roughgarden, H. A., unpublished manuscript, 2000.
49. Booth, D. B., Urbanization and the natural drainage system — impacts, solutions, and prognoses, *Northwest Environ. J.,* 7, 93, 1991.
50. Schueler, T., The importance of imperviousness, *Watershed Prot. Tech.,* 1, 100, 1994.
51. Aubry, K. B. and P. A. Hall, Terrestrial amphibian communities in the Southern Washington Cascade Range, in *Wildlife and Vegetation of Unmanaged Douglas-Fir Forests*, K. B. Aubry and M. H. Brookes, Eds., U.S. Forest Service, Pacific Northwest Research Station, Portland, OR, 1991, p. 327.
52. Gilbert, F. F. and R. Allwine, Terrestrial amphibian communities in the Oregon Cascade Range, in *Wildlife and Vegetation of Unmanaged Douglas-Fir Forests*, K. B. Aubry and M. H. Brookes, Eds., U.S. Forest Service, Pacific Northwest Research Station, Portland, OR, 1991, p. 319.
53. McComb, W. C., C. L. Chambers, and M. Newton, Small mammal and amphibian communities and habitat associations in red alder stands, central Oregon coast range, *Northwest Sci.,* 67, 181, 1993.
54. McComb, W. C., K. McGarigal, and R. G. Anthony, Small mammal and amphibian abundance in streamside and upslope habitats of mature Douglas-fir stands, western Oregon, *Northwest Sci.,* 67, 7, 1993.
55. Gomez, D. M. and R. G. Anthony, Amphibian and reptile abundance in riparian and upslope areas of five forest types in Western Oregon, *Northwest Sci.,* 70, 109, 1996.
56. Fahrig, L., J. H. Pedlar, S. E. Pope, P. D. Taylor, and J. F. Wegner, Effect of road traffic on amphibian density, *Biol. Conserv.,* 73, 177, 1995.
57. Vos, C. C. and J. P. Chardon, Effects of habitat fragmentation and road density on the distribution pattern of the moor frog *Rana arvalis*, *J. Appl. Ecol.,* 35, 44, 1998.
58. Richter, K. O. and H. A. Roughgarden, Wetland characteristics and oviposition site selection by northwestern salamander (*Ambystoma gracile*), Unpublished manuscript, King County Department of Natural Resources, Water and Land Resources Section, Seattle, WA, 2000.
59. Abacus Concepts Inc., D. S. Feldman, Jr., R. Hoffman, J. Gagnon, and J. Simpson, *StatView SE+*, Berkley, CA, 1988.

APPENDIX 5-1
1988 Capture Rates of Amphibians in Various Wetlands (Number of Captures per 100 Trap Nights)

Wetland	Year	Northwestern Salamander *Ambystoma gracile*	Long-Toed Salamander *Ambystoma macrodactylum*	Western Toad *Bufo boreas*	Pacific Giant Salamander *Dicamptodon tenebrosus*	Ensatina *Ensatina eschscholtzii*	Western Red-Backed Salamander *Plethodon vehiculum*	Pacific Treefrog *Hyla regilla*	Red-Legged Frog *Rana aurora*	Rough-Skinned Newt *Taricha granulosa*	Total
AL3	1988										
B3I	1988		1.00								1.00
BBC24	1988	0.67	0.34						0.67		1.68
ELS39	1988								0.33		0.33
ELS61	1988	1.00	0.67						1.00		2.67
ELW1	1988										
FC1	1988		1.33			0.33					1.67
HC13	1988	0.33					0.33		1.36		1.67
JC28	1988	0.33				0.33	0.33			0.33	1.69
LCR93	1988					0.33	0.33	0.33	2.33		3.33
LPS9	1988	0.33	1.00				0.33	0.33	0.69		2.33
MGR36	1988	0.33	0.33			1.00	2.00		1.33		5.36
NFIC12	1988										
PC12	1988				0.33	0.33			0.67		1.33
RR5	1988	0.34		1.68					0.67		2.69
SC4	1988										
SC84	1988										
SR24	1988				0.33					0.67	1.00
TC13	1988										
Total		3.34	4.67	1.68	0.67	2.33	3.33	0.67	9.06	1.00	26.75

APPENDIX 5-2
1989 Capture Rates of Amphibians in Various Wetlands (Number of Captures per 100 Trap Nights)

Wetland	Year	Northwestern Salamander *Ambystoma gracile*	Long-Toed Salamander *Ambystoma macrodactylum*	Western Toad *Bufo boreas*	Pacific Giant Salamander *Dicamptodon tenebrosus*	Ensatina *Ensatina eschscholtzii*	Western Red-Backed Salamander *Plethodon vehiculum*	Pacific Treefrog *Hyla regilla*	Red-Legged Frog *Rana aurora*	Rough-Skinned Newt *Taricha granulosa*	Total
AL3	1989					0.67	0.33	0.33	0.33		1.67
B3I	1989										
BBC24	1989	9.67	0.33			0.33			1.00		11.33
ELS39	1989	0.33				0.67		0.33			1.33
ELS61	1989	0.67	0.33						0.67		1.67
FC1	1989										
HC13	1989			0.33			0.33	0.33	1.67		2.67
JC28	1989						1.00		0.33		1.33
LCR93	1989		0.33				0.33		0.67		1.33
LPS9	1989								0.57		0.57
MGR36	1989	0.67					0.67				1.33
NFIC12	1989	1.34				0.34			1.01		2.69
PC12	1989	0.67		0.67				0.33	2.00		3.67
RR5	1989	0.34									0.34
SC4	1989	1.00				0.67	0.33				2.00
SC84	1989						0.33				0.33
SR24	1989			0.67		1.00		0.67	0.33	0.67	3.33
TC13	1989	0.33	0.33			0.33			0.67		1.67
Total		15.02	1.33	1.67	0.00	4.00	3.33	2.00	9.25	0.67	37.27

APPENDIX 5-3
1993 Capture Rates of Amphibians in Various Wetlands (Number of Captures per 100 Trap Nights)

Wetland	Year	Northwestern Salamander *Ambystoma gracile*	Long-Toed Salamander *Ambystoma macrodactylum*	Western Toad *Bufo boreas*	Pacific Giant Salamander *Dicamptodon tenebrosus*	Ensatina *Ensatina eschscholtzii*	Western Red-Backed Salamander *Plethodon vehiculum*	Pacific Treefrog *Hyla regilla*	Red-Legged Frog *Rana aurora*	Rough-Skinned Newt *Taricha granulosa*	Total
AL3	1993					0.33	0.33	0.33			1.00
B3I	1993		0.33								0.33
BBC24	1993					0.33					0.33
ELS39	1993					0.33					0.33
ELS61	1993										
FC1	1993										
HC13	1993		0.67						0.67		1.33
JC28	1993										
LCR93	1993								2.00		2.00
LPS9	1993										
MGR36	1993										
NFIC12	1993					0.33					0.33
PC12	1993										
RR5	1993								0.33		0.33
SC4	1993		0.33			0.33		0.33			1.00
SC84	1993										
SR24	1993		0.33								0.33
TC13	1993										
Total		0.00	1.67	0.00	0.00	1.67	0.33	0.67	3.00	0.00	7.33

APPENDIX 5-4
1995 Capture Rates of Amphibians in Various Wetlands (Number of Captures per 100 Trap Nights)

Wetland	Year	Northwestern Salamander *Ambystoma gracile*	Long-Toed Salamander *Ambystoma macrodactylum*	Western Toad *Bufo boreas*	Pacific Giant Salamander *Dicamptodon tenebrosus*	Ensatina *Ensatina eschscholtzii*	Western Red-Backed Salamander *Plethodon vehiculum*	Pacific Treefrog *Hyla regilla*	Red-Legged Frog *Rana aurora*	Rough-Skinned Newt *Taricha granulosa*	Total
AL3	1995										
B3I	1995		0.33								0.33
BBC24	1995	0.33				0.33				0.33	1.00
ELS39	1995		0.36			0.36					0.71
ELS61	1995	0.33	1.00								1.33
FC1	1995										
HC13	1995		0.36			0.36	0.36		0.73		1.82
JC28	1995						0.33				0.33
LCR93	1995		0.67			0.33			0.36		1.36
LPS9	1995										
MGR36	1995	0.33					0.67				1.00
NFIC12	1995					1.00	0.33	0.33			1.67
PC12	1995		1.01					0.34			1.35
RR5	1995			0.69							0.69
SC4	1995	0.33									0.33
SC84	1995	0.33							0.39		0.73
SR24	1995										
TC13	1995					0.67		0.33			1.00
Total		1.67	3.73	0.69	0.00	3.05	1.70	1.00	1.48	0.33	13.66

6 Bird Distribution, Abundance, and Habitat Use

Klaus O. Richter and Amanda L. Azous

CONTENTS

Introduction 167
Methods 169
Results 170
 Annual Variation 170
 Species Diversity 171
 Wetland Characteristics and Species Richness 179
 Birds of Social (i.e., Regulatory) Significance 180
 Uncommon, Exotic, and Aggressive Species 182
 Migrants and Residents 182
 Versatility Ratings 185
 Community Assemblages 185
Discussion 186
 Cautionary Points 189
Acknowledgments 190
References 190
Appendix 6-1: Common and Scientific Names of Birds Used in Text 194
Appendix 6-2: Bird Species Detection Rates in Individual Wetlands for All Years Combined 196

INTRODUCTION

Human values and ecological benefits of wetlands gained growing recognition in the 1970s.[1,2] Consequently, wetlands are now considered sensitive and unique ecosystems with diverse functions that are protected at federal, state, and local levels. Of the many benefits that wetlands exhibit, their ability to provide resting, feeding, and breeding habitat for a wide diversity of birds is among the most noticeable and appreciated by the general public. In rural areas waterfowl are important in providing food and hunting opportunities, thereby significantly contributing to the well-being of families and local economies. In fact, it was the commercial value of waterfowl and wading birds that historically accounted for the impetus for wetland protection.[3]

1-56670-386-7/00/$0.00+$.50

In urban and suburban environments, a rich and often highly visible wetland avifauna has increasingly become an important component of passive recreation, enriching the quality of residential life. Despite these benefits, many hectares of marshes, swamps, and other wetlands are converted to uplands by filling and draining.[4-6] Others are destroyed by altering watershed hydrology or from the introduction of pollutants.[7,8] Rates of losses continue at more than 160,000 ha annually.[9] One reason for the ongoing loss and deterioration of wetlands, in part, is attributable to our inadequate knowledge of wetland ecosystems and need for a more complete understanding of their complex ecological functions and human benefits, including their use by avifauna.

Bird distribution and abundances have been intensively studied in upland habitats of deciduous forests of the east-central U.S. and in managed coniferous forests of the Pacific Northwest for many years.[10-16] Riparian transition zones along streams have also been studied and their ecological importance to avifauna recognized. For example, the riparian fringe has been found to be particularly valuable to passerines, woodpeckers, and other nongame species in deserts, shrub steppes, grasslands, and other arid and semi-arid regions of the U.S.[17-19] Birds of prairie marshes have also been especially well studied.[20-23] In these areas ecologists have surveyed waterfowl and marsh birds in potholes and other open landscapes of the Central Flyway.

In contrast, studies of birds in freshwater palustrine wetlands of the U.S. have received less attention, although studies suggest that communities within wetlands may be rich and productive. For example, over 200 nongame species were documented using wetlands throughout the north-central and northeast U.S.[24] Unusually high densities of insectivorous birds were found in woods near wetlands and this disproportionate use was attributed to the abundant food provided by aquatic insects emerging from those wetlands.[25] Songbird communities at wetlands in unfragmented forests were found to include higher species richness of long-distance migrants, forest ground nesters, insectivores, and species raising single broods, but fewer nest predators and nest parasites when compared to songbird communities in mixed and open wetlands of agriculture-dominated landscapes.[26] Different waterfowl and songbird use was also documented among beaver ponds of differing ages and successional stages.[60]

Remarkably, we are aware of only one detailed avifaunal study of freshwater palustrine wetlands in the Central Puget Sound Basin. It describes April through July bird communities at 23 small urban wetlands influenced by development within King (22 wetlands) and Kitsap (1 wetland) counties.[27] Consequently, the purpose of this chapter is to comprehensively describe the distribution and abundance of diurnal birds at 19 palustrine wetlands under differing watershed conditions in the Central Puget Sound Basin. We describe bird species richness and abundances in 1989, 1991, and 1995 within the context of wetland size and habitat characteristics and across a gradient of watershed urbanization. Hence, we describe the avifauna at some large wetlands in relatively pristine watersheds, thereby augmenting the urban bird communities previously described for relatively small (<4.5 ha) wetlands in urban areas and build on the preliminary richness and abundance findings in related studies.[27-29] Finally, we assess the role of landscape characteristics including the breadth of development within watersheds and land adjacent to wetlands in accounting for wetland bird use.

METHODS

The geologic, hydrologic, water quality, and vegetative description of the wetlands in this study are presented Chapters 1, 2, and 3. Here, we determine the distribution and relative abundance of birds from surveys at permanently marked census stations during breeding from mid-May through mid-June 1989, 1991, and 1995. Species were identified by visual sightings, nonterritorial calls, territorial song, pecking, drumming, and wetland-related flyovers during 15-min point counts. Four ornithologists usually surveyed each wetland on different mornings to account for variable survey skills and daily bird activity. Surveys commenced a half-hour after sunup and generally ended no later than 9:00 a.m. Stations were surveyed in alternating sequence to minimize time biases. Surveys were not undertaken at NFIC12 in 1989, RR5 in 1991, and SR24 in 1995. Hence, species data for these wetlands was omitted from analysis during respective years.

We used average detection values for each species observed more than once as a surrogate for abundance. Abundance was calculated by dividing the total number of a species sighted at a wetland by the total number of 15-min observation periods at a wetland. This averaging method represents conservative abundance estimates as it includes recounted individuals during four replicate surveys (pseudoreplication) of a census station rather than the maximum count of breeding individuals observed during just one survey. Nevertheless, it standardizes detection data among surveyors and wetlands with unequal sampling effort (e.g., larger wetlands had more stations and hence more time was spent at large wetlands).

Species were characterized into several groupings based on social/regulatory concern and by ecologically similar behaviors. Avifauna of social/regulatory significance include species listed as endangered, threatened, or of other concern by the U.S. Fish and Wildlife Service, Washington State, and King County.[30-33]

The number of species identified as possible and probable breeders in the Central Puget Sound Basin was determined from previously reported distribution maps.[34] Ecological groupings include uncommon, exotic, and aggressive species and birds with similar migration patterns and similar habitat affinities. Categorization into uncommon, exotic, and aggressive was based to some extent on species' ability to tolerate human intrusion. Birds were also classified into one of three migratory classes depending on wintering location: (1) long-distance migrants wintering in Central and South America (i.e., Neotropical migrants), (2) short-distance migrants wintering south of the Puget Sound Lowlands but north of the tropics, and (3) permanent residents that remain in the vicinity of wetlands.[35,38] We included American Robin, Cedar Waxwing, Coopers Hawk, Golden-crowned Kinglet, and Red-tailed Hawks as residents although small-scale movements occur in these species.

Habitat affinities were quantified from previously reported versatility ratings for bird species.[36] These represent the sum total of the number of plant communities and stand conditions used for breeding and feeding by a species. Correspondingly, we also identified wetland-dependent species as birds commonly recognized as exclusively or partially dependent on wetlands and their immediate buffers for breeding and most other aspects of their life history. We included Red-winged Blackbird within this group despite their expanding distribution across agricultural and other terrestrial landscapes.[37] Wetland-independent species are birds commonly nesting and feeding in upland landscapes including fields and forests and do not require wetlands to meet their life history needs.

Species were characterized by their sensitivity to anthropogenic impacts, including their adaptability to human perturbations.[38] Common names and scientific nomenclature conform to those published in the latest Check-List of North American Birds (see Appendix 6-1).[39-44] Weather data was obtained from NOAA stations monitored at Snoqualmie Falls and Landsburg, Washington.[40] Habitat characterizations were based on National Wetland's Inventory (NWI) definitions and are described in more detail in Chapter 3.[41]

Population changes among species were identified from detection rates. We identified birds as exhibiting an increase or decrease when populations changed by more than 10% between survey years, and considered them unchanged when detection rates did not exceed these limits. When populations both increased and decreased among the study years, no trend was identified.

Land use adjacent to wetlands was identified, categorized, and quantified for the proportion (percent) of favorable native bird habitat within 100, 500, and 1000 m circumscribing each wetland using a combination of satellite imagery, geographic information data, and field surveys. Forests (forests with low-density single-family housing included), lakes, wetlands, shorelines, undeveloped meadow, and shrub land were identified as suitable bird habitats adjacent to wetlands. Commercial developments, multifamily housing, agricultural lands, and cleared areas were considered unsuitable for native birds.

Statistical analysis of correlations and hypothesis testing utilized parametric statistics when assumptions of normality were met and nonparametric statistics when assumptions were violated. We chose $p \leq .05$, and $p > .05$ and $\leq .10$ with $R \geq .4$ as significant and weakly significant, respectively. Nevertheless, significance should be interpreted cautiously because of the high variability of bird data and concomitantly wide confidence intervals for the predictive level of significance attributable to only a maximum of four replicates. Discontinuities in wetland habitat characteristics and unequal representation of all wetland size classes along the watershed-development gradient also required extrapolations across limited data sets.

We used PC-ORD to calculate diversity measures and to run ordination routines.[42] For diversity statistics S = richness and represents the total identified taxa, E = evenness and is calculated as H′/ln (richness) and H′ = Shannon diversity expressed as $-S[p_i \times \ln (P_i)]$. We also used PC-ORD for cluster analysis and to provide dendrograms for identifying similar communities of species at wetlands.

RESULTS

ANNUAL VARIATION

We counted a total of 9709 birds during the three years of breeding surveys. Avifaunal sightings differed greatly between years: 3849 in 1989, 1962 in 1991, and 3898 in 1995, indicating low abundance in 1991 with only 54% of 1989 and 44% of 1995 totals observed. Bird richness was also significantly lower ($p < .05$) among all

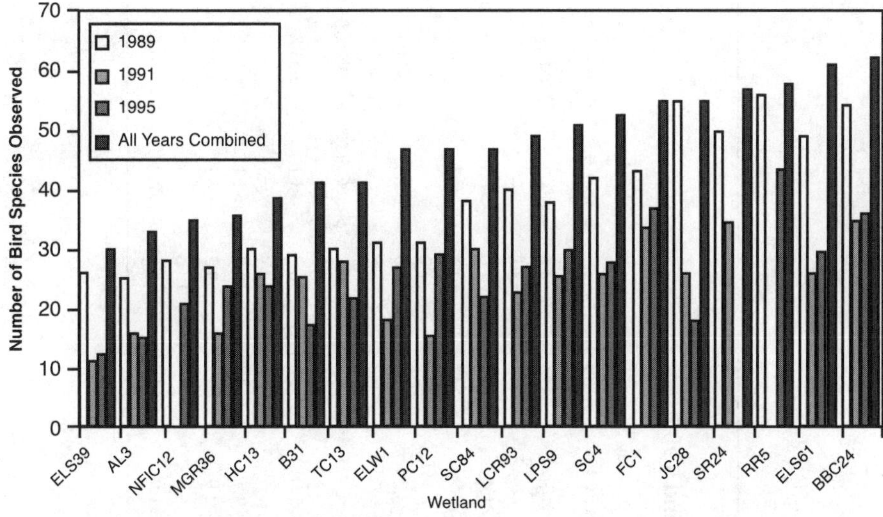

FIGURE 6-1 Number of bird species (richness) observed in wetlands in 1989, 1991, 1995, and all years combined.

wetlands in 1991 (Figure 6-1). Only 68 species were observed among all wetlands as opposed to 83 and 86 species in 1989 and 1995, respectively. Richness within each wetland differed between 1989 and both 1991 ($p = .0001$) and 1995 ($p = .001$), but not between 1991 and 1995 (Wilcoxon Signed Rank test; $p = .35$).

It is unlikely that weather during surveys accounted for the 1991 lows because weather in 1991 was similar to 1995 and not much different than in 1989 (Table 6-1). Specifically, in 1991 surveys were undertaken on 27% clear, 3% partly cloudy, and 70% overcast days. In 1995 surveys occurred on 27% clear, 48% partly cloudy, and 25% on overcast days. In 1989, 39% of surveys were on clear days, 34% partly cloudy days, and 25% on overcast days. Although unlikely, the lower richness and abundance in 1991 could be attributable to cooler average temperatures in 1991 than in either 1989 or 1995. Most importantly, perhaps, may have been the higher total precipitation in April 1991. It was nearly twice that recorded in either April 1989 or 1995. Finally, the previous years' weather, potentially an index of recruitment for the following year, appeared to have little to do with the lower bird sightings in 1991. Temperatures in 1991 were warmer during spring and early summer of 1990 and precipitation was lower in March and April, although not in May and June.

Species Diversity

A total of 90 species were identified on at least two or more occasions during our three years of surveys (Table 6-2). No single wetland exhibited the more than 69% (62) of species found across all wetlands (Figure 6-1) and richness ranged to a low of 33% (30) of total species sighted. The highest count of 62 species was identified at BBC24, a relatively small 2.12-ha, vegetatively diverse, beaver pond wetland

TABLE 6-1
Spring and Early Summer Precipitation and Temperatures During 1989, 1991, and 1995

Factor		March			April			May			June		
Precipitation	**Year**	**cm**	**><cm 91**	**><% 91**	**cm**	**><cm 91**	**><% 91**	**cm**	**><cm 91**	**><% 91**	**cm**	**><cm 91**	**><% 91**
Snoqualmie	89	22.3	>6.0	>36.7	9.8	<9.6	<49.4	9.4	>1.8	>24.2	4.6	<4.2	<47.8
	91	16.3	—	—	19.4	—	—	7.6	—	—	8.8	—	—
	95	13.6	<2.7	<16.7	10.7	<8.7	<44.9	6.2	<1.4	<17.8	7.0	>1.8	>20.0
Landsburg	89	21.1	>5.3	>25.3	11.3	<10.6	<48.2	10.5	>2.9	>37.4	5.8	>0.4	>6.6
	91	15.8	—	—	21.9	—	—	7.7	—	—	5.5	—	—
	95	14.5	<1.3	<8.1	6.8	<15.1	<69.0	4.5	<3.2	<41.3	7.9	>2.5	>44.7
Temperature		**°C**	**>< °C 91**	**>< °C %91**	**°C**	**>< °C 91**	**>< °C %91**	**°C**	**>< °C 91**	**>< °C %91**	**°C**	**>< °C 91**	**>< °C %91**
Snoqualmie	89	5.9	>0.1	>2.1	10.8	>2.2	>25.5	12.3	>1.0	>0.1	16.3	>3.1	>23.8
	91	5.8	—	—	8.6	—	—	11.3	—	—	13.2	—	—
	95	7.3	>1.5	>24.9	9.1	>0.6	>6.6	13.4	>2.1	>18.5	15.2	>2.1	>15.6
Landsburg	89	5.2	>0.01	>0.002	10.3	<1.1	<9.4	8.4	<2.9	<25.7	9.2	<4.1	<30.7
	91	5.2	—	—	11.4	—	—	11.4	—	—	13.3	—	—
	95	7.1	>1.9	>35.8	15.5	>4.1	36.1	13.3	>1.9	>17.0	15.6	>2.2	>16.6

TABLE 6-2
Species and Life History Traits of Birds Sighted at Study Wetlands in Order of Highest Occurrence Among Wetlands Over All Years

Bird Species	Percent of 19 Wetlands Species Observed				Status[a]	Adaptability	Versatility Rating	Population Trend
	1989	1991	1995	Percent of Wetlands (Years 1989, '91, and '95)				
American Robin	100%	100%	100%	100%	PR	Adapter	37	Undetermined
Song Sparrow	100%	100%	100%	100%	PR	Adapter	24	Decreasing
Black-capped Chickadee	100%	94%	94%	100%	PR	Adapter	18	Decreasing
Swainson's Thrush	95%	100%	94%	100%	LDM	Avoider	26	Increasing
Winter Wren	95%	94%	67%	100%	PR	Adapter	27	Decreasing
Pacific-slope Flycatcher	95%	82%	83%	100%	LDM	Adapter	24	Decreasing
Golden-crowned kinglet	95%	71%	33%	100%	PR	Adapter	27	Decreasing
Wilson's Warbler	89%	76%	56%	100%	LDM	Avoider	33	Decreasing
Black-throated Gray Warbler	53%	18%	39%	100%	LDM	Avoider	28	No Change
Spotted Towhee	89%	76%	83%	95%	PR	Adapter	31	No Change
Willow Flycatcher	84%	82%	78%	95%	LDM	Avoider	20	No Change
American Crow	84%	71%	94%	95%	PR	Exploiter	32	Increasing
Bushtit	84%	47%	22%	95%	PR	Adapter	22	No Change
Bewick's Wren	74%	65%	61%	95%	PR	Adapter	22	No Change
Cedar Waxwing	84%	71%	44%	89%	PR	Exploiter	27	Decreasing
Chestnut-backed Chickadee	84%	47%	28%	89%	PR	Adapter	28	Decreasing
Brown-headed Cowbird	58%	18%	56%	89%	SDM	Adapter	34	No Change
Rufous Hummingbird	53%	41%	61%	89%	SDM	Adapter	37	No Change
Hermit Thrush	84%	24%	22%	84%	SDM	Adapter	30	Decreasing
Yellow Warbler	74%	65%	22%	84%	LDM	Adapter	19	Decreasing
Dark-eyed Junco	74%	47%	33%	84%	PR	Adapter	39	Decreasing
Orange-crowned Warbler	74%	41%	33%	84%	LDM	Adapter	31	Decreasing

TABLE 6-2 *(continued)*
Species and Life History Traits of Birds Sighted at Study Wetlands in Order of Highest Occurrence Among Wetlands Over All Years

Bird Species	Percent of 19 Wetlands Species Observed				Status[a]	Adaptability	Versatility Rating	Population Trend
	1989	1991	1995	Percent of Wetlands (Years 1989, '91, and '95)				
Tree Swallow	58%	24%	33%	84%	LDM	Adapter	29	No Change
Hairy Woodpecker	79%	41%	28%	79%	SDM	Adapter	22	Decreasing
Warbling Vireo	68%	12%	22%	79%	LDM	Avoider	26	No Change
Townsend's Warbler	68%	6%	28%	79%	LDM	Avoider	22	Decreasing
Steller's Jay	58%	53%	61%	79%	PR	Adapter	32	No Change
Red-breasted Nuthatch	53%	41%	56%	79%	PR	Adapter	24	No Change
Downy Woodpecker	47%	47%	56%	79%	PR	Adapter	21	No Change
Marsh Wren	74%	12%	17%	74%	PR	Adapter	8	Decreasing
American Goldfinch	63%	47%	61%	74%	PR	Adapter	23	No Change
Ruby-crowned Kinglet	42%	53%	11%	74%	LDM	Adapter	26	No Change
Varied Thrush	63%	41%	11%	68%	PR	Avoider	28	No Change
Purple Finch	63%	29%	28%	68%	PR	Adapter	29	No Change
House Finch	63%	0%	28%	68%	PR	Exploiter	28	Undetermined
Common Yellowthroat	58%	71%	44%	68%	LDM	Adapter	8	Decreasing
Red-winged Blackbird	53%	35%	50%	68%	SDM	Adapter	11	Decreasing
Black-headed Grosbeak	84%	59%	78%	63%	LDM	Avoider	28	No Change
Violet-green Swallow	42%	35%	56%	63%	LDM	Adapter	33	No Change
Wood Duck	32%	24%	39%	63%	SDM	Adapter	25	Undetermined
Western Tanager	47%	29%	22%	58%	LDM	Avoider	30	No Change
Northern Flicker	37%	29%	33%	58%	PR	Adapter	33	No Change
Great Blue Heron	37%	12%	17%	58%	PR	Adapter	14	Undetermined
Mallard	42%	24%	33%	53%	PR	Exploiter	10	No Change

European Starling	42%	18%	17%	53%	PR	Exploiter	27	Undetermined
Cassin's Vireo	21%	41%	11%	53%	LDM	Avoider		Undetermined
Barn Swallow	26%	24%	33%	47%	LDM	Exploiter	17	No Change
Hutton's Vireo	42%	6%	6%	42%	PR	Avoider	28	Undetermined
Western Wood-pewee	32%	6%	33%	42%	LDM	Adapter	29	Undetermined
Belted Kingfisher	26%	12%	11%	42%	PR	Avoider	u	Undetermined
Brown Creeper	26%	12%	17%	42%	PR	Avoider	29	Undetermined
Yellow-rumped Warbler	26%	12%	17%	42%	SDM	Adapter	31	Undetermined
Virginia Rail	32%	12%	17%	37%	PR	Adapter	10	Undetermined
Brewer's Blackbird	26%	12%	11%	37%	PR	Exploiter	24	Undetermined
Hammond's Flycatcher	26%	12%	11%	37%	LDM	Adapter	26	Undetermined
Red-breasted Sapsucker	21%	0%	22%	37%	PR	Adapter	26	Undetermined
White-crowned Sparrow	32%	24%	6%	32%	PR	Adapter	34	Undetermined
Evening Grosbeak	26%	6%	6%	32%	SDM	Adapter	33	Undetermined
Pine Siskin	21%	0%	17%	32%	PR	Adapter	32	Undetermined
Pied-billed Grebe	26%	0%	11%	26%	PR	Adapter	u	Undetermined
Red-Tailed Hawk	21%	6%	11%	26%	PR	Adapter	29	Undetermined
Pileated Woodpecker	21%	0%	11%	26%	PR	Avoider	27	Undetermined
Olive-sided Flycatcher	16%	18%	11%	26%	LDM	Avoider	24	Undetermined
House Sparrow	16%	12%	6%	26%	PR	Exploiter	u	Undetermined
Killdeer	16%	0%	11%	26%	PR	Exploiter	4	Undetermined
MacGillivray's Warbler	11%	0%	22%	26%	LDM	Avoider	27	Undetermined
Gadwall	21%	0%	6%	21%	PR	Adapter	10	Undetermined
Sharp-shinned Hawk	21%	0%	0%	21%	SDM	Adapter	33	Undetermined
Canada Goose	11%	6%	11%	21%	PR	Adapter	8	Undetermined
California Quail	11%	0%	17%	21%	PR	Adapter	22	Undetermined
Hooded Merganser	16%	0%	6%	16%	LDM	Adapter	12	Undetermined
Green Heron	11%	6%	6%	16%	LDM	Avoider	6	Undetermined
Bullock's Oriole	11%	0%	6%	16%	PR	Adapter	20	Undetermined
Chipping Sparrow	11%	0%	6%	16%	LDM	Avoider	36	Undetermined

TABLE 6-2 *(continued)*
Species and Life History Traits of Birds Sighted at Study Wetlands in Order of Highest Occurrence Among Wetlands Over All Years

	Percent of 19 Wetlands Species Observed							
Bird Species	1989	1991	1995	Percent of Wetlands (Years 1989, '91, and '95)	Status[a]	Adaptability	Versatility Rating	Population Trend
Cooper's Hawk	11%	0%	6%	16%	LDM	Adapter	32	Undetermined
Ruffed Grouse	5%	12%	0%	16%	PR	Avoider	28	Undetermined
Fox Sparrow	5%	0%	11%	16%	PR	Adapter	34	Undetermined
Sora	0%	6%	11%	16%	PR	Avoider	10	Undetermined
Anna's Hummingbird	11%	0%	0%	11%	SDM	Exploiter	25	Undetermined
Glaucous-winged Gull	11%	0%	6%	11%	PR	Adapter	u	Undetermined
Vaux's Swift	11%	0%	0%	11%	LDM	Exploiter	34	Undetermined
Cliff Swallow	5%	12%	6%	11%	LDM	Exploiter	12	Undetermined
Band-tailed Pigeon	5%	6%	0%	11%	PR	Adapter	32	Undetermined
Northern Pygmy-owl	5%	6%	0%	11%	PR	Avoider	36	Undetermined
Red-eyed Vireo	5%	6%	6%	11%	LDM	Adapter	14	Undetermined
Bald Eagle	0%	0%	11%	11%	PR	Adapter	19	Undetermined
Blue-winged Teal	0%	0%	11%	11%	LDM	Avoider	10	Undetermined
Caspian Tern	0%	0%	11%	11%	LDM	Avoider	u	Undetermined
American Coot	5%	6%	6%	5%	PR	Adapter	10	Undetermined

[a] LDM = Long Distance Migrant, PR = Permanent Resident, SDM = Short Distance Migrant.

TABLE 6-3
Summary of Bird Diversity Statistics for 19 Wetlands Surveyed in King County, Puget Sound Ecoregion

Wetland Name	Richness (S)	Evenness (E)	Shannon Diversity (H¢)
AL3	33	.887	3.103
B3I	41	.812	3.014
BBC24	62	.765	3.158
ELS39	30	.884	3.008
ELS61	61	.772	3.172
ELW1	47	.824	3.173
FC1	55	.826	3.11
HC13	39	.871	3.191
JC28	55	.845	3.388
LCR93	49	.8	3.112
LPS9	51	.745	2.928
MGR36	36	.791	2.835
NFIC12	35	.864	3.073
PC12	47	.83	3.196
RR5	58	.872	3.542
SC4	53	.752	2.985
SC84	47	.792	3.048
SR24	57	.837	3.385
TC13	41	.863	3.203
Averages	47.2	.823	3.149

system. In contrast, the lowest richness of 30 species was identified at ELS39, a 1.74-ha scrub-shrub wetland dominated by Douglas Spirea (*Spirea douglasii*).

Evenness in wetlands ranged from a high of 0.887 at AL3 to a low of 0.745 at LPS9 (Table 6-3). Evenness increased along with diversity but was high more often in wetlands with lower bird richness. Wetlands with the highest bird richness such as BBC24 (62 species) and ELS61 (61 species) had among the lowest evenness values of 0.765 and 0.772, respectively, although there were exceptions such as RR5, which had both high richness of 58 species and high evenness of 0.872.

Shannon diversity ranged from a high of 3.542 at RR5 to a low of 2.835 at MGR36. Wetlands in pristine watersheds often exhibited a higher Shannon diversity than those in wetlands surrounded by development, but the relationship was not consistent among all wetlands. Where diversity was low in wetlands of developed watersheds such as LPS9, SC4, and B3I, low diversity resulted from fewer species and with an abundance of Song Sparrows, American Robins, and three to four other species.

Two species (Song Sparrow and Swainson's Thrush) were sighted at every wetland every year. Seven additional species (8% of total), were sighted at every wetland sometime during the three years surveyed. In contrast, 24 species (27%)

TABLE 6-4
Bird Species and the Number of Wetlands in Which They Were Sighted during the Three Years of Surveys

Most Widely Distributed	Most Restrictively Distributed
19 Wetlands	1 Wetland
American Robin	American Coot
Black Capped Chickadee	Sora
Black-headed Grosbeak	2 Wetlands
Golden-crowned Kinglet	Anna's Hummingbird
Pacific-slope Flycatcher	Bald Eagle
Song Sparrow	Band-tailed Pigeon
Swainson's Thrush	Blue-winged Teal
Wilson's Warbler	Caspian Tern
Winter Wren	Cliff Swallow
18 Wetlands	Glaucous-winged Gull
American Crow	Northern Pygmy-owl
Bewick's Wren	Red-eyed Vireo
Spotted Towhee	Vaux's Swift
Willow Flycatcher	

were found in fewer than five wetlands or 26% of all wetlands. In 1989, 36%, and in 1991 and 1995, 61 and 53% or more of bird species, respectively, were identified at less than 26% of total wetlands.

Species we expected but rarely found at surveyed wetlands include Sora and Virginia Rail. Although secretive birds of emergent zones, they nevertheless appear not to be widely distributed. Birds breeding or feeding at lakes such as Bald Eagles, Glaucous-winged Gull, and American Coot were only sighted at the two wetlands adjacent to Lake Washington. Finally, birds of dense upland forests including the Northern Pygmy-owl were presumably uncommon (Table 6-2).

The most widely distributed species are also the most abundant species, although a few geographically restricted species were found in relatively large numbers (Appendix 6-1). For instance, of the 9 species identified at all 19 wetlands, 7 ranked among the top 5 species in abundance at wetlands in which they were found (Table 6-4). An example is the Song Sparrow, which is the most widely distributed and abundant species at our wetlands. It ranks among the top five in sightings at all wetlands and exhibits an average ranking of 1.5, indicating it is either the first or second most abundant species at each wetland (Table 6-5). Swainson's Thrush, the next most widely distributed species, is also found at every wetland, but ranks among the top five in abundance at only 12 wetlands and exhibits an average abundance ranking of 3.0. The American Robin, sighted at 18 of 19 wetlands exhibits the third highest abundance ranking of 3.2. Conversely, the Red-winged Blackbird is among the top five most abundant species in 6 of the 13 wetlands at which it was sighted. However, it is the second most abundant species after the Song Sparrow at select wetlands. Other species, also sighted at numerous wetlands, are abundant at only a

TABLE 6-5
The Five Most Abundant Bird Species at Wetlands, the Number of Wetlands in Which They Ranked in the Top Five, and Their Average Rank

Total Number of Wetlands and Species Found	Number of Wetlands Species In Top 5 Abundance Ranking	Average Top 5 Abundance Ranking for Species
19 Wetlands		
Song Sparrow	19	1.5
Swainson's Thrush	12	3.0
American Robin	18	3.2
Winter Wren	4	3.6
Pacific-slope Flycatcher	7	4.1
Wilson's Warbler	3	4.5
Black Capped Chickadee	8	4.7
Black-headed Grosbeak	0	—
Golden-crowned Kinglet	0	—
18 Wetlands		
American Crow	4	3.9
Spotted Towhee	8	3.9
Willow Flycatcher	4	4.1
Bewick's Wren	0	—
17 Wetlands		
Cedar Waxwing	2	5.0
16 Wetlands		
Tree Swallow	1	5.0
14 Wetlands		
Marsh Wren	1	3.0
13 Wetlands		
Red-winged Blackbird	6	1.8
Common Yellowthroat	2	4.0
10 Wetlands		
Mallard	1	4.0

few. These include the Tree Swallow, Marsh Wren, and Mallard, which are ranked in the top five in abundance at only 1 wetland and are sighted in at least 10 of the 18 remaining wetlands.

Wetland Characteristics and Species Richness

Bird richness increased with increasing wetland area (Fisher's r to z, r = .53, p = .018). Among the 6 wetlands with areas greater than 4 ha, 5 had more than 50 species (83%). Among the remaining 13 wetlands with less than 4 ha, only 5 had richness of greater than 50 species (38%) (Figure 6-2). Wetland size, although a factor accounting for richness of bird communities, is clearly not the only important trait as evidenced by high richness of 53 species at SC4, a small 1.62-ha wetland.

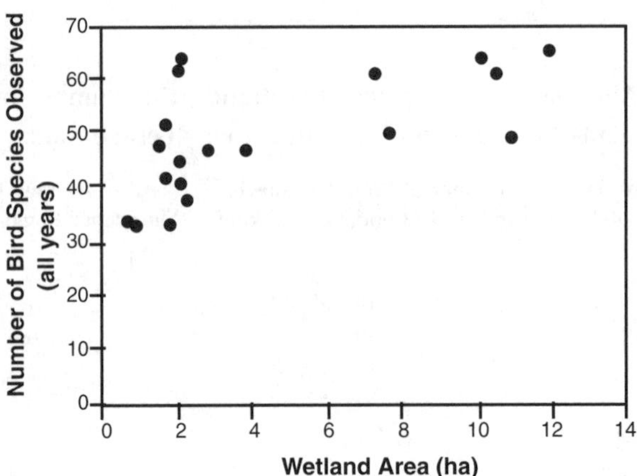

FIGURE 6-2 Relationship between bird species richness and wetland size.

The proximity to lakes and an open water component at wetlands also increases bird richness. Waterfowl bolstered species richness at FC1 and ELW1, wetlands adjacent to Lake Washington. Both SR28 and BBC24 also had high richness because of waterfowl using their open water areas. Most frequently observed waterfowl were Mallard (99 observations), Pied-billed Grebe (26), Canada goose (10), Hooded Merganser (9), and Gadwall (7), with only occasional sightings of Blue-winged Teal. Although not considered waterfowl, Glaucous-winged Gulls and Caspian Terns were also sighted at wetlands with large open bodies of water.

The structural complexity of a wetland characterized by the number of NWI habitat classes, was found to be a contributing factor accounting for species richness (Figure 6-3) (Fisher's r to z, r = .62, p = .004). When three or more NWI habitat classes were present, richness was significantly higher in most wetlands but not all. The exceptions were two wetlands of very different character, LPS9 with 51 species, a large wetland in a highly urbanized area and SC4 (53 species), a small forested wetland in a much more rural area.

Birds of Social (i.e., Regulatory) Significance

Twelve avian species listed by federal, state, and county agencies were detected (Table 6-6). Bald Eagles, a federally and state-listed threatened species within the range of surveyed wetlands, were sighted at ELW1 and FC1. Both wetlands are contiguous with Lake Washington a 72-km^2 lake that provides fish and waterfowl for food. Our few sightings most likely are attributable to our survey technique that inadequately censuses raptors and other species with large territories.

The Olive-sided Flycatcher, a federally listed species of concern, was sighted at five wetlands. It was found in low abundances at ELW1, JC28, TC13, SC4, and SC84, with detection values of 0.4, 0.6, 0.8, 0.8, and 0.15, respectively.

Pileated Woodpecker and Vaux's Swift are two state candidate species we sighted. Pileated Woodpeckers were observed at BBC24, SR24, PC12, and RR5, with abundance values of 0.09, 0.01, 0.04, and 0.15, respectively. These are all open

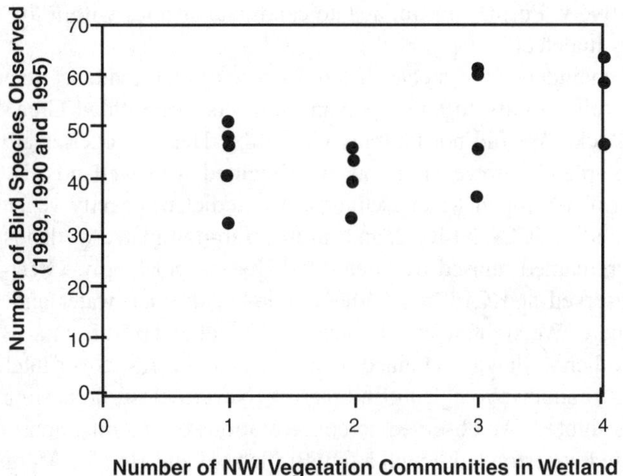

FIGURE 6-3 Relationship between bird species richness and vegetation community complexity.

TABLE 6-6
Listed Species (During the Study Period) Sighted Within Study Wetlands

Species	Federal	State Listing and/or Description of Areas
Great Blue Heron		2 — Aggregate Breeding Areas
Wood Duck		3 — Breeding Areas
Common Goldeneye		3 — Breeding Areas
Bufflehead		3 — Breeding Areas
Hooded Mergansers		3 — Breeding Areas
Bald Eagle	T	T — Breeding areas, communal roosts,
Osprey		3 — Breeding Areas
Band-tailed Pigeon		3 — Breeding areas, regular concentrations
Vaux's Swift		C — Also breeding areas, communal roosts
Pileated Woodpecker		C — Also breeding areas
Purple Martin (not sighted)		C — Candidate, breeding areas
Olive-sided Flycatcher	SC	

Note: SC = species of concern; T = threatened; C = candidate; No. = code for priority habitats of species.

water wetlands. Pileated Woodpeckers, however, were also sighted at JC28, a forested wetland were abundance was calculated at 0.03. Regardless of water, all four wetlands contained snags attributable to flooding, windthrow, or competition. Pileated Woodpecker nesting activity and sightings were especially common at RR5, a series of beaver ponds in which many trees had died from flooding. Vaux's Swift were sighted only at FC1 and SR24. Their abundance values were low at 0.08 and

0.01, respectively. Purple Martin, a state candidate species within the range of our surveys, was undetected.

Of the remainder of the species listed because of their tendency to aggregate and hence their vulnerability to anthropogenic impacts, we sighted Great-blue Herons and Wood Ducks. We did not find any Great-blue Heron rookeries at our wetlands and adjacent uplands. However, herons were sighted at 11 wetlands (58%) including all large, permanent open water wetlands. As predicted, heavily vegetated forested wetlands including JC28, NFIC12, and shrub-scrub-dominated wetlands such as AL3 and ELS39 remained unused by Great-blue Herons. Strikingly, Great-blue Herons were not observed at HC13, a 1.62-ha wetland with open water and an extensive emergent fringe. We sighted Wood Ducks at 12 wetlands. Buffers at BBC24, RR5, and several other wetlands contained many snags and forest cover interspersed with water habitat conducive to nesting and rearing. Nevertheless, no nesting was noticed or fledglings sighted. We observed no concentrations or breeding areas for Washington State-listed waterfowl although a few Bufflehead and Hooded Mergansers rested and fed at wetlands during their migrations to breeding areas elsewhere.

Uncommon, Exotic, and Aggressive Species

Uncommon birds sighted at our wetlands included Anna's Hummingbird at ELS61 (0.03) and BBC24 (0.01); Fox Sparrows at AL3 (0.08), B3I (0.08), BBC24 (0.01); and LCR93 (0.06), and Chipping Sparrows at JC28 (0.03), LPS9 (0.01), and PC12 (0.04). White-crowned Sparrows were sighted at six wetlands.

Exotic birds we encountered at numerous wetlands included European Starling, House Sparrow, and Brown-headed Cowbirds. We found European Starling at 10 wetlands. At FC1 they were found nesting in the holes of snags, displacing Tree Swallows. At ELW1 they were found competing for nesting cavities with Violet Green Swallows, and Black-capped Chickadees. They are also known to displace Northern Flickers from nesting trees (from observation). Brown-headed cowbirds are well-recognized brood parasites and were sighted at 17 wetlands. Finally, the American Crow, an aggressive native species, was found at 18 wetlands. It is a well-known predator of eggs and chicks of passerines and smaller birds.

The observations of birds known to avoid suburban and urban development both declined and increased depending on species, but few changes were significant or consistent enough to warrant categorization as a trend. Three avoiders, the Orange-crowned Warbler, Varied Thrush, and Willow Flycatcher, declined while the Black-throated Gray Warbler and Swainson's Thrush increased (Table 6-7). There were 14 species generally thought to be adaptable to urbanization that declined, another 14 showed no change, and 26 changed but with no determinable trend.

Migrants and Residents

We identified 31 long-distance migrants (LDM), 10 short-distance migrants (SDM), and 49 permanent residents (PR) representing 34, 11, and 54% of total species, respectively (Table 6-2). We observed as few as 12 LDM at AL3 and ELS39 and as many as 24 LDM at BBC24, ELS61, JC28, and RR5. The proportional abundance

TABLE 6-7
Bird Species Abundances and Detection Rates

Bird Species	Number of Individuals				Detection Rate			
	1989	1991	1995	All Years	1989	1991	1995	All Years
Song Sparrow	492	270	307	1069	1.757	1.164	1.163	1.378
American Robin	290	160	213	663	1.036	0.690	0.807	0.854
Red-winged Blackbird	302	94	153	549	1.079	0.405	0.580	0.707
Swainson's Thrush	154	122	212	488	0.550	0.526	0.803	0.629
Black-capped Chickadee	162	82	119	363	0.579	0.353	0.451	0.468
Pacific-slope Flycatcher	127	93	90	310	0.454	0.401	0.341	0.399
Spotted Towhee	108	72	115	295	0.386	0.310	0.436	0.380
Willow Flycatcher	114	56	102	272	0.407	0.241	0.386	0.351
Winter Wren	131	57	74	262	0.468	0.246	0.280	0.338
American Crow	55	48	109	212	0.196	0.207	0.413	0.273
Wilson's Warbler	115	46	40	201	0.411	0.198	0.152	0.259
Common Yellowthroat	96	41	47	184	0.343	0.177	0.178	0.237
Black-headed Grosbeak	56	24	49	129	0.200	0.103	0.186	0.166
Bewick's Wren	50	18	53	121	0.179	0.078	0.201	0.156
Yellow Warbler	67	37	11	115	0.239	0.159	0.042	0.148
American Goldfinch	45	30	37	112	0.161	0.129	0.140	0.144
Steller's Jay	37	25	48	110	0.132	0.108	0.182	0.142
Cedar Waxwing	56	26	25	107	0.200	0.112	0.095	0.138
Golden-crowned kinglet	71	20	14	105	0.254	0.086	0.053	0.135
Hermit Thrush	84	8	8	100	0.300	0.034	0.030	0.129
Bushtit	55	24	13	92	0.196	0.103	0.049	0.119
Marsh Wren	57	4	23	84	0.204	0.017	0.087	0.108
Mallard	30	11	42	83	0.107	0.047	0.159	0.107
Tree Swallow	42	11	26	79	0.150	0.047	0.098	0.102
Dark-eyed Junco	44	13	9	66	0.157	0.056	0.034	0.085
Chestnut-backed Chickadee	43	13	10	66	0.154	0.056	0.038	0.085
Violet-green Swallow	18	10	37	65	0.064	0.043	0.140	0.084
Orange-crowned Warbler	38	17	9	64	0.136	0.073	0.034	0.082
Purple Finch	24	18	17	59	0.086	0.078	0.064	0.076
Hairy Woodpecker	39	10	8	57	0.139	0.043	0.030	0.073
Red-breasted Nuthatch	15	21	20	56	0.054	0.091	0.076	0.072
Warbling Vireo	38	2	13	53	0.136	0.009	0.049	0.068
Black-throated Gray Warbler	25	7	17	49	0.089	0.030	0.064	0.063
Townsend's Warbler	39	2	6	47	0.139	0.009	0.023	0.061
Barn Swallow	12	7	27	46	0.043	0.030	0.102	0.059
Downy Woodpecker	18	10	16	44	0.064	0.043	0.061	0.057
Brown-headed Cowbird	22	4	17	43	0.079	0.017	0.064	0.055
Rufous Hummingbird	19	8	15	42	0.068	0.034	0.057	0.054
European Starling	25	5	9	39	0.089	0.022	0.034	0.050
Northern Flicker	10	8	21	39	0.036	0.034	0.080	0.050
Ruby-crowned Kinglet	15	18	3	36	0.054	0.078	0.011	0.046
Varied Thrush	19	11	4	34	0.068	0.047	0.015	0.044
Western Tanager	17	6	7	30	0.061	0.026	0.027	0.039
Great Blue Heron	14	2	14	30	0.050	0.009	0.053	0.039
House Finch	24		5	29	0.086	0.000	0.019	0.037

TABLE 6-7 *(continued)*
Bird Species Abundances and Detection Rates

	Number of Individuals				Detection Rate			
Bird Species	1989	1991	1995	All Years	1989	1991	1995	All Years
Western Wood-pewee	11	4	11	26	0.039	0.017	0.042	0.034
Pied-billed Grebe	8		16	24	0.029	0.000	0.061	0.031
White-crowned Sparrow	14	6	1	21	0.050	0.026	0.004	0.027
Hutton's Vireo	18	1	1	20	0.064	0.004	0.004	0.026
Wood Duck	9	4	7	20	0.032	0.017	0.027	0.026
Brown Creeper	10	4	5	19	0.036	0.017	0.019	0.024
Hammond's Flycatcher	10	5	3	18	0.036	0.022	0.011	0.023
Cassin's Vireo	5	11	2	18	0.018	0.047	0.008	0.023
Pileated Woodpecker	13		4	17	0.046	0.000	0.015	0.022
Belted Kingfisher	7	2	8	17	0.025	0.009	0.030	0.022
Virginia Rail	9	2	5	16	0.032	0.009	0.019	0.021
Brewer's Blackbird	7	2	6	15	0.025	0.009	0.023	0.019
Olive-sided Flycatcher	5	7	2	14	0.018	0.030	0.008	0.018
Yellow-rumped Warbler	7	2	3	12	0.025	0.009	0.011	0.015
Pine Siskin	6		6	12	0.021	0.000	0.023	0.015
Green Heron	9	1	1	11	0.032	0.004	0.004	0.014
Hooded Merganser	8		3	11	0.029	0.000	0.011	0.014
Red-Tailed Hawk	7	2	2	11	0.025	0.009	0.008	0.014
American Coot	4	1	6	11	0.014	0.004	0.023	0.014
Cliff Swallow	4	3	2	9	0.014	0.013	0.008	0.012
Canada Goose	2	1	6	9	0.007	0.004	0.023	0.012
Killdeer	5		3	8	0.018	0.000	0.011	0.010
Red-breasted Sapsucker	4		4	8	0.014	0.000	0.015	0.010
Gadwall	5		2	7	0.018	0.000	0.008	0.009
Evening Grosbeak	5	1	1	7	0.018	0.004	0.004	0.009
House Sparrow	4	2	1	7	0.014	0.009	0.004	0.009
MacGillivray's Warbler	2		5	7	0.007	0.000	0.019	0.009
Cooper's Hawk	2		4	6	0.007	0.000	0.015	0.008
Fox Sparrow	1		5	6	0.004	0.000	0.019	0.008
Vaux's Swift	5			5	0.018	0.000	0.000	0.006
California Quail	2		3	5	0.007	0.000	0.011	0.006
Red-eyed Vireo	2	1	2	5	0.007	0.004	0.008	0.006
Sharp-shinned Hawk	4			4	0.014	0.000	0.000	0.005
Chipping Sparrow	3		1	4	0.011	0.000	0.004	0.005
Glaucous-winged Gull	2		2	4	0.007	0.000	0.008	0.005
Caspian Tern			4	4	0.000	0.000	0.015	0.005
Sora		1	3	4	0.000	0.004	0.011	0.005
Bullock's Oriole	2		1	3	0.007	0.000	0.004	0.004
Band-tailed Pigeon	2	1		3	0.007	0.004	0.000	0.004
Ruffed Grouse	1	2		3	0.004	0.009	0.000	0.004
Anna's Hummingbird	2			2	0.007	0.000	0.000	0.003
Spotted Sandpiper	2			2	0.007	0.000	0.000	0.003
Northern Pigmy-Owl	1	1		2	0.004	0.004	0.000	0.003
Bald Eagle			2	2	0.000	0.000	0.008	0.003
Blue-winged Teal			2	2	0.000	0.000	0.008	0.003

of LDM ranged from a low of 33% of total species at LC93 and SR24 to a high of 58% of total species at RR5. The twofold differences between low and high richness among LDM at these wetlands is most likely attributable to the small size and simple plant communities of the former wetlands.

We sighted eight SDM (9% of total species). A maximum of five were observed at each of seven wetlands. Five SDM also represented 14% of all species found at MGR36, the greatest proportional abundance of SDM among all wetlands. Six LDM, three SDM, and eight PR species declined in abundance over the study period. Only one LDM, Swainson's Thrush, and one PR species, American Crow, increased, while the remaining species either showed no change or both an increase and decrease in the years surveyed. Many PR species, including both Black-capped and Chestnut-backed Chickadees, were more abundant at forested wetlands, whereas others, such as Song Sparrow, were more abundant in scrub-shrub wetlands.

Versatility Ratings

Birds at wetlands may be found breeding and feeding in a large number of plant communities and stand conditions, as suggested by the wide range of versatility ratings of birds sighted (Table 6-2): 47 species, slightly more than half (57%), are characterized by ratings of 26 or higher with 25 species, approximately half of these (30%) exhibiting ratings between 26 and 30. We only found 16 species, representing 19% of all species with ratings below 16. Clearly, the lowest rankings were among wetland obligates and in our surveys included Marsh Wren, Green Heron, Canada Goose, and other waterfowl. Both Killdeer and Yellowthroat, also with specific and few breeding and feeding habitats, were observed. Species with the highest ratings included the American Robin, Rufous Hummingbird, and Chipping Sparrow.

Community Assemblages

Cluster analysis defined two major and several minor groups of similar bird associations within wetlands (Figure 6-4). The two major clusters generally identified species associated with open water and those not associated with open water. The bottom group, although extending from AL3 through ELS61, includes wetlands ELW1 through ELS61 — wetlands adjacent to or characterized by proportionately large areas of permanent open water. The top clusters are mostly forested and scrub-shrub-dominated wetlands with proportionately small areas of semipermanent and shallow open water. Clearly, some wetlands are uncharacteristic for these respective clusters as a whole. RR5, for example, is a 10.52-ha permanent open water wetland and is clustered between ELS39, a 1.74-ha scrub-shrub wetland, and TC13, a small 2.06-ha forested wetland, both without permanent water. Likewise, B3I, a 1.98-ha scrub-shrub wetland with a small creek running through it is sandwiched between ELW1 and FC1, which are both wetlands adjacent to a large lake (Lake Washington) and is clustered among large open water wetlands.

Smaller groupings of interest include the clusters established from similar bird communities identified at TC13, HC13, and SC84, which all have forested habitat adjacent to scrub-shrub habitats. In addition, ELW1 fuses with LPS9 and LCR93

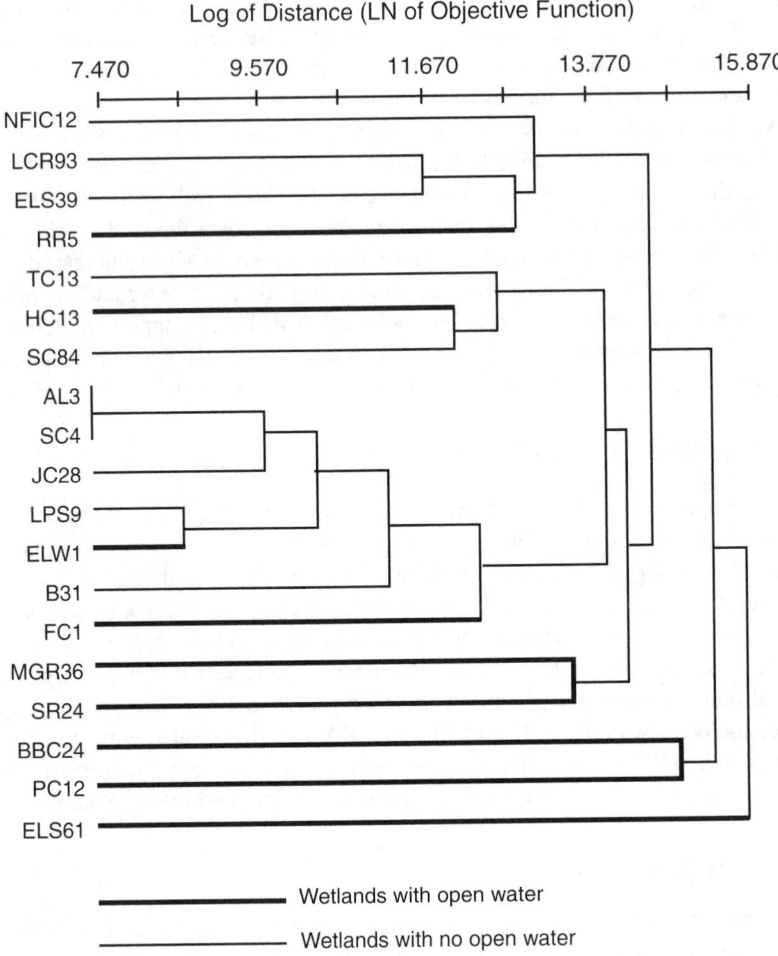

FIGURE 6-4 Cluster diagram showing two major and several minor groups of similar bird associations within study wetlands.

with ELS39 at the lower, left end of the dendrogram identifying similar bird communities in the three wetlands, all of which are dominated by scrub-shrub habitat. Despite some unexplained discrepancies, our 33.64% chaining for clusters suggest a reasonably good differentiation of bird communities at wetlands.

DISCUSSION

We identified 90 bird species within freshwater wetlands of the Central Puget Sound Basin. When compared with diversities of birds found in upland areas, our sightings suggest a disproportionately greater use at wetlands by avifauna, and indicate that wetlands are probably the single most productive ecosystems for birds in the Puget Sound Basin. Most likely, this is because wetlands provide a wide diversity of cover and nesting microhabitats that overlap with water and foraging microhabitats. Specifically, our high bird richness is attributable to diversity of bird groups that utilize the many different habitat niches often characteristic of palustrine wetlands. We found waterfowl (e.g., Hooded Merganser, Blue-winged Teal), wading birds (Great-blue Heron), shorebirds (Spotted Sandpiper), and other obligate wetland birds at larger open water wetlands with shallow, exposed shorelines and emergent vegetation. Raptors including Cooper's, Sharp-shinned, and Red-tailed Hawks were observed hunting for birds and small mammals at smaller scrub-shrub, emergent, and wet meadow-dominated wetlands. We observed some passerines, including swallows and flycatchers, feeding on flying invertebrates above open water, emergent zones, and low vegetation. We found cavity-nesting birds such as Black-capped Chickadees and Pileated Woodpeckers utilizing trees killed by wetland flooding. Finally, we found many other passerines (i.e., songbirds) including sparrows (Song Sparrow, White-crowned Sparrow), thrushes (Swainson's and Varied), and warblers (Orange-crowned, Yellow and Black-throated Gray), all with no special adaptations for wetland characteristics, drawn to wetlands by the diverse vegetation and its associated food and cover.

Our wetland richness values are 19% higher than the 73 species collectively found across a range of managed western Washington Douglas-fir forests, 74% higher than the 23 species and 38% higher than the 56 species in rural upland second-growth forests, and 47% higher than 48 species identified in urban parks.[14-16,43] Although, structurally simple wetlands, such as ELS39 with its Spirea-dominated scrub-shrub community also exhibit low richness overall, our findings confirm the historical literature that wetlands are valuable bird habitats to many more species than are uplands.

Avian richness was distributed among all wetlands, with no wetland exhibiting the full range of diversity found across all wetlands. Not surprisingly, the greatest richness was detected at wetlands adjacent to large lakes and at wetlands with multiple habitats that included open water, emergent edges, and complex vegetation communities as well as a mosaic of scrub-shrub and forested stands of deciduous and evergreen trees. Structurally simple wetlands such as ELW1 and FC1 likely exhibited high richness because of their adjacency to Lake Washington, whereas SR24, PC12, and RR5, wetlands with a more complex habitat, had many species because of open water adjacent to several habitat types. These wetlands provide cover and foraging areas for truly aquatic species such as waterfowl — most notably Mallard, American Coots, Pied-billed Grebes, and Hooded Mergansers. Moreover, Wood Ducks utilize the surrounding buffer and make incursions into wetlands for cover, feeding, and other life functions.

Wetland size helped to account for species richness. Consequently, our results are consistent with previous findings supporting the well-documented relationships between area and bird species richness.[18,27,44-46] Bird species richness correlated with wetland area primarily because larger wetlands exhibited a greater diversity of habitats. Our total richness is 30% higher than the 56 species identified at smaller urban wetlands and associated uplands.[27] Specifically, waterfowl and wading birds including Wood duck, Common Mallard, Gadwall, Canada Goose, and Great blue Heron were constrained to larger wetlands characterized by open water. In woodlot studies, area was also found to be highly correlated with richness and abundance of birds, although the length of woodlot perimeter was found to be an even better predictor of bird richness. It accounted for 82% of richness and 96% of abundance, whereas woodlot area provided considerably less powerful predictions.[47] Unfortunately, our studies did not evaluate wetland perimeter as a factor related to species richness. Nevertheless, we suspect that beyond a minimum wetland area, increasing wetland perimeter may similarly be related to diversity as woodlot species. So far, however, we found wetland area and habitat diversity to be critical factors in maintaining high biodiversity in wetland bird communities. This finding was expected in lieu of findings by other investigators, demonstrating that 10 of 25 species did not occur in wetlands smaller than 5 ha and that bird species richness is greater in wetland complexes than in isolated wetlands of equal or larger size.[18] Our largest wetlands exhibited the greatest richness and density when standardized for size, however that did not preclude smaller wetlands from providing habitat to diverse, sometimes uncommon species, some with considerable densities.

Our avian abundance within individual wetlands was somewhat lower than that found in terrestrial studies, although many similar species dominate. For example, in forests, shrubs, and meadows of the Metropolitan Greenspace Natural Area of Portland, Oregon the five most abundant species were Song Sparrow (6.9%), European Starling (6.2%), American Robin (5.9%), Bushtit (4.6%), and American Goldfinch (4.3%), with Spotted Towhee (4.1%), Cedar Waxwing (3.9%), Red-winged Blackbird (2.8%), Black-capped Chickadee (2.7%), and Brown-headed Cowbird (2.6%) also being common.[48]

Our bird observations also reflected the vegetation community and adjacent landscape habitat within which respective wetlands were situated. For example, we found MacGillivray's Warbler, Willow Flycatcher, Song Sparrow, American Goldfinch, and White-crowned Sparrow at wetlands of open landscapes including meadows, logged fields, and shrub areas. We found Chestnut-backed Chickadee, Golden-crowned Kinglet, and Winter Wren within forested wetlands. Because some wetlands are habitats within a larger forest matrix, we found both groups of birds at some wetlands.

As expected, species richness was lowest at forested wetlands (e.g., AL3, HC13, NFIC12, TC13, but see exception at JC28) as such communities typically exhibit lower avian richness.[9] Nevertheless, forested and small, monotypic wetlands contributed to total species richness by providing food and shelter, and therefore should not be undervalued as bird habitat as is happening in some areas.[50]

For the most part, wetland avifauna is an extension of the upland fauna with few species being restricted to lentic-water habitats. It is clear from our studies that species breeding beyond the immediate wetland use wetlands. Many of the species we identified may be ecologically classified as edge species. Consequently, buffer conditions are important to determining wetland bird communities.

Richness of birds at palustrine wetlands in relatively undisturbed watersheds did not vary widely between wetlands, or between wetlands and their adjacent habitats, with the exception of open water obligates such as waterfowl and shorebirds. The most widely distributed and abundant birds at wetlands are the same species identified at uplands. Many of the species we sighted at wetlands are common breeding species within the Pacific Northwest that can exploit a variety of habitats. These species may have been inhabiting adjacent terrestrial habitats, coming to wetlands to drink, augment their diet, and support their young. Highly mobile upland avifauna of adjacent habitats (e.g., American Robin), although not depending on wetlands for nesting, forage widely on readily accessible and abundant food sources available at wetlands and within their buffers.

We identified Anna's Hummingbird at two wetlands. This species was first recorded locally in 1987 at the Montlake wetlands in Seattle but is now observed more frequently within urban parks and gardens.[35] We identified six records for Chipping Sparrows. This is a rare bird within the Central Puget Sound Basin, although 30 to 40 years ago it was much more common. Individuals were identified by visual sightings and vocalizations at three wetlands by two different observers, so it most likely was not mistaken for Dark-eyed Junco, which has similar vocalizations. We also found Warbling and Red-eyed Vireos, and the Yellow Warbler, thought to be in decline since 1970. Yellow Warblers are especially sensitive to wetland disturbance and deterioration and their numbers are often impacted by nest predation from Brown-headed Cowbirds.[35] Interestingly, no breeding Wood Ducks were observed at any of our wetlands over the duration of the study. This is surprising given Wood Duck preferences for snags and the prevalence of these snags in beaver ponds, forested wetlands, and wooded swamps that characterized some of our sites.

The presence of exotics (e.g., European Starling) and aggressive natives (e.g., American Crow) at wetlands in our more pristine watersheds is disturbing. American Crows are known nest predators, potentially reducing nesting success at wetlands, and Starlings often aggressively take over nesting cavities from native species.[51,52]

Most resident Pacific Northwest species are reported to be maintaining their populations despite increasing urbanization.[53] Our study results generally corroborate this finding though we did not have sufficient data to assess trends in all species we observed. Declines were observed among some migrating species and some adapters. Infrequent sightings of several other species may have been attributable to either their overall rareness or their use of wetlands for only part of their life history. Clearly, to firmly establish population trends will require more targeted and longer-term studies than those undertaken in our work.

Cautionary Points

Bird wetland use varies by species, age, social structure, breeding stage, and season.[20,54] Our work in this chapter describes species usage from May through June, which is generally considered the breeding season, with habitat requirements during this time being critical. Nevertheless, different habitat use may be expected during migration and overwintering. Several factors need greater attention in accounting for the distribution of species. These include habitat fragmentation, the proximity of wetlands to each other and the role of metapopulation ecology in structuring wetland avifauna.[18,55] Others are the determination of breeding success within palustrine wetlands of varying habitat characteristics as well as identifying other habitat uses (feeding, cover, defense) of birds at wetlands and their buffers.

The diverse wetland and habitat classification schemes, and the spatial and structural complexity of wetlands make it difficult to determine whether an existing avifaunal community is predictable within an area being studied. Birds unique to freshwater marshes, wet coniferous forests, and other habitats identified by ornithologists when studying bird distribution and behavior, may not fit well within current classification schemes. Consequently, determining the relative paucity or heightened richness across regional scales is difficult. Bird species diversity and abundances are known to vary yearly and our studies are no exception. Bird and nesting mortality may be affected by weather over several years, the preceding year, the survey year and during breeding, migration, or over-wintering period.[29,56-59,61] There are also biases in our study associated with overall wetland size and the diverse wetland habitats surveyed in only 19 wetlands. Finally, it is well recognized that bird identification by vocalizations is difficult in urban areas with ongoing traffic and other activities.

ACKNOWLEDGMENTS

We would like to thank Dr. Gordon H. Orians for helping to establish our monitoring protocols. We also thank Michael Gratz, Michael Emers, Vanessa L. Artman, Ellen Martin-Yanny, Claus R. Svendsen, James M. Shields, Christopher Chappell, Timothy Quinn, and other graduate students who carried out bird censuses over the years. Paul R. Adamus, Kenneth R. Brunner, David A. Manuwal, and Dennis R. Paulson helped in the preparation of our species list and bird classifications. We appreciate the editorial assistance of Drew Kerr during the final draft of the manuscript.

REFERENCES

1. Good, R. E., D. F. Whigham, R. L. Simpson, and J. J. Crawford, Eds., *Freshwater Wetlands: Ecological Processes and Management Potential*, Academic Press, San Diego, 1978.
2. Greeson, P. E., J. R. Clark, and J. E. Clark, Eds., *Wetland Functions and Values: The State of Our Understanding*, American Water Resources Association, Minneapolis, MN, 1979.
3. Casalena M. J., Birds as indicators of wetland ecosystems, in *Ecology of Wetlands and Associated Systems,* S. K. Majumdar, E. W. Miller, and F. J. Brenner, Eds., Pennsylvania Academy of Science, Easton, PA, 1998, p. 313.
4. Shaw, S. P. and C. G. Fredine, Wetlands of the United States, Circular 39, U.S. Fish and Wildlife Service, Washington, D.C., 1956.
5. Tiner, R. W., Jr., Wetlands of the United States: current status and recent trends, National Wetlands Inventory, Washington, D.C., 1984.
6. Dahl, T. E., Wetland Losses in the United States: 1780's to 1980's, U.S. Department of Interior, Fish and Wildlife Service, Washington, D.C., 1990.
7. Loucks, O. L., Restoration of the pulse control function of wetlands and its relationship to water quality objectives, in *Wetland Creation and Restoration: The Status of the Science*, Vol. 1, J. A. Kusler and M. E. Kentula, Eds., USEPA/7600/3-89/038, U.S. Environmental Protection Agency, Environmental Research Laboratory, Corvallis, OR, 1989, p. 273.
8. Kusler, J. J. and T. Opheim, *Our National Wetland Heritage: A Protection Guide*, 2nd ed., Environmental Law Institute, Washington, D.C., 1996.

9. Eddleman, W. R., Knopf, F. L., Meanley, B., Reid, F. A., and Zembal, R., Conservation of North American rallids, *Wilson Bull.*, 100, p. 458, 1988.
10. Blake, J. G. and J. R. Karr, Species composition of bird communities and the conservation benefit of large versus small forests, *Biol. Conserv.*, 30, 173, 1984.
11. Blake, J. G., Species-area relationship of migrants in isolated woodlots in east-central Illinois, *Wilson Bull.*, 98, 291, 1986.
12. Blake, J. G. and J. R. Karr, Breeding birds of isolated woodlots: area and habitat relationships, *Ecology*, 68, 1724, 1987.
13. Morse, S. F. and S. K. Robinson, Nesting success of a neotropical LDM in a multi-use forested landscape, *Biol. Conserv.*, 13, 327, 1999.
14. Manuwal, D. A. and S. Pearson, Bird populations in managed forests in the western Cascade mountains, Washington, in *Wildlife Use of Managed Forests: A Landscape Perspective*, Vol. 2, Aubry, K. B., J. G. Hallett, S. D. West, M. A. O'Connell, and D. A. Manuwal, Eds., Sect. 5, West-side Studies, Research Results, Timber Fish and Wildlife, TFW-WL4-98-002, 1997, p. 1.
15. Stofel, J. L., *Evaluating Wildlife Responses to Alternative Silvicultural Practices*, University of Washington Press, Seattle, 1993.
16. Artman, V. L., *Breeding Bird Population and Vegetation Characteristics in Commercially Thinned and Unthinned Western Hemlock Forests of Washington*, University of Washington Press, Seattle, 1990.
17. Anderson, B. W. and R. D. Ohmart, Vegetation structure and bird use in the lower Colorado River Valley, in Importance, Preservation and Management of Riparian Habitat: A Symposium, General Tech. Rep. RM-166, U.S. Forest Service, Rocky Mountain Forest and Range Experiment Station, Fort Collins, CO, 1977.
18. Brown, M. and J. J. Dinsmore, Implications of marsh size and isolation for marsh bird management, *J. Wildl. Manage.*, 50, 392, 1986.
19. Knopf, F. L. and F. B. Samson, Scale perspectives on avian diversity in western riparian ecosystems, *Biol. Conserv.*, 8, 669, 1994.
20. Weller, M. W. and C. S. Spatcher, *Role of Habitat in the Distribution and Abundance of Marsh Birds*, Iowa State University Press, Ames, 1965.
21. Weller, M. W. and L. H. Fredrickson, Avian ecology of a managed glacial marsh, *Living Bird*, 12, 269, 1974.
22. Weller, M. W., Birds of some Iowa wetlands in relation to concepts of faunal preservation, *Proc. Iowa Acad. Sci.*, 86, 81, 1979.
23. Weller, M. W., Bird-habitat relationships in a Texas estuarine marsh during summer, *Wetlands*, 14, 293, 1994.
24. Landin, M. C., The importance of wetlands in the north central and northeast United States to non-game birds, in *Management of North Central and Northeastern Forest for Nongame Birds*, M. DeGraff and K. E. Evans, Eds., USDA GTR NC-51, U.S. Department of Agriculture, Washington, D.C., 1979, p. 179.
25. Busby, D. G. and S. G. Sealy, Feeding ecology of a population of nesting Yellow Warblers, *Can. J. Zool.*, 57, 1670, 1979.
26. Gaudette, M. T., Modeling Wetland Songbird Community Integrity In Central Pennsylvania, M.S. thesis, Pennsylvania State University, University Park, PA, 1998.
27. Milligan, D. A., The Ecology of Avian Use of Urban Freshwater Wetlands in King County, Washington, M.S. thesis, University of Washington, Seattle, 1985.
28. Azous, A. L., An Analysis of Urbanization Effects on Wetland Biological Communities, M.S. thesis, University of Washington, Seattle, 1991.
29. Martin-Yanny, E., The Impacts of Urbanization on Wetland Bird Communities, M.S. thesis, University of Washington, Seattle, 1992.

30. U.S. Fish and Wildlife Service, Endangered, threatened, proposed and candidate species, critical habitat and species of concern in the western portion of Washington State, North Pacific Coast Ecoregion, Rev. April 5/99, U.S. Fish and Wildlife Service, Western Washington Office, Olympia, WA, 1999.
31. Washington Department of Fish and Wildlife, *Species of Concern*, Effective 16 March, 1999, URL: http://www.wa.gov/wdfw/wlm/diversity/soc/soc.htm.
32. Washington Department of Fish and Wildlife, *Priority Habitats and Species*, 1999, URL: http://www.wa.gov/wdfw/hab/phsvert.htm.
33. King County Comprehensive Plan, December 1997 Update, Department of Development and Environmental Services, King County, WA, 1997.
34. Smith, M. R., P. W. Mattocks, Jr., and K. M. Cassidy, Breeding Birds of Washington State, Vol. 4, in *Washington State Gap Analysis — Final Report* , K. M. Cassidy, C. E. Grue, M. R. Smith, and K. M. Dvornich, Eds., Seattle Audubon Society, Seattle, 1997.
35. Hunn, E. S., *Birding in Seattle and King County*, Seattle Audubon Society, Seattle, WA, 1982.
36. Brown, R. E., Management of Wildlife and Fish Habitats in the Forests of Western Oregon and Washington, U. S. Forest Service, Pacific Northwest Region, Portland, OR, USA, 1985.
37. Beletsky, L., *The Red-Winged Blackbird*, Academic Press, New York, 1996.
38. Manuwal, D. A. and P. Adamus, Personal communication, 2000.
39. AOU, *Check-List of North American Birds,* 7th ed., American Ornithologists' Union, Washington, D.C., 1983.
40. NOAA, Washington Climate Summaries: Snoqualmie Falls (No. 82) and Landsburg (No. 83), Washington, 1999, http://www.wrcc.dri.edu/summary/mapwa.html.
41. Cowardin, L. M., V. Carter, F. C. Goulet, and E. T. LaRoe, Classification of Wetlands and Deepwater Habitat of the United States, U.S. Fish and Wildlife Service, Washington, D.C., 1979.
42. McCune, B. and M. J. Mefford, *PC-ORD Multivariate Analysis of Ecological Data*, Ver. 3.0. MjM Software Design, Gleneden Beach, Oregon, 1997.
43. Gavareski, C. A., Relation of park size and vegetation to urban bird populations in Seattle, Washington, *Condor,* 78, 375, 1976.
44. MacArthur, R. H. and E. O. Wilson, An equilibrium theory of insular biogeography, *Evolution,* 17, 373, 1963.
45. Müller, W., Die Besiedlung der Eichenwälder im Kanton Zürich durch den Mittelspecht, *Dendrocopus medius*, *Orn. Beob.*, 79, 105, 1982.
46. Burger, J., Habitat selection in temperate marsh-nesting birds, in *Habitat Selection in Birds*, M. L. Cody, Ed., Academic Press, London, 1985, p. 253.
47. Gotfryd, A. and R. I. C. Hansell, Prediction of bird-community metrics in urban woodlots, in *Wildlife 2000: Modeling Habitat Relationships of Terrestrial Vertebrates*, J. Verner, M. L. Morrison, and C. J. Ralph, Eds., University of Wisconsin Press, Madison, WI, 1986, p. 321.
48. Poracsky, J., L. Sharp, E. Lev, and M. Scott, Metropolitan Greenspaces Program Data Analysis, Part 1: Field-Based Biological Data, Final Report, Metro, 600 NE Grand Ave, Portland, OR, 1992.
49. Aubry, K. B., J. G. Hallett, S. D. West, M. A. O'Connell, and D. A. Manuwal, *Wildlife Use of Managed Forests: A Landscape Perspective*, Volume 1, Executive Summaries and Introduction and Technical Approach, Timber Fish and Wildlife, TFW-WL4-98-001, 1997.

50. Evans, C. D. and K. E. Black, Duck production studies on prairie potholes of South Dakota, Special Sci. Rep. (Wildlife) No. 20, U.S. Fish and Wildlife Service, Washington, D.C., 1956.
51. Ehrlich, P. R., D. S. Dobkin, and D. Wheye, *The Birder's Handbook: A Field Guide To The Natural History Of North American Birds*, Simon & Shuster, New York, 1988.
52. Vander Haegen, W. M. and R. M. Degraaf, Predation on artificial nests in forested riparian buffer strips, *J. Wildl. Manage.,* 60, 542, 1996.
53. Paulson, D. R., Northwest bird diversity: From extravagant past and changing present to precarious future, *Northwest Environ. J.,* 8, 71, 1992.
54. Orians, G. H., Some Adaptations of Marsh-nesting Blackbirds, *Monographs in Population Biology 14,* Princeton University Press, Princeton, NJ, 1980.
55. Opdam, P., Metapopulation theory and habitat fragmentation: a review of holarctic breeding bird studies, *Landscape Ecol.,* 5, 93, 1991.
56. Holmes, R. T., T. W. Sherry, and F. W. Sturges, Bird community dynamics in a deciduous forest: Long-term trends at Hubbard Brook, *Ecol. Monogr.*, 50, 201, 1986.
57. Blake, J. G., G. J. Niemi, and J. M. Hanowski, Drought and annual variation in bird populations, in *Ecology and Conservation of Neotropical LDM Landbirds*, J. M. Hagan, III and D.W. Johnston, Eds., Smithsonian Institution Press, Washington, D.C., 1989.
58. Graber, J. W. and R. R. Graber, Severe winter weather and bird populations in southern Illinois, *Wilson Bull.,* 91, 88, 1979.
59. Duffield, J. M., Waterbird use of a urban stormwater wetland system in Central California, *U.S. Colon. Waterbirds,* 9, 227, 1986.
60. Prosser, D. J., Avian Use of Different Successional Stage Beaver Ponds in Pennsylvania, M.S. thesis, Pennsylvania State University, University Park, PA, 1998.
61. Holmes, R. T. and T. W. Sherry, Assessing population trends of New Hampshire forest birds: local vs. regional patterns, *Auk*, 105, 756, 1988.

APPENDIX 6-1
Common and Scientific Names of Birds Used in Text

Bird Species	Genus	Species	Family
American Coot	*Fulica*	*americana*	
American Crow	*Corvus*	*brachyrhynchos*	
American Goldfinch	*Carduelis*	*tristis*	
American Robin	*Turdus*	*migratorius*	
Anna's Hummingbird	*Calypte*	*anna*	
Bald Eagle	*Haliaeetus*	*leucocephalus*	
Band-tailed Pigeon	*Columba*	*fasciata*	
Barn Swallow	*Hirundo*	*rustica*	
Belted Kingfisher	*Ceryle*	*alcyon*	
Bewick's Wren	*Thryomanes*	*bewickii*	
Black Headed Grosbeak	*Pheucticus*	*melanocephalus*	
Black-capped Chickadee	*Parus*	*atricapillus*	
Black-throated Gray Warbler	*Dendroica*	*nigrescens*	
Blue-winged Teal	*Anas*	*discors*	
Brewer's Blackbird	*Euphagus*	*cyanocephalus*	
Brown Creeper	*Certhia*	*americana*	
Brown-headed Cow Bird	*Molothrus*	*ater*	
Bullock's Oriole	*Icterus*	*bullockii*	
Bushtit	*Psaltriparus*	*minimus*	
California Quail	*Callipepla*	*californica*	
Canada Goose	*Branta*	*canadensis*	
Caspian Tern	*Sterna*	*caspia*	
Cassin's Vireo	*Vireo*	*solitarius*	
Cedar Waxwing	*Bombycilla*	*cedrorum*	
Chestnut-backed Chickadee	*Parus*	*rufescens*	
Chipping Sparrow	*Spizella*	*passerina*	
Cliff Swallow	*Hirundo*	*pyrrhonota*	
Common Yellow-throat	*Geothlypis*	*trichas*	
Cooper's Hawk	*Accipiter*	*cooperii*	
Dark-eyed Junco	*Junco*	*hyemalis*	
Downy Woodpecker	*Picoides*	*pubescens*	
European Starling	*Sturnus*	*vulgaris*	
Evening Grosbeak	*Coccothraustes*	*vespertinus*	
Fox Sparrow	*Passerella*	*iliaca*	
Gadwall	*Anas*	*strepera*	
Glaucous-winged Gull	*Larus*	*glaucescens*	
Golden-crowned kinglet	*Regulus*	*satrapa*	
Great Blue Heron	*Ardea*	*herodias*	
Green Heron	*Butorides*	*virescens*	
Green-winged Teal	*Anas*	*crecca*	
Hairy Woodpecker	*Picoides*	*villosus*	
Hammond's Flycatcher	*Empidonax*	*hammondii*	
Hermit Thrush	*Catharus*	*guttatus*	
Hooded Merganser	*Lophodytes*	*cucullatus*	
House Finch	*Carpodacus*	*mexicanus*	
House Sparrow	*Passer*	*domesticus*	

APPENDIX 6-1 *(continued)*
Common and Scientific Names of Birds Used in Text

Bird Species	Genus	Species	Family
Hutton's Vireo	*Vireo*	*huttoni*	
Killdeer	*Charadrius*	*vociferus*	
MacGillivray's Warbler	*Oporornis*	*tolmiei*	
Mallard	*Anas*	*platyrhynchos*	
Marsh Wren	*Cistothorus*	*palustris*	
Northern Flicker	*Colaptes*	*auratus*	
Northern Pigmy-owl	*Glaucidium*	*gnoma*	
Olive-sided Flycatcher	*Contopus*	*borealis*	
Orange-crowned Warbler	*Vermivora*	*celata*	
Pacific-slope Flycatcher	*Empidonax*	*difficilis*	
Pied-billed Grebe	*Podilymbus*	*podiceps*	
Pileated Woodpecker	*Dryocopus*	*pileatus*	
Pine Siskin	*Carduelis*	*pinus*	
Purple Finch	*Carpodacus*	*purpureus*	
Red-breasted Nuthatch	*Sitta*	*canadensis*	
Red-breasted Sapsucker	*Sphyrapicus*	*ruber*	
Red-eyed Vireo	*Vireo*	*olivaceus*	
Red-Tailed Hawk	*Buteo*	*jamaicensis*	
Red-winged Blackbird	*Agelaius*	*phoeniceus*	
Ring-billed Gull	*Larus*	*delawarensis*	
Ruby Crowned Kinglet	*Regulus*	*calendula*	
Ruffed Grouse	*Bonasa*	*umbellus*	
Rufous Hummingbird	*Selasphorus*	*rufus*	
Sharp-shinned Hawk	*Accipiter*	*striatus*	
Song Sparrow	*Melospiza*	*melodia*	
Sora	*Porzana*	*carolina*	
Spotted Sandpiper	*Actitis*	*macularia*	
Spotted Towhee	*Pipilo*	*maculatus*	
Steller's Jay	*Cyanocitta*	*stelleri*	
Swainson's Thrush	*Catharus*	*ustulatus*	
Townsend's Warbler	*Dendroica*	*townsendi*	
Tree Swallow	*Tachycineta*	*bicolor*	
Varied Thrush	*Ixoreus*	*naevius*	
Vaux's Swift	*Chaetura*	*vauxi*	
Violet-green Swallow	*Tachycineta*	*thalassina*	
Virginia Rail	*Rallus*	*limicola*	
Warbling Vireo	*Vireo*	*gilvus*	
Western Tanager	*Piranga*	*ludoviciana*	
Western Wood-pewee	*Contopus*	*sordidulus*	
White-crowned Sparrow	*Zonotrichia*	*leucophrys*	
Willow Flycatcher	*Empidonax*	*traillii*	
Wilson's Warbler	*Wilsonia*	*pusilla*	
Winter Wren	*Troglodytes*	*troglodytes*	
Wood Duck	*Aix*	*sponsa*	
Yellow Warbler	*Dendroica*	*petechia*	
Yellow-rumped Warbler	*Dendroica*	*coronata*	

APPENDIX 6-2
Bird Species Detection Rates in Individual Wetlands for All Years Combined

Bird Species	AL3	B3I	BBC24	ELS39	ELS61	ELW1	FC1	HC13	JC28	LCR93	LPS9	MGR36	NFIC12	PC12	RR5	SC4	SC84	SR24	TC13	All Wetlands All Years (1989, '91 and '95)
American Coot	0.000	0.000	0.000	0.000	0.000	0.000	0.229	0.000	0.000	0.000	0.000	0.000	0.000	0.000	0.000	0.000	0.000	0.000	0.000	0.014
American Crow	0.250	0.583	0.059	0.500	0.100	0.214	0.229	0.125	0.328	0.083	0.389	0.278	0.250	0.000	0.333	0.479	1.025	0.033	0.375	0.273
American Goldfinch	0.000	0.208	0.059	0.083	0.100	0.000	0.104	0.125	0.375	0.083	0.611	0.000	0.083	0.000	0.104	0.083	0.075	0.043	0.000	0.144
American Robin	0.417	1.125	0.868	0.750	0.725	1.036	0.521	0.625	0.844	0.792	0.972	0.833	0.500	0.917	0.729	2.083	1.575	0.272	0.667	0.854
Anna's Hummingbird	0.000	0.000	0.015	0.000	0.025	0.000	0.000	0.000	0.000	0.000	0.000	0.000	0.000	0.000	0.000	0.000	0.000	0.000	0.000	0.003
Bald Eagle	0.000	0.000	0.000	0.000	0.000	0.036	0.021	0.000	0.000	0.000	0.000	0.000	0.000	0.000	0.000	0.000	0.000	0.000	0.000	0.003
Band-tailed Pigeon	0.000	0.000	0.000	0.000	0.000	0.000	0.000	0.083	0.000	0.000	0.000	0.000	0.000	0.000	0.000	0.021	0.000	0.000	0.000	0.004
Barn Swallow	0.000	0.042	0.088	0.000	0.075	0.071	0.458	0.000	0.031	0.000	0.056	0.000	0.000	0.000	0.083	0.042	0.000	0.000	0.000	0.059
Belted Kingfisher	0.000	0.000	0.103	0.000	0.000	0.071	0.063	0.042	0.000	0.021	0.000	0.028	0.000	0.000	0.000	0.021	0.000	0.011	0.000	0.022
Bewick's Wren	0.083	0.167	0.029	0.083	0.100	0.357	0.313	0.083	0.078	0.083	0.361	0.000	0.083	0.083	0.063	0.479	0.350	0.011	0.083	0.156
Black-capped Chickadee	0.417	1.042	0.235	0.167	0.250	0.821	0.313	0.500	0.359	0.250	0.708	0.444	0.292	0.625	0.625	0.813	0.575	0.293	0.500	0.468
Black-throated Gray Warbler	0.000	0.000	0.015	0.000	0.025	0.000	0.000	0.250	0.094	0.104	0.028	0.028	0.208	0.000	0.125	0.146	0.150	0.000	0.125	0.063
Black Headed Grosbeak	0.083	0.083	0.265	0.083	0.275	0.250	0.188	0.125	0.109	0.188	0.208	0.056	0.042	0.250	0.146	0.167	0.050	0.185	0.125	0.166
Blue-winged Teal	0.000	0.000	0.000	0.000	0.025	0.000	0.021	0.000	0.000	0.000	0.000	0.000	0.000	0.000	0.000	0.000	0.000	0.000	0.000	0.003
Brewer's Blackbird	0.000	0.167	0.000	0.000	0.025	0.000	0.125	0.000	0.016	0.000	0.014	0.000	0.000	0.000	0.000	0.021	0.000	0.011	0.000	0.019
Brown-headed Cow Bird	0.167	0.208	0.015	0.083	0.125	0.036	0.000	0.042	0.031	0.063	0.083	0.028	0.000	0.167	0.104	0.021	0.050	0.022	0.042	0.055
Brown Creeper	0.000	0.000	0.059	0.000	0.050	0.000	0.000	0.000	0.000	0.021	0.000	0.000	0.000	0.042	0.104	0.042	0.000	0.022	0.083	0.024
Bullock's Oriole	0.000	0.000	0.015	0.000	0.000	0.000	0.021	0.000	0.000	0.000	0.000	0.000	0.000	0.000	0.021	0.000	0.000	0.000	0.000	0.004
Bushtit	0.000	0.208	0.015	0.167	0.100	0.071	0.104	0.083	0.063	0.083	0.375	0.111	0.125	0.083	0.063	0.250	0.025	0.087	0.125	0.119
California Quail	0.000	0.000	0.000	0.083	0.050	0.000	0.000	0.000	0.016	0.000	0.000	0.000	0.000	0.000	0.000	0.021	0.000	0.000	0.000	0.006
Canada Goose	0.000	0.000	0.015	0.000	0.000	0.179	0.042	0.000	0.000	0.000	0.000	0.000	0.000	0.000	0.021	0.000	0.000	0.000	0.000	0.012
Caspian Tern	0.000	0.000	0.000	0.000	0.000	0.036	0.063	0.000	0.000	0.000	0.000	0.000	0.000	0.000	0.000	0.000	0.000	0.000	0.000	0.005

Cassin's Vireo	0.000	0.000	0.029	0.000	0.025	0.000	0.021	0.000	0.047	0.000	0.014	0.000	0.000	0.083	0.000	0.083	0.025	0.011	0.083	0.023
Cedar Waxwing	0.083	0.500	0.147	0.000	0.100	0.071	0.250	0.167	0.125	0.104	0.028	0.000	0.042	0.542	0.271	0.167	0.100	0.065	0.083	0.138
Chestnut-backed Chickadee	0.083	0.083	0.044	0.000	0.025	0.000	0.021	0.083	0.094	0.063	0.069	0.083	0.042	0.167	0.250	0.229	0.125	0.043	0.083	0.085
Chipping Sparrow	0.000	0.000	0.000	0.000	0.000	0.000	0.000	0.000	0.031	0.000	0.014	0.000	0.000	0.042	0.000	0.000	0.000	0.000	0.000	0.005
Cliff Swallow	0.000	0.042	0.000	0.000	0.000	0.000	0.167	0.000	0.000	0.000	0.000	0.000	0.000	0.000	0.000	0.000	0.000	0.000	0.000	0.012
Common Yellow-throat	0.000	0.083	0.471	0.000	0.325	0.036	0.208	0.208	0.266	0.542	0.222	0.389	0.000	0.958	0.438	0.000	0.000	0.043	0.000	0.237
Cooper's Hawk	0.083	0.000	0.000	0.000	0.000	0.000	0.000	0.000	0.000	0.083	0.000	0.000	0.000	0.000	0.000	0.000	0.000	0.011	0.000	0.008
Dark-eyed Junco	0.167	0.000	0.044	0.167	0.050	0.000	0.000	0.125	0.219	0.021	0.056	0.028	0.125	0.042	0.083	0.188	0.075	0.087	0.250	0.085
Downy Woodpecker	0.083	0.042	0.074	0.000	0.025	0.000	0.063	0.083	0.000	0.042	0.097	0.167	0.000	0.083	0.083	0.063	0.100	0.022	0.042	0.057
European Starling	0.000	0.333	0.000	0.000	0.025	0.071	0.292	0.000	0.078	0.042	0.042	0.028	0.000	0.000	0.000	0.042	0.025	0.000	0.000	0.050
Evening Grosbeak	0.083	0.000	0.015	0.000	0.000	0.000	0.000	0.000	0.031	0.000	0.014	0.000	0.000	0.000	0.021	0.000	0.000	0.011	0.000	0.009
Fox Sparrow	0.000	0.083	0.015	0.000	0.000	0.000	0.000	0.000	0.000	0.063	0.000	0.000	0.000	0.000	0.000	0.000	0.000	0.000	0.000	0.008
Gadwall	0.000	0.000	0.000	0.000	0.025	0.000	0.083	0.000	0.000	0.000	0.000	0.000	0.000	0.042	0.021	0.000	0.000	0.000	0.000	0.009
Glaucous-winged Gull	0.000	0.000	0.000	0.000	0.000	0.036	0.063	0.000	0.000	0.000	0.000	0.000	0.000	0.000	0.000	0.000	0.000	0.000	0.000	0.005
Golden-crowned kinglet	0.250	0.125	0.118	0.250	0.050	0.071	0.021	0.333	0.172	0.104	0.042	0.139	0.292	0.083	0.250	0.208	0.175	0.065	0.292	0.135
Great Blue Heron	0.000	0.042	0.044	0.000	0.050	0.107	0.229	0.000	0.000	0.000	0.014	0.028	0.000	0.125	0.042	0.021	0.000	0.022	0.000	0.039
Green Heron	0.000	0.083	0.000	0.000	0.000	0.000	0.167	0.000	0.000	0.000	0.000	0.028	0.000	0.000	0.000	0.000	0.000	0.000	0.000	0.014
Hairy Woodpecker	0.000	0.042	0.103	0.000	0.050	0.036	0.000	0.125	0.016	0.083	0.000	0.083	0.125	0.083	0.167	0.063	0.150	0.098	0.167	0.073
Hammond's Flycatcher	0.000	0.000	0.000	0.000	0.025	0.000	0.000	0.125	0.000	0.000	0.000	0.056	0.000	0.083	0.104	0.000	0.100	0.011	0.000	0.023
Hermit Thrush	0.250	0.000	0.088	0.000	0.125	0.000	0.021	0.333	0.031	0.229	0.014	0.167	0.208	0.250	0.146	0.104	0.275	0.207	0.167	0.129
Hooded Merganser	0.000	0.000	0.015	0.000	0.025	0.000	0.000	0.000	0.000	0.000	0.000	0.000	0.000	0.000	0.188	0.000	0.000	0.000	0.000	0.014
House Finch	0.167	0.125	0.059	0.167	0.125	0.071	0.042	0.000	0.000	0.000	0.028	0.000	0.000	0.000	0.021	0.021	0.050	0.022	0.042	0.037
House Sparrow	0.000	0.042	0.000	0.000	0.025	0.000	0.063	0.000	0.016	0.000	0.000	0.000	0.000	0.042	0.000	0.000	0.000	0.000	0.000	0.009
Hutton's Vireo	0.000	0.000	0.015	0.000	0.050	0.000	0.021	0.000	0.000	0.000	0.000	0.000	0.042	0.000	0.104	0.125	0.050	0.000	0.083	0.026
Killdeer	0.000	0.000	0.000	0.000	0.000	0.071	0.000	0.000	0.047	0.000	0.014	0.000	0.000	0.000	0.000	0.021	0.000	0.011	0.000	0.010
MacGillivray's Warbler	0.000	0.000	0.000	0.000	0.000	0.000	0.000	0.000	0.031	0.042	0.014	0.000	0.000	0.000	0.021	0.000	0.000	0.000	0.042	0.009
Mallard	0.000	0.042	0.221	0.000	0.175	0.071	0.771	0.000	0.000	0.000	0.028	0.167	0.000	0.083	0.167	0.000	0.000	0.033	0.000	0.107
Marsh Wren	0.000	0.000	0.059	0.083	0.025	0.250	0.854	0.000	0.047	0.104	0.069	0.056	0.000	0.083	0.042	0.021	0.025	0.098	0.000	0.108
Northern Flicker	0.000	0.167	0.147	0.000	0.075	0.000	0.000	0.000	0.016	0.021	0.014	0.083	0.000	0.000	0.104	0.083	0.125	0.022	0.000	0.050
Northern Pigmy-owl	0.000	0.000	0.000	0.000	0.000	0.000	0.000	0.000	0.000	0.021	0.000	0.000	0.000	0.000	0.000	0.000	0.025	0.000	0.000	0.003
Olive-sided Flycatcher	0.083	0.000	0.000	0.000	0.000	0.036	0.000	0.000	0.063	0.000	0.000	0.000	0.000	0.000	0.000	0.000	0.150	0.000	0.083	0.018

APPENDIX 6-2 *(continued)*
Bird Species Detection Rates in Individual Wetlands for All Years Combined

Bird Species	AL3	B3I	BBC24	ELS39	ELS61	ELW1	FC1	HC13	JC28	LCR93	LPS9	MGR36	NFIC12	PC12	RR5	SC4	SC84	SR24	TC13	All Wetlands All Years (1989, '91 and '95)
Orange-crowned Warbler	0.000	0.083	0.147	0.333	0.175	0.071	0.063	0.083	0.031	0.021	0.028	0.000	0.000	0.083	0.021	0.063	0.125	0.130	0.250	0.082
Pacific-slope Flycatcher	0.917	0.125	0.456	0.167	0.125	0.071	0.063	0.708	0.391	0.500	0.167	0.722	0.667	0.625	0.563	0.479	0.250	0.435	0.750	0.399
Pied-billed Grebe	0.000	0.000	0.000	0.000	0.025	0.179	0.333	0.000	0.000	0.000	0.000	0.000	0.000	0.000	0.021	0.000	0.000	0.011	0.000	0.031
Pileated Woodpecker	0.000	0.000	0.088	0.000	0.000	0.000	0.000	0.000	0.031	0.000	0.000	0.000	0.000	0.042	0.146	0.000	0.000	0.011	0.000	0.022
Pine Siskin	0.000	0.000	0.015	0.083	0.000	0.000	0.000	0.000	0.063	0.000	0.000	0.000	0.042	0.000	0.000	0.021	0.100	0.000	0.000	0.015
Purple Finch	0.000	0.083	0.000	0.083	0.075	0.036	0.000	0.000	0.141	0.000	0.097	0.000	0.125	0.125	0.292	0.208	0.100	0.011	0.042	0.076
Red-breasted Nuthatch	0.083	0.000	0.088	0.083	0.050	0.036	0.000	0.000	0.063	0.042	0.014	0.000	0.042	0.083	0.208	0.042	0.300	0.065	0.208	0.072
Red-breasted Sapsucker	0.000	0.000	0.015	0.000	0.025	0.000	0.000	0.000	0.000	0.021	0.000	0.028	0.000	0.042	0.042	0.021	0.000	0.000	0.000	0.010
Red-eyed Vireo	0.000	0.000	0.000	0.000	0.000	0.000	0.000	0.083	0.000	0.063	0.000	0.000	0.000	0.000	0.000	0.000	0.000	0.000	0.000	0.006
Red-Tailed Hawk	0.000	0.000	0.088	0.000	0.050	0.000	0.021	0.000	0.000	0.000	0.000	0.000	0.000	0.042	0.000	0.000	0.000	0.011	0.000	0.014
Red-winged Blackbird	0.000	0.042	2.353	0.000	2.450	0.429	2.000	0.125	0.016	0.604	1.056	1.750	0.000	0.000	0.104	0.000	0.000	0.033	0.083	0.707
Ruby Crowned Kinglet	0.000	0.000	0.059	0.000	0.025	0.071	0.000	0.125	0.063	0.021	0.042	0.000	0.042	0.042	0.083	0.083	0.100	0.022	0.083	0.046
Ruffed Grouse	0.000	0.000	0.000	0.000	0.000	0.000	0.000	0.000	0.000	0.021	0.000	0.000	0.000	0.042	0.000	0.000	0.000	0.011	0.000	0.004
Rufous Hummingbird	0.167	0.042	0.044	0.167	0.050	0.000	0.000	0.208	0.047	0.042	0.028	0.083	0.167	0.042	0.083	0.021	0.075	0.033	0.042	0.054
Sharp-shinned Hawk	0.000	0.000	0.000	0.083	0.000	0.036	0.000	0.000	0.000	0.021	0.014	0.000	0.000	0.000	0.000	0.000	0.000	0.000	0.000	0.005
Song Sparrow	1.000	1.917	1.941	1.083	0.725	1.536	1.167	1.542	0.922	1.146	2.153	1.139	0.708	1.625	1.146	2.375	2.050	0.696	0.833	1.378
Sora	0.000	0.000	0.029	0.000	0.000	0.000	0.021	0.000	0.000	0.021	0.000	0.000	0.000	0.000	0.000	0.000	0.000	0.000	0.000	0.005
Spotted Sandpiper	0.000	0.000	0.000	0.000	0.000	0.000	0.000	0.000	0.031	0.000	0.000	0.000	0.000	0.000	0.000	0.000	0.000	0.000	0.000	0.003
Spotted Towhee	0.250	0.375	0.059	0.917	0.350	0.607	0.042	0.167	0.328	0.042	0.722	0.000	0.792	0.583	0.188	1.354	0.775	0.087	0.417	0.380
Steller's Jay	0.083	0.042	0.368	0.000	0.000	0.036	0.000	0.083	0.234	0.146	0.028	0.000	0.125	0.333	0.250	0.167	0.225	0.109	0.250	0.142

Swainson's Thrush	0.917	0.333	0.324	0.333	0.250	0.036	0.104	1.000	0.641	1.333	0.333	0.917	0.875	1.083	1.042	1.042	1.000	0.239	1.333	0.629
Townsend's Warbler	0.333	0.000	0.059	0.167	0.025	0.036	0.000	0.083	0.094	0.042	0.000	0.028	0.167	0.000	0.167	0.063	0.125	0.033	0.042	0.061
Tree Swallow	0.083	0.083	0.088	0.167	0.025	0.179	0.646	0.000	0.031	0.000	0.042	0.028	0.042	0.042	0.313	0.042	0.025	0.054	0.000	0.102
Varied Thrush	0.250	0.000	0.044	0.000	0.050	0.107	0.042	0.167	0.031	0.000	0.014	0.000	0.042	0.000	0.083	0.021	0.000	0.054	0.125	0.044
Vaux's Swift	0.000	0.000	0.000	0.000	0.000	0.000	0.083	0.000	0.000	0.000	0.000	0.000	0.000	0.000	0.000	0.000	0.000	0.011	0.000	0.006
Violet-green Swallow	0.000	0.000	0.235	0.000	0.100	0.179	0.271	0.000	0.094	0.000	0.083	0.000	0.042	0.125	0.042	0.083	0.075	0.022	0.000	0.084
Virginia Rail	0.000	0.000	0.088	0.000	0.050	0.000	0.063	0.000	0.016	0.021	0.000	0.056	0.000	0.042	0.000	0.000	0.000	0.000	0.000	0.021
Warbling Vireo	0.000	0.042	0.118	0.083	0.050	0.036	0.063	0.042	0.016	0.208	0.097	0.028	0.000	0.000	0.146	0.042	0.075	0.054	0.000	0.068
Western Tanager	0.167	0.000	0.132	0.000	0.000	0.000	0.000	0.125	0.031	0.000	0.014	0.000	0.000	0.083	0.063	0.042	0.025	0.033	0.083	0.039
Western Wood-pewee	0.250	0.000	0.044	0.000	0.075	0.071	0.000	0.000	0.000	0.000	0.000	0.000	0.000	0.125	0.125	0.000	0.025	0.000	0.208	0.034
White-crowned Sparrow	0.083	0.000	0.000	0.333	0.125	0.000	0.021	0.000	0.109	0.000	0.000	0.000	0.000	0.000	0.000	0.063	0.000	0.000	0.000	0.027
Willow Flycatcher	0.167	0.125	0.426	0.667	0.475	0.250	0.354	0.583	0.172	0.833	0.528	0.528	0.083	0.292	0.458	0.000	0.175	0.141	0.583	0.351
Wilson's Warbler	0.583	0.083	0.235	0.167	0.050	0.071	0.188	0.583	0.203	0.688	0.014	0.500	0.250	0.292	0.771	0.167	0.075	0.141	0.333	0.259
Winter Wren	1.167	0.083	0.794	0.083	0.075	0.250	0.125	0.542	0.438	0.146	0.069	0.333	0.167	0.417	0.604	0.229	0.125	0.272	1.083	0.338
Wood Duck	0.000	0.000	0.059	0.000	0.050	0.071	0.021	0.000	0.000	0.021	0.000	0.028	0.042	0.042	0.021	0.000	0.050	0.033	0.042	0.026
Yellow-rumped Warbler	0.000	0.042	0.015	0.000	0.000	0.000	0.000	0.125	0.000	0.000	0.014	0.000	0.042	0.042	0.000	0.063	0.000	0.011	0.000	0.015
Yellow Warbler	0.250	0.250	0.162	0.000	0.075	0.321	0.250	0.083	0.125	0.458	0.083	0.000	0.125	0.000	0.063	0.146	0.050	0.174	0.083	0.148

7 Terrestrial Small Mammal Distribution, Abundance, and Habitat Use

Klaus O. Richter and Amanda L. Azous

CONTENTS

Introduction 201
Methods 202
Results 203
Discussion 209
Acknowledgments 213
References 213
Appendix 7-1: Capture Rates of Small Mammals by Wetland — 1988 215
Appendix 7-2: Capture Rates of Small Mammals by Wetland — 1989 216
Appendix 7-3: Capture Rates of Small Mammals by Wetland — 1993 217
Appendix 7-4: Capture Rates of Small Mammals by Wetland — 1995 218

INTRODUCTION

The importance of large mammals including beaver (*Castor canadensis*), muskrat (*Ondatra zibethicus*), and river otter (*Lutra canadensis*) to wetland ecosystems is well known. Beaver especially are recognized as keystone species because of their extensive pre-Columbian North American distribution, abundance, and resultant impacts on the creation and maintenance of complex wetland ecosystems.[1-3] More subtle and less well understood is the ecological role of small mammals.

Small mammals are widespread and abundant animals of most landscapes, yet little appears to be known about their distribution at wetlands. Some, such as the water shrew, are wetland obligate species. Others may be especially abundant at wetlands where food and shelter are readily available. Insectivores such as shrews (Soricidae) and moles (Talpidae) are prolific feeders on mollusks and arthropods. Shrews additionally feed on small amphibians at wetlands and some, such as the wandering shrew, eat dead birds and mammals when available. [4] Rodents (Rodentia) such as voles primarily eat plants. Some, such as the bushy-tailed woodrat, consume meat when found, whereas other rodents such as deer mice are omnivorous, feeding on vegetation, seeds, and invertebrates alike. The southern red-backed vole, creeping vole, and other

1-56670-386-7/00/$0.00+$.50

voles as well as the bushy-tailed woodrat and deer mouse, are avid foragers on truffles and other mushrooms.[5,6] Consequently, they disperse mycorrhizal fungi, facilitating nutrient cycling between soil, litter, and vegetation. Clearly, diverse and abundant small mammals with widely varying foraging modes may contribute to wetland biodiversity and presumably in structuring wetland communities.

In turn, many small mammals are prey items for aquatic, avian, and terrestrial carnivores. Nocturnal small mammals (e.g., Townsend mole, bushy-tailed woodrat) are important prey for owls. Diurnal and crepuscular species (e.g., Townsend chipmunk, shrew mole) may be eaten by hawks, other avian raptors, and great blue herons. Reptilian predators including garter snakes may also eat them as well as mammals such as coyotes, foxes, bobcats, raccoons, weasels, ermines, and skunks. Some small mammal species are active at any time (e.g., Townsend voles) and hence may be eaten by all of the above predators whenever available.

In the Northwest, the regional distribution of small mammals and small mammal habitat associations within unmanaged old-growth Douglas-fir forests have been described by numerous biologists.[5-8] Terrestrial small mammals in second-growth forests under several cutting practices have been described. More recently, mammal richness and abundance in managed forest stands of 2- to 3-year-old clearcuts, precommercial stands, commercially thinned stands of 30 to 40 years of age, and in 50- to 70-year-old harvest-age forest stands have been studied.[9,10] Small mammal species within three large conifer-dominated urban parks of 64 to 113 ha in Seattle have also been described.[11]

For riparian areas in general, some qualitative descriptions of small mammals at wetlands have been published.[12-14] Comprehensive, quantitative studies of the richness and abundance of terrestrial small mammals in stream-side corridors have been documented for the central Oregon coast.[15-18] We are unaware, however, of any terrestrial small mammal studies at wetlands and their buffers, and particularly within wetlands of the Central Puget Sound Basin. Consequently, our objectives are to describe the biodiversity, relative distribution, and abundance of terrestrial small mammals across these wetlands. We also examine wetland traits such as size, vegetation habitats, and the presence of nonnative mammals to gain insight into wetland characteristics important in accounting for diversity and maintaining endemic and unique species. Finally, we assess watershed land use coverage and urbanization influences adjacent to wetlands to see if such large-scale landscape features can explain small mammal richness.

We believe our description and quantification of the distribution and abundance of small mammals (similar to that of macroinvertebrates, amphibians, and birds in Chapters 4, 5, and 6) may be useful in assessing the environmental health of wetlands and their associated watersheds. Moreover, small mammals through their high reproductive rate, active burrowing, and extensive foraging exhibit the ability to shape wetlands through their influence on soil, water, and plants.

METHODS

We used mid-October to mid-November Sherman and pitfall trap captures in 1988, 1989, and 1995 to identify terrestrial small mammals at wetlands. We augmented these data with just pitfall captures in 1993 as 1993's program targeted only amphibians and therefore did not use Sherman traps. We installed traps along two, 250-m transects on opposite sides of each wetland. A combination of 25 Sherman traps at 10-m intervals, with the central 10 traps paired with pitfall traps (without drift fences), were used as a trap line. To minimize the hydrostatic ejection of pitfalls, transects were located above yearly high-ground water levels. Pitfalls were operated for a total of 14 days. All traps were baited with rolled oats and ground beef. Moreover, polyester was provided in each for cover and insulation. Shermans were operated for total of six days by moving traps between paired wetlands after three consecutive trapping days at each wetland. Trapping was done on consecutive days, although we closed or removed traps disturbed by dogs, cats, raccoons, and other mammals and continued trapping when disturbance was no longer expected. We relocated flooded traps to nearby higher ground. At wetlands in which trap nights were less than the total required because of ongoing problems, we subtracted one half the number of traps unavailable from the total number of functioning traps and extrapolated to the full monitoring period using previously published methods.[19]

Traps were checked daily and all small mammals identified to species. Tail lengths distinguished deer mice from forest deer mice. Adults with tails exceeding 96 mm were identified as deer mice and those with tails less than or equal to 96 mm classified as forest deer mice. We marked captured mice by snipping the hairs from the tip of tails, allowing us to determine recaptures and hence obtain rough indices of abundances. The nearly total mortality of shrews enabled us to obtain their numbers by direct count.

We compared the number of National Wetlands Inventory habitats with the richness of mammal communities at wetlands.[20] We also compared small mammal richness to wetland size, land use, and forest land within 1000 m of the wetland, using the same methods described for amphibians in Chapter 5. The number, length, width, and characteristics of woody debris exceeding 15 cm diameter was inventoried at four parallel line transects, one along the shore and three within the buffer at four representative wetlands. The relative quantity of woody debris at each wetland was then grouped into one of four classes ranging from low to high when ranked against these surveyed wetlands. These categorical values were subsequently used in our analysis.

For our correlates of richness against wetland traits we omitted the northern flying squirrel, Douglas squirrel, and ermine. They are not well captured by our trapping methods and would unrealistically bias the richness at the few wetlands in which they were caught. Although the Townsend chipmunk and shrew mole may also not be readily captured by pitfall and Sherman traps, we nevertheless included them in our analysis as they were regularly captured and sighted at numerous wetlands, indicating a higher trapability than generally assumed. Our taxonomy follows Wilson and Ruff.[6]

RESULTS

We captured a total of 22 small mammal species, 19 of which are native to the natural habitats of the Central Puget Sound Basin (Table 7-1). Our richness included 14 species of rodents (Rodentia), seven species of insectivores (Insectivora), and one species of carnivore (Carnivora). We captured three nonnative rodents including the Norway rat, black rat, and house mouse. Several wetlands were visited by dogs (BBC24, PC12), opossums (ELW1, FC1), raccoons (LPS9), bears (RR5), and cougars (RR5). Free-ranging dogs, raccoons, and unidentified species sometimes disrupted our trapping program.

The range of terrestrial small mammal species at wetlands varied widely from a low of one species, the Norway rat, at ELW1, to a high of 13 species at LCR93 (Table 7-1). This maximum richness represented 68% of the total observed native species among local wetlands. We identified a mean of 7.2 native species including an average of 3.8 rodents, 3.2 insectivores, and 0.2 carnivores at wetlands.

The masked shrew was the most unexpected small mammal captured. It is a fairly uncommon species in this area because its primary distribution is in the boreal forests of northern Canada and Alaska. One individual was trapped at LCR93. The most unusual capture was that of a northern flying squirrel. Mostly an arboreal species, a single flying squirrel was captured in a Sherman trap at BBC24.

A bushy-tailed woodrat was captured at wetland LPS9 and a water shrew was observed but not captured at TC13. Species identified at more than one wetland, but less than six, included the Pacific jumping mouse at two wetlands (11%), long-tailed vole at four wetlands (21%), southern red-backed vole at four wetlands (21%), and the Townsend vole at five (26%) wetlands (Table 7-1).

Native mice were the most widespread small mammals, the deer mouse being found at 18 of 19 or 95% of our wetlands and the forest deer mouse at 17 or 89% of total wetlands. Nonnative rodents were identified at four or 21% of the wetlands. These species were found at B3I, ELS61, ELW1, and LPS9, wetlands close to human habitation. Surprisingly, B3I, a small 1.9-ha urban wetland with black rats also had six native mammal species. ELW1, with the Norway rat present, had no other mammal captures. Other widespread species in decreasing order of their distribution at wetlands include the Trowbridge shrew, vagrant shrew, creeping vole, and dusky shrew.

Native small mammal richness ranged widely within wetlands and between years (Figure 7-1). For example, LCR93, which had the highest number of native species over the entire study period, exhibited ten species the first year in 1988, yet only five species in 1995 under comparable trapping protocols. At HC13, we identified eight, nine, and seven species, respectively, in 1988, 1989, and 1995. Interestingly, during our amphibian pitfall trapping in 1993 we captured a greater number of mammal species at numerous wetlands (e.g., PC12, MGR36) than captured using both Sherman and pitfall traps in 1995 (Figure 7-1).

The most widely distributed species are also the most abundant in that the top five most widespread species, with the exception of the Townsend vole, were also most often captured (Figure 7-2). The deer mouse, for example, was the most widely distributed species and by far the most abundant mammal captured in all years over all wetlands. Nevertheless, we failed to detect the deer mouse during all trapping years at some of the wetlands even though it was abundant in other years (Figure 7-3). The dusky shrew and forest deer mouse were the next most abundant species but were captured in far fewer numbers compared to the deer mouse (Figure 7-3). Like the deer mouse, dusky shrew and forest deer mouse were found at wetlands

TABLE 7-1
Small Mammals Captured and Observed in Palustrine Wetlands of the Puget Sound Basin

Common Name	Scientific Name	AL3	B31	BBC24	ELS39	ELS61	ELW1	FC1	HC13	JC28	LCR93	LPS9	MGR36	NFIC12	PC12	RR5	SC4	SC84	SR24	TC13	Sum of Wetlands	Percent of Total Wetlands
Rodent	Rodentia																					
Black Rat	Rattus rattus		■									■									2	0.11
Bushy-tailed Woodrat	Neotoma cinerea											■									1	0.05
Creeping Vole	Microtus oregoni	■	■	■	■	■		■	■		■		■	■	■	■		■		■	14	0.74
Deer Mouse	Peromyscus maniculatus	■	■	■	■	■		■	■	■	■	■	■	■	■	■	■	■	■	■	18	0.95
Douglas Squirrel	Tamiasciurus douglasii								■				■								2	0.11
Forest Deer Mouse	Peromyscus oreas	■	■	■	■	■			■	■	■	■	■	■	■	■	■	■	■	■	17	0.89
House Mouse	Mus musculus					■															1	0.05
Long-tailed Vole	Microtus longicaudus			■							■	■		■							4	0.21
Northern Flying Squirrel	Glaucomys sabrinus			■																	1	0.05
Norway Rat	Rattus norvegicus						■														1	0.05
Pacific Jumping Mouse	Zapus trinotatus			■												■					2	0.11
Southern Red-backed Vole	Clethrionomys gapperi		■		■						■								■		4	0.21
Townsend Chipmunk	Tamias townsendi				■				■		■								■	■	5	0.26
Townsend Vole	Microtus townsendi				■	■						■			■			■			5	0.26
Insectivore	Insectivora																					
Dusky Shrew	Sorex monticolus				■	■			■		■	■	■	■	■	■			■		10	0.53
Marsh Shrew	Sorex bendirei			■					■		■		■			■			■	■	7	0.37
Masked Shrew	Sorex cinereus										■										1	0.05

TABLE 7-1 *(continued)*
Small Mammals Captured and Observed in Palustrine Wetlands of the Puget Sound Basin

Common Name	Scientific Name	AL3	B31	BBC24	ELS39	ELS61	ELW1	FC1	HC13	JC28	LCR93	LPS9	MGR36	NFIC12	PC12	RR5	SC4	SC84	SR24	TC13	Sum of Wetlands	Percent of Total Wetlands
Shrew-mole	Neurotricus gibbsii			X					X		X		X		X	X	X			X	8	0.42
Trowbridge Shrew	Sorex trowbridgii	X	X	X	X	X			X	X	X	X	X	X	X	X	X	X	X	X	17	0.89
Vagrant Shrew	Sorex vagrans	X	X	X	X	X		X	X		X	X	X	X	X	X		X	X	X	16	0.84
Water Shrew	Sorex palustris																			X	1	0.05
Carnivore	Carnivora																					
Ermine	Mustela erminea								X		X		X								3	0.16
Total Number of Species in Wetland:		5	7	10	9	8	1	3	11	3	13	9	10	7	8	9	4	6	8	9	19	1
Number of Native Species:		5	6	10	9	7	0	3	11	3	13	9	9	7	8	9	4	6	8	8	Mean:	7.2
Number of Rodents:		3	4	6	6	4	0	2	5	2	6	5	4	4	4	4	2	4	4	4	Mean:	3.8
Number of Insectivores:		2	2	4	3	3	0	1	5	1	6	3	5	3	4	5	2	2	4	5	Mean:	3.2
Number of Carnivores:		0	0	0	0	0	0	0	1	0	1	0	1	0	0	0	0	0	0	0	Mean:	0.2

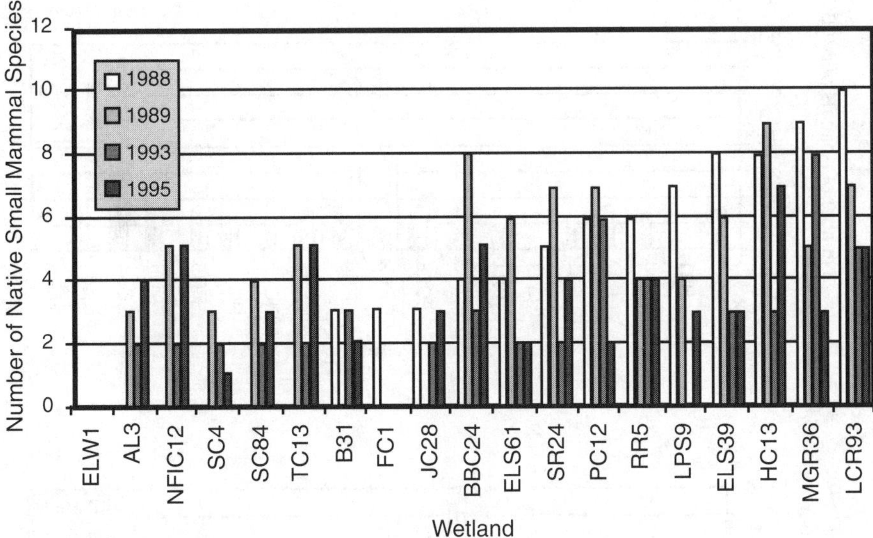

FIGURE 7-1 Native small mammal richness in 1988, 1989, 1993, and 1995 in study wetlands (1993 data is from pitfall traps only).

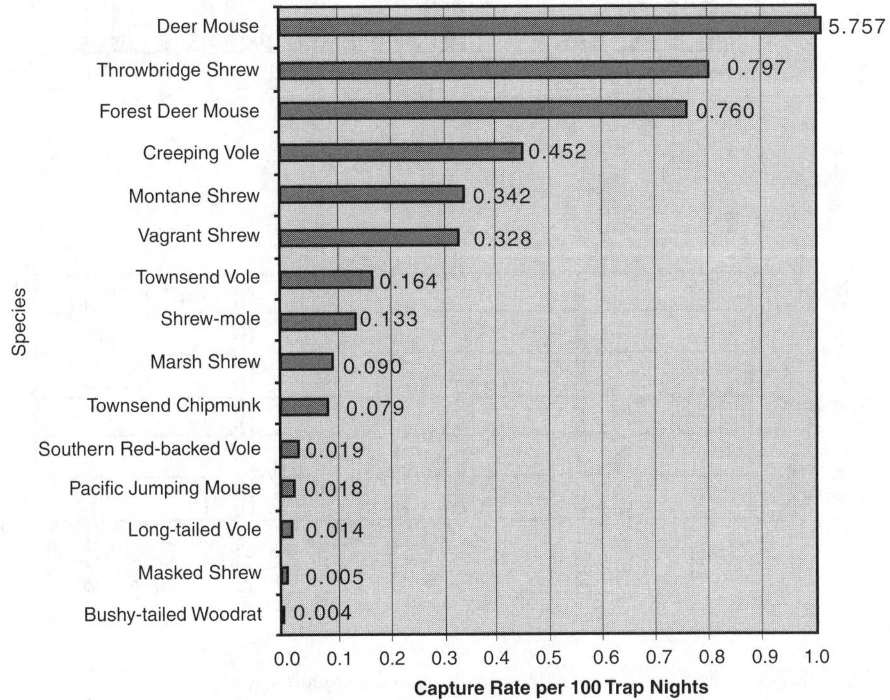

FIGURE 7-2 Capture rates of terrestrial small mammals within 19 wetlands. Data from a four-year study.

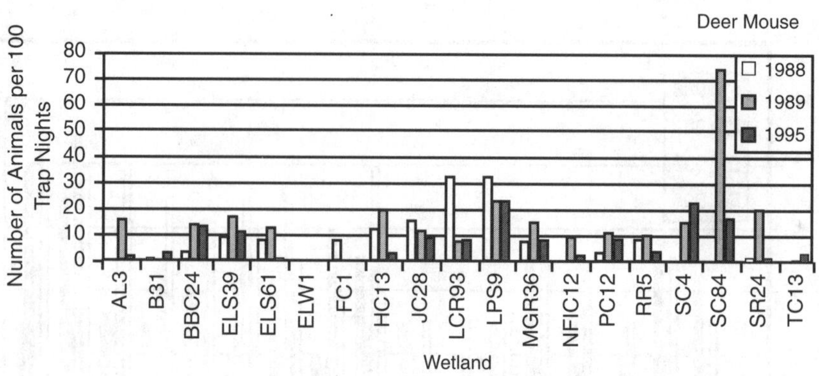

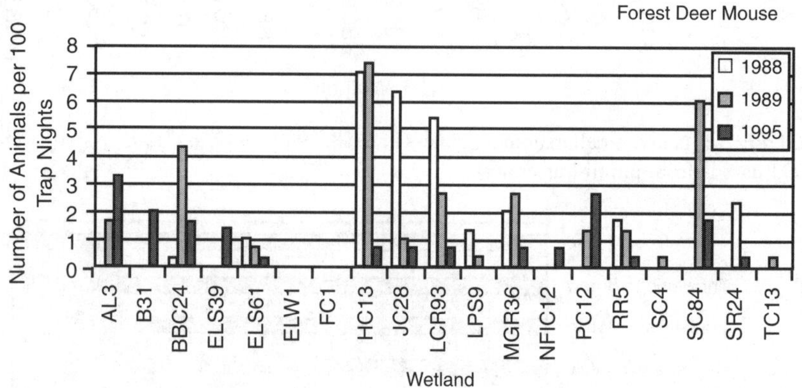

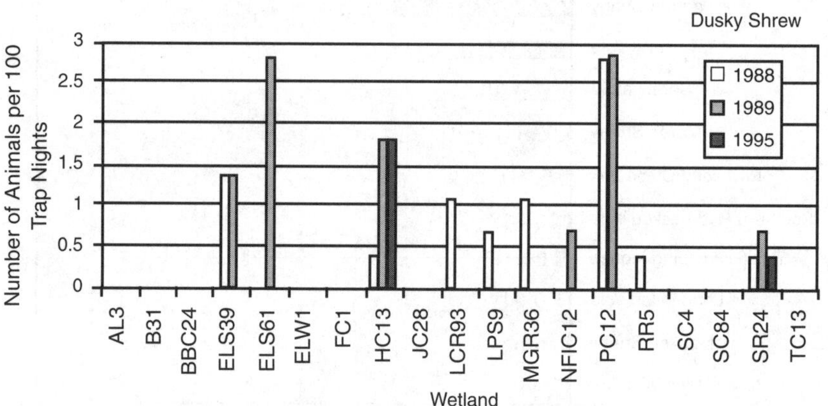

FIGURE 7-3 Yearly variations in three small mammal captures.

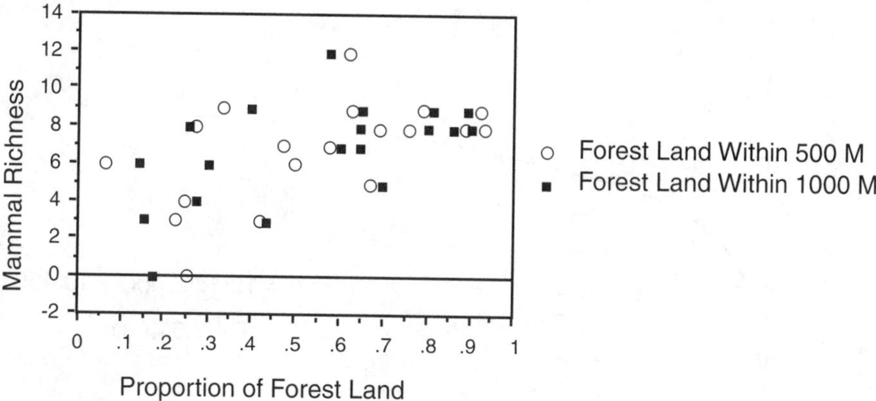

FIGURE 7-4 Relationship between small mammal richness and forest cover within 500 and 1000 m of wetland.

during some years but not others. Clearly, small mammal abundance varied widely between species, wetlands and between years (Appendices 1 through 4).

Wetland size was not found to be a significant correlate of either mammal richness or abundance (Fisher's r to z, r = –.13, $p = .6$). The total number of habitat classes was also not found to be correlated with small mammal richness (Fisher's r to z, r = –.14, $p = .57$).

We found that the total area of undeveloped land (including forest, shrub communities, agricultural fields, meadows, etc.) adjacent to the wetland weakly correlated with mammal richness (Fisher's r to z, r = .36, $p = .13$) within the first 500 m. The percent of forest land alone, with at most one single-family dwelling, within the first 500 m of a wetland was much more strongly correlated to richness (Fisher's r to z, r = .55, $p = .014$ and Figure 7-4) than undeveloped land in general. From 500 to 1000 m adjacent to a wetland, proportions of forest land were positively correlated with small mammal richness (Fisher's r to z, r = .44, $p = .06$ and R = .59, $p = .007$, respectively).

Finally, large, coarse woody debris at a wetland is related to small mammal richness, with an increase in wood correlated to an increase in small mammal species richness. Most importantly, however, small mammal richness was best associated with the combined factors of wetland size, adjacent forest retention, and the quantity of large, coarse woody debris within wetland buffers (Figure 7-5).

DISCUSSION

Our study shows that the small mammal communities at some wetlands may be among the most diverse when compared to that found in other terrestrial communities in the Central Puget Sound Basin. We captured 22 species, 19 of which were native. Excluding the Douglas and northern flying squirrel and the nonnative Murinae, we sighted six more species than were found in second-growth alder and Douglas-fir retention forests.[9] In contrast, we did not find the coast mole (*Scapanus orarius*).

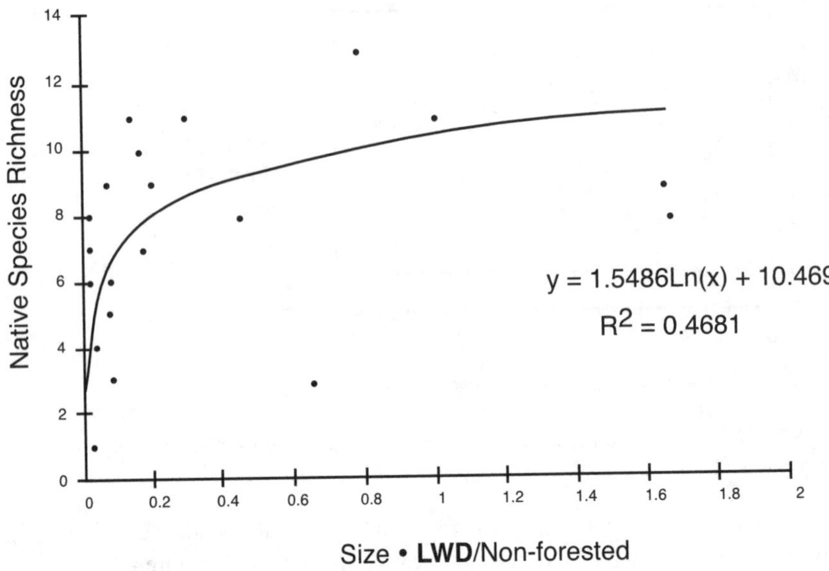

FIGURE 7-5 Relationship between small mammal richness and habitat variables of wetland size, land use, cover, and large woody debris.

Our richness, however, is comparable to the total richness of 18 native species identified in three years of trapping in four forest structural classes including clearcuts, pre- and postcommercially thinned stands, and 50- to 70-year-old harvest-age stands.[10] We found 16 species (89%) in common. In upland forests, coast and Townsend moles were exclusively captured whereas at wetlands, the bushy-tailed woodrat and Pacific jumping mouse were unique.

As expected, we sighted more native species than the seven found at three large upland urban parks.[11] This may be due to the wide range of urbanization encompassed by our study wetlands, from near pristine watersheds to metropolitan areas. Surprisingly, we captured rats and mice (Murinae) that were not captured in urban Seattle, suggesting that they perhaps do better adjacent to aquatic environments. In contrast, we did not capture or observe nonnative eastern cottontail, (*Sylvilagus floridanus*), fox squirrel (*Sciurus niger*), and eastern gray squirrel (*S. carolinensis*) even in wetlands adjacent to dense urbanization. Despite these current differences in richness, we expect wetland small mammal communities to become increasingly depauperate, monotypic, and cosmopolitan with ongoing urbanization and habitat fragmentation.

We identified several endemic Pacific Northwest/northern California species such as the Trowbridge shrew, shrew-mole, western red-backed vole, and creeping vole. Unexpectedly, we captured a masked shrew, which is at the southwestern edge of its continental distribution in Western Washington.[10] It has been captured in a wide range of successional habitats with no particular association to wetlands.

Native mice were the most widely distributed and abundant taxa, presumably because of their broad habitat niches. Both the deer mouse and forest deer mouse breed, feed, and rest in all types of habitats including wetlands and their edges. They have been identified within the greatest diversity of plant communities and stand conditions of any small mammal in the Northwest.[21] Nevertheless, it may be that native mice preferentially select wetlands. Riparian-associated upland red alder communities are often dominated by deer mice, whereas in upland forests they are only the third most abundant species after the Trowbridge shrew and creeping vole.[10,18] Interestingly, other species we captured such as the Townsend vole, Pacific jumping mouse, marsh shrew, and long-tailed vole, were found by other investigators to be more abundant along riparian corridors than in adjacent upland areas of coniferous stands.[16, 17]

Native mice most likely play a disproportionately important role in the trophic dynamics of palustrine wetlands because we found them at almost all sites, capturing them on a regular basis and in large numbers when compared to other small mammals. Although other species breed throughout the year (e.g., Townsend vole, creeping vole) their abundances were never as great, thus their ecological role may not be as important.

Several species of voles were captured. Townsend vole populations reach highest abundances in the wet field and meadow environments of the Puget Sound lowlands.[10] They are often found in grassy areas interspersed with sedges and rushes and, in fact, sometimes live in lawns adjacent to such areas.[6] Consequently, their limited distribution and low abundances at our surveyed wetlands is disturbing since they inhabit areas with high water tables.

Although our captures are few, it is interesting that we did not capture long-tailed voles at wetlands where we captured the Townsend vole. Some believe the Townsend vole may be more aggressive and hence may not be sympatric with long-tailed voles.[6] Creeping voles may be found in a variety of environments but often are found at wetlands because their preferred breeding, feeding, and resting habitats are riparian plant communities and especially wet meadows. However, they are also associated with forests and are abundant where forests have been cut.[6,10] Long-tailed voles, although arboreal, were captured in relatively large numbers at our wetlands. Long-tailed voles are found in grass and forb communities of wetlands and have been captured in riparian zones.[6] Southern red-backed voles are generally found in coniferous forests with substantial amounts of large fallen trees and around openings in such forests.[10] We found southern-red-backed voles at only one wetland that exhibited these traits, but not in the seven other wetlands that generally exhibited such forest structure. Its absence from these potential sites remains unexplored.

The richness and abundance of shrews at wetlands may be attributable to their small size, high metabolic rates, and correspondingly high water losses, requiring them to remain within high moisture environments.[21-23] In fact, research shows that hydric habitats supported 2.7 times as many shrew species ($\bar{x} = 4.7$) as xeric habitats ($\bar{x} = 1.8$ species).[24] In the Pacific Northwest, water and moisture may not be a limiting factor with the exception the of summer drought periods. Consequently, our shrew richness may be similar to that identified for uplands.[10] Also, in contrast to expectations, the vagrant shrew reaches its greatest abundances in exposed environments and, in fact, increases in number with clearcut area and mean clearcut patch size.[10] Moreover, the Trowbridge shrew is an omnivorous shrew that occupies the greatest variety of habitats, including the range from the wettest to the driest areas.[4] Interestingly, we found both species at almost all wetlands.

Clearly, water shrews are strictly riparian within deciduous, coniferous, and mixed forest types and are the only species that may be considered wetland obligated. However, we only sighted one in our surveys, possibly because our traps were beyond the hyporheic zone in which this wetland obligate may be more regularly found. In contrast, the marsh shrew (a wetland-associated species) was captured at seven wetlands and is more often found further from sources of water. Both species swim under water, feeding on aquatic insects, most likely dragonfly naiads, nymphal mayflies, and alderflies, and consequently may be expected to be abundant at wetlands. [4]

Pacific jumping mice are associated with grass, herbs, and moist or marshy areas with skunk cabbage. [4] Although our sightings of this species were expected, we nevertheless found them at only two wetlands.

We believe exotic rats at wetlands are detrimental to native, less assertive small mammals. Norway rats particularly may be more aggressive and detrimental to native mammals than black rats, since wetlands with Norway rats did not have any native species. Norway rats presumably out-competed native species at ELW1 and a second wetland in Snohomish County, where only Norway rats were captured but no native species.[25] Moreover, these wetlands in urban areas are also visited by domestic cats, which are known to kill vagrant and Trowbridge shrew, and likely most other small mammals, but possibly avoid Norway rats.[4] Clearly, pet predation on small mammals additionally reduces species distribution, richness, and abundance in urban areas.

We expected mammal richness to be strongly related to wetland size, since intuitively one would expect larger wetlands to have more niches and habitat complexity, but did not find this to be the case. The total number of habitat classes, an indirect measure of habitat niches, also was not found to be correlated with small mammal richness.

Perhaps one of our more significant findings is the importance of forest land adjacent to wetlands to maintain diversified mammal populations. The highest small mammal richness occurred in wetlands that had greater than 60% of the first 500 m of buffer in forest land. Undeveloped land can provide habitat for some species, but more important is forest land, and its associated component of large woody debris, in accounting for small mammal richness within the wetland buffer.

Our earliest models attempted to account for small mammal richness only through wetland size, the presence of vegetation types, and the presence of development (with its associated human and animal impacts). However, they failed to show the strong relationship exhibited by utilizing the joint characteristics of forest land and large, woody debris to account for small mammal populations. Consequently, our work confirms that of others who have shown Trowbridge shrew, shrew mole, deer mice, and southern red-backed vole increase in abundance with increasing quantities of forest floor woody debris.[26] Our findings, however, additionally suggest that a certain amount of development can occur and nonnative mammals such as stray cats and dogs may be tolerated at wetlands as long as forest land and large, coarse woody debris remains available for cover, food, shelter, and microclimatic relief. It is forest land that produces a continuous supply of trees, large logs, and stumps that provide small mammal habitat over time. Finally, our findings also point out the value of conserving and maintaining whatever large woody debris exists at wetlands and within wetland buffers to maintain small mammal habitat.

ACKNOWLEDGMENTS

Our thanks to Drew Kerr, a.k.a. Dave, for helping to edit this manuscript and to Stephen D. West for setting up monitoring protocols. We gratefully acknowledge the many graduate students at the University of Washington who aided in trapping and data collection including, but not limited to, Vanessa Artman, Sang-Don Lee, Timothy Quinn, Mark E. Rector, Heather Roughgarden, Claus Svendsen, and Julie Stoffel.

REFERENCES

1. Paine, R. T., Food web complexity and species diversity, *Am. Nat.,* 100, 65, 1966.
2. Neiman, R. J., J. M. Melillo, and J. E. Hobbie, Ecosystem alteration of boreal forest streams by beaver (*Castor canadensis*), *Ecology,* 67, 1254, 1986.
3. Neiman, R. J., C. A. Johnston, and J. C. Kelley, Alteration of North American streams by beaver, *BioScience,* 38, 289, 1988.
4. Maser, C., Land Mammals, in *Natural History of Oregon Coast Mammals*, C. Maser, B. R. Mate, J. F. Franklin, and C. T. Dyrness, Eds., USDA Forest Service Gen. Tech. Rep. 133, U.S. Department of Agriculture, Pacific Northwest Forest and Range Experiment Station, Portland, OR, 1981, p. 35.
5. Maser, C., J. M. Trappe, and R. A. Nussbaum, Fungal-small mammal interrelationships with emphasis on Oregon coniferous forests, *Ecology,* 59, 799, 1978.
6. Wilson, D. E. and S. Ruff, Eds., *The Smithsonian Book of North American Mammals,* Smithsonian Institute Press, Washington and London, 1999.
7. Ingles, L. G., *Mammals of the Pacific States,* Stanford University Press, Stanford, CA, 1965.
8. Aubry, K. B. and M. H. Brookes, Eds., *Wildlife and Vegetation of Unmanaged Douglas-Fir Forests*, USDA, Forest Service, Gen. Tech. Rep. PNW-GTR-28, U.S. Department of Agriculture, Pacific Northwest Research Station, Portland, OR, 1991.
9. Stofel, J. L., Evaluating Wildlife Responses to Alternative Silvicultural Practices, M.S. thesis, University of Washington, Seattle, 1993.
10. West, S. D., Terrestrial Small Mammals, in *Wildlife Use of Managed Forests: A Landscape Perspective*, K. B. Aubry, S. D. West, D. A. Manuwal, A. B. Stringer, J. L. Erickson, and S. Pearson, Vol. 2, Section 3, West-side Studies Res. Results, Timber Fish and Wildlife, Olympia, WA, 1997.
11. Gavareski, C. A., Relation of Park Size and Vegetation to Bird and Mammal Populations in Seattle, Washington, M.S. thesis, University of Washington, Seattle, 1972.
12. Fritzell, E. K., Mammals and Wetlands, in *The Ecology and Management of Wetlands*, Vol. 1, D. D. Hook, W. H. McKee, Jr., H. K. Smith, J. Gregory, V. Burrell, Jr., M. R. DeVoe, R. E. Sojka, S. Gilbert, R. Bani, L. H. Stolzy, C. Brooks, T. D. Matthews, and T. H. She, Eds., Timber Press, Portland, OR, 1988, p. 213.
13. Brooks, R. P. and M. J. Croonquist, Wetland habitat and trophic response guilds for wildlife species in Pennsylvania, *J. Pa. Acad. Sci.,* 64, 93, 1990.
14. Croonquist, M. J. and R. P. Brooks, Use of avian and mammal guilds as indicators of cumulative impacts in riparian-wetland areas, *Environ. Manage.,* 15, 701, 1991.

15. Doyle, A. T., Use of riparian and upland habitats by small mammals, *J. Mammal.*, 71, 14, 1990.
16. Gomez, A. T., Small Mammal and Herpetofauna Abundance in Riparian and Upslope Areas of Five Forested Conditions, M.S. thesis, Oregon State University, Corvallis, 1992.
17. McComb, W. C., K. McGarigal, and R. G. Anthony, Small mammal and amphibian abundance in streamside and upslope habitats of mature Douglas-fir stands, western Oregon, *Northwest Sci.*, 67, 7, 1993.
18. McComb, W. C., C. L. Chambers, and M. Newton, Small mammal and amphibian communities and habitat associations in red alder stands, central Oregon coast range, *Northwest Sci.*, 67, 181, 1993.
19. Nelson, L., Jr. and F. W. Clark, Correction for sprung traps in catch/effort calculations of trapping results, *J. Mammal.*, 54, 295, 1973.
20. Cowardin, L. M., V. Carter, F. C. Goulet, and E. T. LaRoe, Classification of Wetlands and Deepwater Habitat of the United States, U.S. Fish and Wildlife Service, FWS/OBS-79/31, Washington, D.C., 1979.
21. Brown, R., Management of Wildlife and Fish Habitats in Forests of Western Oregon and Washington: Part 2-Appendices, USDA Forest Service, Publ. No. R6-F&WL-192-1985, Pacific Northwest Region, Portland, OR, 1983.
22. Manville, R. H., A study of small mammal populations in northern Michigan, *Misc. Publ. Mus. Zol. Univ. Mich.*, 73, 1, 1949.
23. Getz, L. L., Factors influencing the local distribution of shrews, *Am. Midl. Nat.*, 65, 67, 1961.
24. Kirkland, G. L., Jr. and T. L. Serfass, Wetland mammals of Pennsylvania, in *Wetlands Ecology and Conservation: Emphasis in Pennsylvania*, S. K. Majumdar, R. P. Brooks, F. J. Brenner, and R. W. Tiner, Jr., Eds., Pennsylvania Academy of Science, Easton, PA, 1989, p. 216.
25. Wrigley, W. E., J. E. Dubois, and H. W. R. Copeland, Habitat, abundance, and distribution of six species of shrews in Manitoba, *J. Mammal.*, 60, 505, 1979.
26. Svendsen, C. R. and K. O. Richter, Small mammals in a restored urban wetland in response to biotic and landscape characteristics, abstr. presented at Ecological Summit 96, Copenhagen, Denmark, 1996.
27. Carey, A. B. and M. L. Johnson, Small mammals in managed, naturally young, and old-growth forests, *Ecol. Appl.*, 5, 336, 1995.

APPENDIX 7-1
Capture Rates of Small Mammals by Wetland — 1988

1988 SITE_ID	Southern Red-backed Vole		Townsend Chipmunk	Long-tailed Vole		Creeping Vole		Townsend Vole		Ermine	Shrew-mole	Deer Mouse		Forest Deer Mouse	Black Rat	Marsh Shrew	Masked Shrew	Dusky Shrew		Trowbridge Shrew		Vagrant Shrew		Douglas Squirrel	Total Pitfall	Total Sherman
	CLGA	CLGA	TATO	MILO	MILO	MIOR	MIOR	MITO	MITO	MUER	NEGI	PEMA	PEMA	PEOR	RARA	SOBE	SOCI	SOMO	SOMO	SOTR	SOTR	SOVA	SOVA	TADO		
	pitfall	Sherman	Sherman	pitfall	Sherman	pitfall	Sherman	pitfall	Sherman	Sherman	pitfall	pitfall	Sherman	Sherman	Sherman	pitfall	pitfall	pitfall	Sherman	pitfall	Sherman	pitfall	Sherman	Sherman		
B3I	0.36												0.67		1.00							0.34			0.69	1.67
BBC24						0.35	0.67						2.02							0.35	0.34				0.69	3.03
ELS39	0.36		0.33			2.41	1.67	2.41	2.67			0.34	8.33					0.69	0.67	0.34		1.72	0.33		8.28	14.00
ELS61						0.34	1.00						7.33	1.00						0.69	0.33				1.03	9.67
ELW1																										1.33
FC1						2.76	3.67						7.33										0.33		2.76	12.00
HC13							2.33			0.33		0.69	7.33	5.00		0.69		0.34		0.69		0.34			2.76	15.33
JC28												0.34	7.33	4.00							0.67				0.69	12.00
LCR93		0.33	2.00	0.34		0.34						0.34	15.00	2.00		0.69	0.34	1.03		1.03					4.14	19.33
LPS9					0.33			0.69	0.33			1.38	18.33	1.00				0.34	0.33	1.72	0.67	1.03	0.33		5.17	21.33
MGR36						1.72	1.67			0.33	0.34	1.38	4.00	1.33				1.03		0.34	0.67	0.34	0.33	0.33	5.17	8.67
PC12						0.34	1.00	0.34	1.33			0.34	3.33					1.72	1.00	0.34	0.33	1.03			4.14	7.00
RR5						1.03	0.67						5.04	0.67		0.69			0.34	3.10					4.83	6.72
SR24													1.67	1.00				0.34		0.69		0.34			2.07	2.67
Grand Total	0.71	0.33	2.33	0.34	0.33	9.31	12.68	3.45	4.33	0.67	0.34	4.83	87.73	16.01	1.00	2.07	0.34	5.52	2.34	9.31	3.00	5.17	1.33	0.33	42.41	134.75

APPENDIX 7-2
Capture Rates of Small Mammals by Wetland — 1989

1989 SITE_ID	Townsend Chipmunk	Northern Flying Squirrel	Creeping Vole		Townsend Vole		Ermine	Bushy-tailed Woodrat	Shrew-mole	Deer Mouse		Forest Deer Mouse		Norway Rat	Black Rat	Marsh Shrew	Dusky Shrew		Trowbridge Shrew		Vagran Shrew		Pacific Jumping Mouse		Total Pitfall	Total Sherman
	Sherman TATO	Sherman GLSA	pitfall MIOR	Sherman MIOR	pitfall MITO	Sherman MITO	Sherman MUER	Sherman NECI	pitfall NEGI	pitfall PEMA	Sherman PEMA	pitfall PEOR	Sherman PEOR	Sherman RANO	Sherman RARA	pitfall SOBE	pitfall SOMO	Sherman SOMO	pitfall SOTR	Sherman SOTR	pitfall SOVA	Sherman SOVA	pitfall ZATR	Sherman ZATR		
AL3										2.14	7.59		1.38								1.43	0.34			3.57	9.31
BBC24		0.35		0.70						0.36	8.70		3.13			0.36			1.79	0.35	0.72		0.36	0.35	3.58	13.57
ELS39	0.34			1.03	1.07	0.69				0.36	13.45						0.71	0.69			0.36	0.34			2.50	16.55
ELS61				0.34		0.34				0.71	10.34		0.69				1.43	1.03				0.34			2.14	13.10
HC13	1.03			0.69					0.71		12.07		4.83			0.36	1.79		0.36		0.36				3.57	18.62
JC28											8.28		1.03													9.31
LCR93	0.34						0.34		0.36		6.55		2.07			1.07			1.79						3.21	9.31
LPS9								0.34		0.36	12.76		0.34						1.43	1.03					1.79	14.48
MGR36			0.36	0.34						0.71	8.62		0.69			0.36			1.07						2.50	9.66
NFIC12			0.71							3.57	2.76						0.36	0.34	0.71	0.34	0.36				5.71	3.45
PC12				1.38		1.38				0.36	9.66		1.03				2.14	0.34	2.86	0.34	1.43	0.34			6.79	14.48
RR5										0.36	7.30		1.39						0.71			0.35			1.07	9.04
SC4										1.43	7.93		0.34						0.36						1.79	8.28
SC84			0.36		0.36					3.93	30.69	0.36	3.79												5.00	34.48
SR24	0.34									0.36	10.34		1.72			0.36	0.36		2.50		1.79				6.07	12.41
TC13			0.36	0.69							0.69		0.34			0.36			0.36						1.07	1.72
Grand Total	2.07	0.35	1.79	5.18	1.43	2.41	0.34	0.34	1.07	14.64	157.72	0.36	22.80			2.86	6.79	2.41	13.93	2.07	6.43	1.73	0.36	0.35	50.36	197.78

APPENDIX 7-3
Capture Rates of Small Mammals by Wetland — 1993

1993 SITE_ID	Southern Red-backed Vole	Creeping Vole	Townsend Vole	Shrew-mole	Deer Mouse	Forest Deer Mouse	Marsh Shrew	Dusky Shrew	Trowbridge Shrew	Vagrant Shrew	Total Pitfall
	CLGA	MIOR	MITO	NEGI	PEMA	PEOR	SOBE	SOMO	SOTR	SOVA	
	pitfall	pitfall	pitfall	pitfall	pitfall	pitfall	pitfall	pitfall	pitfall	pitfall	
AL3					2.50				1.43		3.93
B3I		0.36			0.71				1.07		2.14
BBC24					0.36						0.36
ELS39			0.36		0.71				0.36		1.43
ELS61									0.71	0.36	1.07
HC13					0.36			0.36	0.36		1.07
JC28					0.36				0.71		1.07
LCR93					2.14	0.36		0.36	3.57	0.36	6.79
MGR36		0.36		0.36	4.29		1.43	1.43	1.43	1.43	10.71
NFIC12					0.36				0.36		0.71
PC12				1.07	4.29	0.36		0.36	2.86	2.14	11.07
RR5				3.21	1.79				2.14	0.71	7.86
SC4				0.36	0.36						0.71
SR24	0.36								0.71		1.07
TC13					0.36						0.36
Grand Total	0.36	0.71	0.36	5.00	18.57	0.71	1.43	2.50	15.71	5.00	50.36

APPENDIX 7-4
Capture Rates of Small Mammals by Wetland — 1995

1995	Townsend Chipmunk	Long-tailed Vole	Creeping Vole		Shrew-mole	Deer Mouse		Forest Deer Mouse		Black Rat	Dusky Shrew	Trowbridge Shrew	Vagrant Shrew		Douglas Squirrel	Pacific Jumping Mouse	Total	Total
	TATO	MILO	MIOR	MIOR	NEGI	PEMA	PEMA	PEOR	PEOR	RARA	SOMO	SOTR	SOVA	SOVA	TADO	ZATR		
SITE_ID	Sherman	pitfall	pitfall	Sherman	pitfall	pitfall	Sherman	pitfall	Sherman	Sherman	pitfall	pitfall	pitfall	Sherman	Sherman	Sherman	Pitfall	Sherman
AL3				0.34			1.72		3.10			1.07					1.07	5.17
B3I							3.45		1.72									5.17
BBC24			2.15		1.43	3.94	9.39		1.74			1.79					9.66	11.13
ELS39	1.38						11.72		1.38									14.83
ELS61						0.36	1.72		0.34								0.36	3.45
HC13					0.71	0.36	3.10	0.36	0.34		1.79	1.07	0.36	0.34	0.34		4.64	4.14
JC28						0.36	8.62		0.69			0.36					0.71	9.31
LCR93			0.71				8.97		0.69			0.36		0.34			1.07	1
LPS9						0.36	23.45			1.72		0.36	0.71				1.43	25.17
MGR36						0.71	8.62		0.69			2.86					3.57	9.31
NFIC12		0.36	1.07				2.41		0.69			0.36					1.79	3.10
PC12						0.36	8.62		2.41								0.36	11.03
RR5					0.71	0.36	3.48		0.35							0.70	1.07	4.52
SC4							23.79											23.79
SC84							16.21		1.72			0.71					0.71	17.93
SR24							1.72		0.34		0.36	1.43					1.79	2.07
TC13	0.34				0.36	0.71	2.76					1.79	1.07				3.93	3.10
Grand Total	1.72	0.36	3.93	0.34	3.22	7.51	139.77	0.36	16.22	1.72	2.14	12.15	2.14	0.69	0.34	0.70	32.16	163.24

Section III

Watershed Development Effects on Freshwater Wetlands in the Central Puget Sound Basin

8 Effects of Watershed Development on Hydrology

Lorin E. Reinelt and Brian L. Taylor

CONTENTS

Introduction....................221
Puget Sound Wetlands and Stormwater Management Research Program....................222
Research Objectives....................222
Data Collection and Analysis Methods....................223
Statistical Analysis of Development Impacts on Wetland Hydrology....................223
Data Collection and Analysis for the Wetland Water Balance....................224
Results and Discussion....................225
Wetland Hydrology and Water Level Fluctuation....................225
Threshold Level Analysis....................225
Multiple Regression Analyses....................229
Dry Period....................230
Hydrologic Characteristics of Two Intensively Studied Wetlands....................230
Wetland Hydrology by Season and Wetland....................232
Urbanization and Other Factors Affecting Wetland Hydrology....................233
Hydrologic Components Error Analysis....................233
Conclusions....................234
Acknowledgments....................235
References....................235

INTRODUCTION

In urbanizing areas, the quantity (peak flow rate and volume) of stormwater can change significantly as a result of changing land use in a watershed. Increases in stormwater may result from new impervious surfaces, removal of forest cover, and installation of constructed drainage systems. Watershed development can also reduce recharge of groundwater, compromising baseflow to streams and result in less evapotranspiration.

1-56670-386-7/00/$0.00+$.50

Changes in hydrology, whether brought about intentionally or incidentally, have an influence on wetland systems. Wetlands will likely have a positive effect on downstream areas by dampening storm flows before discharging to streams and lakes. However, these same higher peak flows and volumes may also adversely impact wetlands. For cases where wetlands are the primary receiving water for urban stormwater from new developments, it is hypothesized that the effects of watershed changes will be manifested through changes in the hydrology of wetlands.

Wetland hydrology is often described in terms of its hydroperiod, the pattern of fluctuating water levels resulting from the balance between inflows and outflows of water, landscape topography and subsurface soil, geology, and groundwater conditions.[1] Hydroperiod alterations are the most common effect of watershed development on wetland hydrology. This usually involves increases in the magnitude, frequency, and duration of wetland water levels. In other words, increased stormwater flows tend to cause higher wetland water levels on more occasions during the wet season and for longer periods of time. These changes in wetland hydroperiod then result in impacts to plant and animal communities that were adapted to the pre-existing hydrologic conditions.

Puget Sound Wetlands and Stormwater Management Research Program

Palustrine wetland hydrology was studied as a part of both components of the research program: (1) the study of the long-term effects of urban stormwater on wetlands, and (2) the study of the water-quality benefits to downstream receiving waters as urban stormwater flows through wetlands. This chapter presents results from the statistical analysis of 19 study wetlands from the long-term effects study and from the water balance of two wetlands from the water-quality benefit study.

Research Objectives

The primary objective of this portion of the research program was to examine the effects of urban stormwater on wetland hydrology. However, there were also a variety of specific hydrologic questions addressed throughout the research, which developed into the following specific objectives:

- Identify the wetland and watershed hydrologic processes and the factors governing these processes.
- Determine how urban catchments behave differently from forested catchments.
- Determine the percent contribution of wetland hydrologic inputs and outputs.
- Relate wetland hydrologic conditions to wetland and watershed characteristics.
- Characterize wetland hydroperiods and develop a set of dependent variables for analysis.

DATA COLLECTION AND ANALYSIS METHODS

As noted in Chapter 1, a conceptual model was used to show the relationship between factors influencing wetland and watershed hydrologic processes and the wetland hydroperiod (Figure 1-4). In the conceptual model, some of the key factors thought to influence wetland water level fluctuation included: (1) forested area, (2) impervious area, (3) wetland morphology, (4) outlet constriction, (5) wetland-to-watershed area ratio, and (6) watershed soils. Statistical analyses were carried out to determine which factors were most important.

Statistical Analysis of Development Impacts on Wetland Hydrology

A variety of graphical and statistical techniques were used in identifying relationships between the watershed or wetland characteristics and wetland hydroperiod.[2] Microsoft EXCEL® was used in processing the data and SYSTAT® was used for statistical analyses.

Graphical analysis was used to identify trends and threshold levels that could then be statistically tested to determine which statistical methods (parametric or nonparametric) were appropriate. Graphical analysis provided insights into which factors correlated to specific aspects of the hydroperiod; however, it failed to show the effects of multiple factors or simultaneously show varying importance.

In order to determine which statistical tests were appropriate for a given hypothesis, the normality of the data was assessed. The Kolmogorov-Smirnoff test was used to compare the maximum difference between two cumulative distributions. The Lilliefors test was used when the mean and variance of the distribution were unknown, in order to automatically standardize the variables and test whether the standardized distributions were normally distributed.[3] The Lilliefors test was used to assess the distribution of water level fluctuation measurements. The significance level used in testing normality was alpha equal to .05.

Threshold testing was done when a scatter plot suggested one or more threshold levels in the response of wetland water level fluctuations to a specific watershed or wetland characteristic. The data were grouped categorically based on thresholds suggested in the scatter plots. These groups were compared in a test of the null hypothesis that all groups were from equivalent distributions.

Because the water level fluctuation measurements were not normally distributed for all of the study sites, nonparametric tests were used: the Mann-Whitney test for two groups and the Kruskal-Wallis test for more than two groups. These two tests are analogous to the independent groups *t*-test for normally distributed data, but are based on data ranks rather than the data values.[3,4] The Kruskal-Wallis test will reject the null hypothesis if any of the groups are significantly different; nonparametric multiple comparisons were done to identify which groups were significantly different.[4] The significance level used in evaluating thresholds was $p \leq .05$.

Multivariate, least-squares, linear regression models were calibrated to the study data to show how various wetland and watershed factors combine to effect wetland hydroperiod.[2] Models were developed by:

1. Using step regression to identify factors important to the aspect of wetland hydroperiod being investigated
2. Determining the best way to quantify or express this factor
3. Evaluating model fit
4. Examining the sensitivity to the predictor variables

The data for each wetland were weighted by sample size when appropriate; mean water level fluctuation was weighted by the total number of observations used in its calculation, while the length of the dry period and seasonal water level fluctuations were weighted by the number of years used in their calculation.

The fit of the regression models was evaluated through various methods: the coefficient of determination (r^2) and the F-ratio, which compares the explanation provided by each predictor to the residual associated with each observation. The final step in the generation of the multiregression models was to examine the sensitivity of each predictor variable. The standardized coefficient of each predictor variable provides a way to compare the significance of the variables.[3] Additionally, variables were removed from the final model one at a time to determine their effect on the model r^2 and the standard error of the estimate.

Data Collection and Analysis for the Wetland Water Balance

In the detailed study of two wetlands (Bellevue 3I and Patterson Creek 12), a complete water balance was performed.[5] This consisted of independent measurements of the following components: precipitation, evapotranspiration, surface inflow, surface outflow, groundwater exchange, and change in wetland storage. Precipitation was measured using an event recorder connected to a tipping-bucket gauge that recorded each .25 mm of rainfall. Continuous water flow measurements were taken at the inlet and outlet of the two wetlands using a variety of different techniques.[6]

Shallow (1.2 to 4 m) and deep (6 to 18 m) piezometers were installed at both wetlands to aid in the estimation of groundwater flow using Darcy's Law (see Chapter 1). The hydraulic conductivity (K) of the underlying aquifer at both wetlands was determined using variable head pump and slug tests as described in previous studies.[7,8] Piezometric head measurements were taken regularly to determine the hydraulic gradient.[9] Control volumes were defined around each wetland to facilitate estimation of the horizontal and vertical components of groundwater flow.

Evapotranspiration was estimated from pan evaporation data from the Washington State University Extension Service Puyallup Station representing the Puget Sound lowlands region. Adjustments were made for differences between pan evaporation, open-water evaporation, and evapotranspiration by plants. Daily changes in wetland water depth (and corresponding storage volume) were estimated by correlating daily outflow data with regular gauge (water depth) readings. Storage volumes were determined for different water levels by multiplying the area of water coverage by water depth.

Identifying and describing seasonal differences in the hydrologic balance of the two wetlands was one objective of the study. Seasons were defined and analyzed by two classification methods. The first method included simply wet (October to March) and dry (April to September) seasons. The second method defined four seasons based on the climate of the Puget Sound region: wet (November to February), dry (June to September), and two transition seasons (March to May and October). The division of data by season allowed for comparison of changes in the relative contributions of different inputs and outputs.

RESULTS AND DISCUSSION

WETLAND HYDROLOGY AND WATER LEVEL FLUCTUATION

Three parameters were used to examine hydrologic conditions in the wetlands: water depth, water level fluctuation (WLF), and length of summer dry period. The minimum, maximum, and range of water depths at the gauges are given in Table 8-1. Also given are the mean (according to Equation 1-4 of Chapter 1) and maximum WLF and days of summer drying in the wetland. Water depth and WLF varied widely for the 19 wetlands.

The largest range of water levels, as well as mean and maximum WLFs, were found at B3I and LPS9, where the basins have among the highest percent of impervious area of any of the study sites and the wetland outlets are constricted (see B3I and LPS9 in Figure 8-1). Those wetlands with 90% or more forested cover and less than 3% impervious surfaces generally exhibited Hanower water depth ranges and low WLFs (see BBC24 and SR24 in Figure 8-1). As can be seen from Figure 8-1, these trends of low or high WLF are independent of whether the base level condition in the wetland is stable or fluctuating. Wetland JC28 was an exception to the normal relationship between high impervious area and high WLF; this was because the watershed soils are predominantly glacial outwash (highly permeable soils), thus reducing runoff volumes.

THRESHOLD LEVEL ANALYSIS

Scatter plots of the event water level fluctuation data were plotted against the various wetland and watershed morphological parameters. Some of these plots showed apparent thresholds that signify a range of the hydrologic parameter where the event fluctuation data are similarly distributed. Within these ranges, characteristics such as the mean and variance of the data were approximately equal. Table 8-2 shows significant threshold levels ($p < .05$ for all thresholds) and characterizes the water level fluctuation data within each range.

A key index relating urbanization to WLF was basin imperviousness. Two thresholds were identified in the relationship between event WLF and impervious area (Figure 8-2). The first threshold (3.5% impervious area) may represent the level of urbanization where scattered clearing of forests is added to by larger developments, and where storm drainage systems that route runoff to the wetland are developed. Development within the first range was usually below 15% low-density residential (LDR), whereas the second range begins around 24% LDR. Wetlands HC13 and LCR93 (in the second range) were exceptions to this tendency, because

TABLE 8-1
Wetland Watershed (Land Use), Outlet and Hydrologic Characteristics

Wetland Name	Forest (%)	Impervious Area (%)	Outlet Constricted 0 = no 1 = yes	Range of Water Depth (m)	Mean WLF (m)	Max. WLF (m)	Mean Dry Period (days)	Calculated Mean WLF (m) Using Multiple Regression
AL3	73.9	3.4	1	0.00–0.62	0.07	0.31	101	0.21
MGR36	88.8	2.7	0	0.13–0.74	0.07	0.26	0	0.08
JC28	34.4	19.3	0	0.00–0.32	0.08	0.17	74	0.14
RR5	62.4	3.2	0	0.02–0.52	0.09	0.24	0	0.11
SC4	46.1	11.8	0	0.00–0.30	0.10	0.15	125	0.13
SR24	100.0	2.0	0	0.00–0.67	0.11	0.23	32	0.07
NFIC12	100.0	2.0	1	0.00–0.53	0.13	0.30	189	0.17
ELS61	0.0	3.9	0	0.05–0.84	0.14	0.33	0	0.19
PC12	75.2	3.9	1	0.20–1.19	0.14	0.84	0	0.20
BBC24	89.5	2.8	0	0.07–0.60	0.14	0.20	0	0.08
TC13	100.0	2.0	0	0.00–0.72	0.16	0.31	156	0.07
ELW1	0.0	19.9	0	0.00–0.66	0.22	0.44	19	0.19
HC13	76.6	3.6	1	0.09–1.56	0.24	0.41	0	0.20
SC84	20.1	15.9	0	0.00–1.08	0.26	0.53	62	0.16
FC1	14.7	30.8	0	0.11–1.01	0.28	0.62	0	0.38
LCR93	44.1	3.9	1	0.00–0.81	0.28	0.57	61	0.24
ELS39	0.0	28.0	1	0.00–1.61	0.46	1.29	151	0.51
B3I	0.0	54.9	1	0.63–2.37	0.57	1.54	0	0.51
LPS9	0.0	21.8	1	0.00–1.72	0.60	1.47	85	0.51

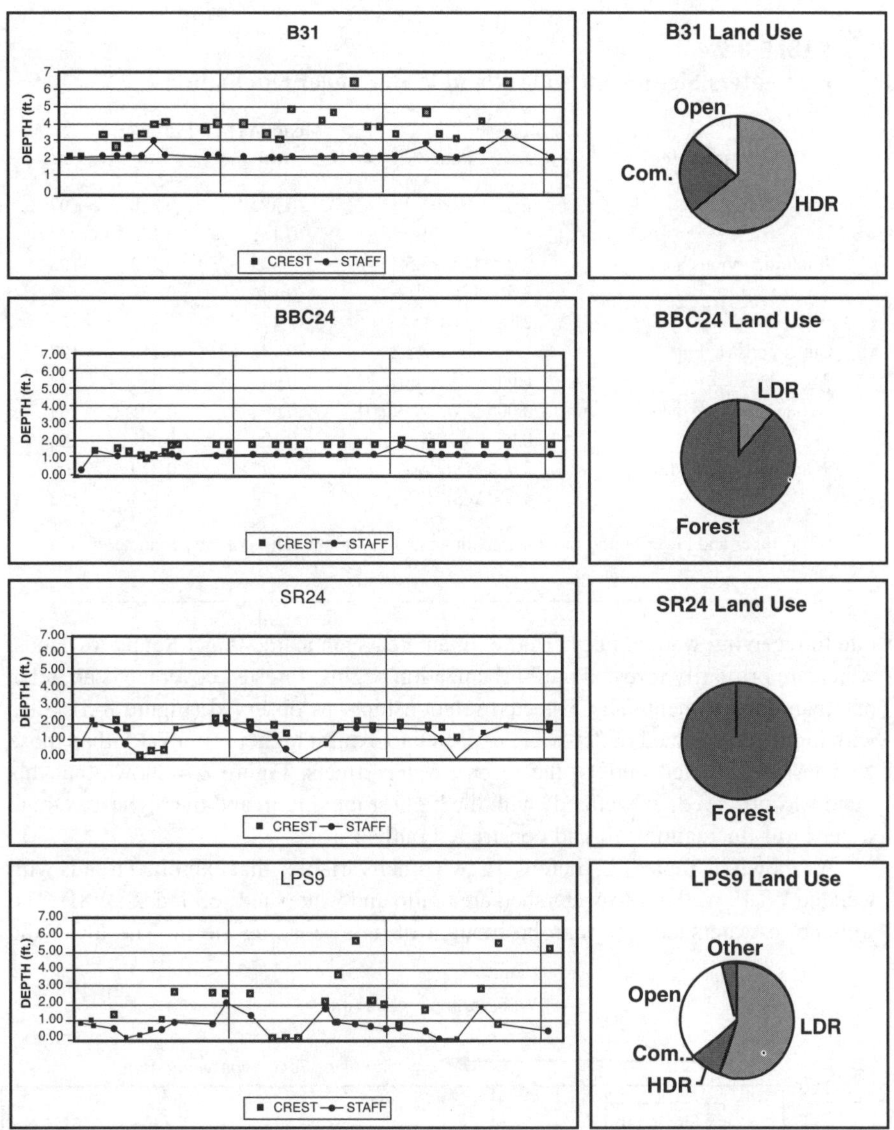

FIGURE 8-1 Wetland hydrographs (base and crest levels) and land use.

of the large proportion of their watersheds that were clear-cut. The second threshold (20% impervious area) may represent the point that changes in storm runoff caused by urbanization (e.g., flow volumes, flashiness) become dominant over the other factors that influence wetland hydroperiod.

The amount of forested area in a watershed was expected to be inversely related to event WLF. Forests store rainwater in the canopy, return water to the atmosphere through evapotranspiration, and typically have a highly permeable litter zone on the soil surface, all of which act to reduce storm runoff volumes and reduce the delivery

TABLE 8-2
Parameters Significant to Wetland Water Level Fluctuation

Parameter	Range[a]	Mean WLF (m)	Std. Dev. (m)	n
Forested area	Forest = 0%	0.384	0.338	97
	Forest ≥14.7%	0.151	0.138	224
Total impervious area	2.0 ≤ TIA ≤3.5%	0.105	0.072	105
	3.5 < TIA ≤20%	0.176	0.151	143
	21.8 < TIA ≤54.9%	0.478	0.348	73
Outlet constriction	Low to moderate	0.148	0.119	198
	High	0.34	0.33	123
Wetland-to-watershed area ratio	0.005 ≤ W/Ws ≤0.04	0.304	0.301	169
	0.05 < W/Ws ≤0.44	0.129	0.091	152
Watershed soils index	3.9 ≤ WSI ≤4.1	0.247	0.279	209
	4.2 < WSI ≤5.8	0.174	0.143	112

[a] The upper and lower bounds are the maximum and minimum values of the parameter within the range.

rate to receiving waters. Furthermore, in an area such as the Puget Sound lowlands, which are primarily forested until urbanization begins, forested coverage is an index of urban development. The expected relationship was observed (Figure 8-3). Sites with highly constricted outlets were expected to exhibit higher event WLF than those with less constricted outlets due to backwater effects. Figure 8-4 shows that this trend was observed, as wetlands with the highest maximum and the highest average water level fluctuations all had constricted outlets.

As shown in Table 8-2, there were two other variables that exhibited trends with wetland WLF: wetland-to-watershed area ratio and watershed soil index (WSI). The wetland-to-watershed ratio can be thought of as a "loading" term. The lower the

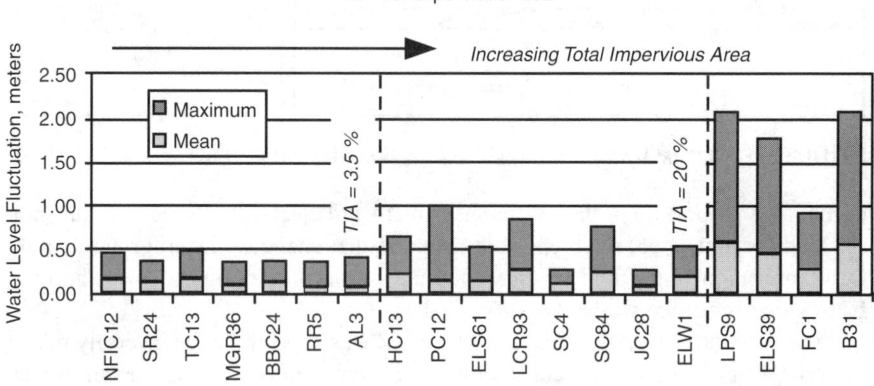

FIGURE 8-2 Relationship between event WLF and impervious area.

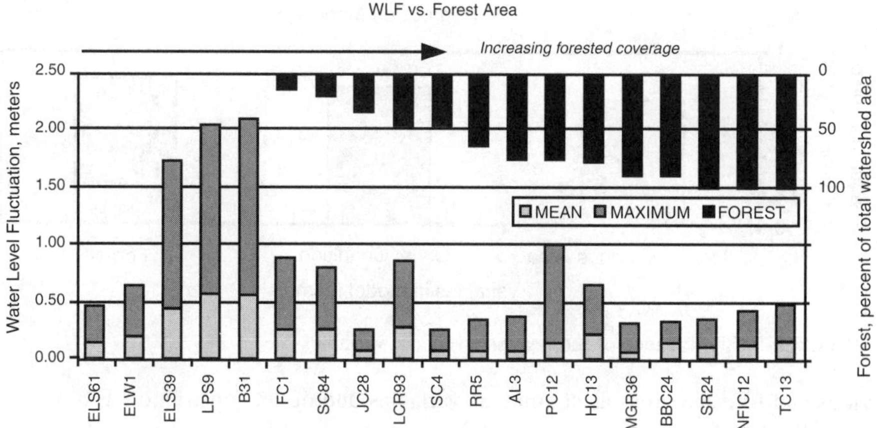

FIGURE 8-3 Relationship between event WLF and forest coverage.

ratio, the less area available to store storm runoff, resulting in higher event WLF. The threshold observed (ratio = 0.045) corresponds with the recommended ratio for stormwater detention ponds, which is five percent.[10] The WSI was developed to quantify the soil drainage characteristics; since higher values indicate soils with high infiltration capacity, these values were expected and found to be associated with low event WLF.

Multiple Regression Analyses

Multiple regression analyses were done on the mean event WLF data from 1988 through 1991. The mean WLF data were weighted by the sample size, with the size of the weighted data set consisting of 321 observations. The best model fit was found using three variables: impervious area, outlet constriction, and forested area (see Figure 8-5). The following equation produced the best fit when using percent imper-

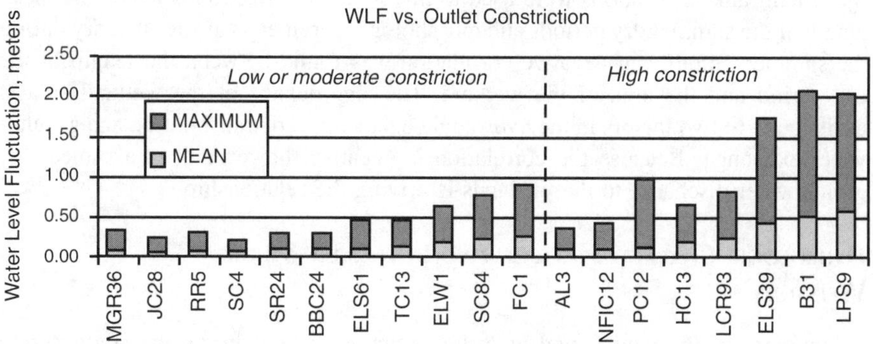

FIGURE 8-4 Relationship between event WLF and outlet condition.

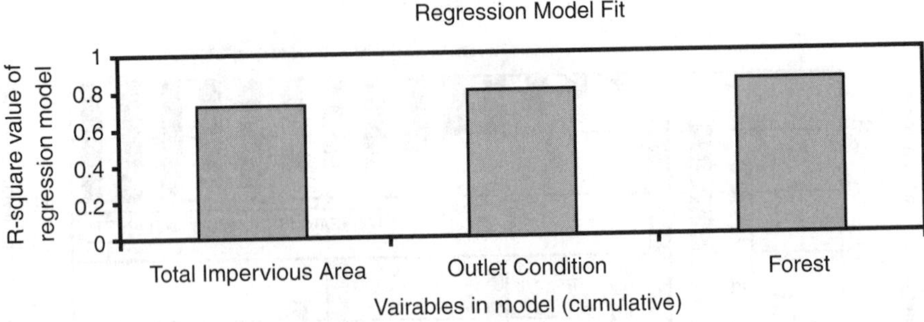

FIGURE 8-5 Significance of the regression model variables.

vious and forested areas as continuous variables and outlet constriction as a binary variable (0 or 1):

$$\text{Mean WLF (m)} = 0.145 + 0.0052 * (\text{Impervious}) + 0.141 * (\text{OC}) - 0.0011 * (\text{Forest}) \quad (8\text{-}1)$$

where $r^2 = 0.790$ and SE = 0.08 m.

The model fit explained 79% of the variation in mean event WLF between sites. Residual analysis showed no deviations from the model assumptions. All the parameter coefficients were of the sign (positive or negative) expected. This model was tested in later years using data from 1993 through 1995 and not confirmed, however, there were some significant differences in the assumptions guiding the selection of data between the two analyses which likely account for the different results.[11]

Dry Period

The length of the summer dry period for the study sites ranged from zero for the sites with stable base flow to nearly 200 days (Table 8-1). A variety of approaches were used to evaluate which factors are important in determining the permanence of a site and the length of the dry period for those sites that dry in the summer. Spearman rank correlations were used to investigate the relation between the mean length of the summer dry periods and morphologic parameters at sites that dry during the summer. Significant negative correlation was found between the length of the dry period and the area of the wetland. The significance of the wetland area is attributable to two factors in the hydrologic balance: evapotranspiration and groundwater exchange. Because the correlation is negative, however, it is assumed that groundwater discharge to the wetlands is driving the relationship.

Hydrologic Characteristics of Two Intensively Studied Wetlands

A summary of the natural and hydrologic characteristics during the study period (1988–1990) for the B3I and PC12 wetlands is given in Table 8-3. The hydrologic reactions to storms exhibited by the two wetlands are typical of the respective

TABLE 8-3
Natural and Hydrologic Characteristics of Two Wetlands

Variable (unit)	B3I Wetland	PC12 Wetland
Dominant land type	Urban	Forest
Watershed area (ha)	187	87
Wetland area (ha)	2	1.5
Wetland-to-watershed ratio	0.011	0.017
Total precipitation (mm)	1813	1934
Precip. volume (m^3) in drainage area	3.4×10^6	1.7×10^6
Mean daily inlet flow (m^3/s)	0.042	0.021
Maximum daily inlet flow (m^3/s)	0.75	0.22
Days with measurable flow during study	730	493
Total flow during study (m^3)	2.7×10^6	0.9×10^6
Wetland storage volume (m^3)[a]	400–5000	600–7000
Runoff/precipitation ratio	0.80	0.53

Note: Study period was two years for B3I and 20 months for PC12.

[a] Wetland storage volume varies depending on season and flow conditions.

watershed land uses. The reaction of B3I inlet flows to storms is fast and dramatic. Flows increase almost immediately because of the large impervious land area and piped storm-drain system. Similarly, when storms end, the flow recedes quickly to near baseflow conditions. The PC12 inlet flow, on the other hand, reacts relatively slowly to storms, with the receding limb of the hydrograph extending much longer than at B3I. Significant inflows occurred at PC12 only from October to June; however, there was water in the wetland year round.

Nearly 80% of the annual precipitation occurred between October and March. The maximum daily precipitation occurred on January 9, 1990 (approximately 80 mm at both sites). Pan evaporation data from the Puyallup station were used for ET estimates at the wetlands. The measured pan evaporation was greatest from May to August (exceeding 100 mm per month) and least from November to March. The maximum monthly and daily evaporation rates during the study were 160 mm (July 1989) and 16 mm (July 30, 1989), respectively.

Water storage volumes varied from 400 to 5000 m^3 at B3I and from 600 to 7000 m^3 at PC12. Generally, changes in storage volume at B3I were short term (on the order of hours) and directly related to storm events. Baseflow rates and water storage were comparable during the wet and dry seasons. At PC12, on the other hand, storage volumes changed during storm events and by season. Water volumes were greatest during large storms or groups of storms during the late wet season.

The results of the groundwater investigation indicate that both wetlands are discharge zones under most conditions, meaning that groundwater discharges to the wetland and becomes surface water. Recharge wetlands, in contrast, replenish groundwater through infiltration of surface water. This was determined by the piezometric head measurements and given the fact that groundwater flows from areas of high to low head. The head measurements in both wetlands generally increase with depth below the water table (as measured by the deep piezometer clusters) and distance

TABLE 8-4
Summary of Hydrologic Inputs and Outputs by Season (All Values are in 1000 m^3; Percent of Total Input or Output in Parentheses)

Season[a]	Inputs Precipitation	Inputs Inflow	Inputs Groundwater[b]	Outputs Outflow	Outputs Evaporation	Error
B3I Wetland[c]						
Dry 88	2 (0.6)	289 (80.8)	66 (18.6)	319 (97.0)	10 (3.0)	28 (8.8)
Wet 88-89	12 (1.6)	639 (85.4)	99 (13.2)	762 (99.9)	1 (0.1)	–12 (–1.6)
Dry 89	6 (0.7)	668 (84.5)	116 (14.8)	627 (98.1)	12 (1.9)	150 (23.5)
Wet 89-90	14 (1.4)	863 (90.0)	82 (8.6)	989 (99.9)	0 (0.1)	–29 (–3.0)
Dry 90	2 (0.7)	239 (87.1)	33 (12.1)	231 (99.2)	2 (0.8)	40 (17.5)
Total	36 (1.2)	2697 (86.1)	398 (12.7)	2928 (99.2)	25 (0.8)	178 (6.0)
PC12 Wetland[d]						
Wet 88-89	12 (2.1)	445 (79.5)	103 (18.4)	535 (99.9)	0 (0.1)	23 (4.4)
Dry 89	5 (3.9)	97 (72.4)	32 (23.8)	136 (93.6)	9 (6.4)	–9 (–6.4)
Wet 89-90	11 (2.5)	312 (74.1)	99 (23.4)	373 (99.9)	0 (0.1)	48 (13.0)
Dry 90	1 (2.5)	49 (82.3)	9 (15.2)	62 (97.7)	1 (2.3)	–4 (–6.4)
Total	29 (2.5)	904 (76.9)	243 (20.7)	1105 (99.0)	11 (1.0)	58 (5.2)

[a] Dry season = April-September; wet season = October–March.
[b] Positive groundwater values indicate groundwater discharge to wetlands.
[c] B3I study period: June 1988–May 1990.
[d] PC12 study period: October 1988–May 1990.

from the wetland, indicating the groundwater flows both vertically and laterally to each wetland. Discharging wetlands have also been documented by other authors.[12,13]

Wetland Hydrology by Season and Wetland

Table 8-4 summarizes the hydrologic inputs and outputs by season for the two wetlands. For both wetlands, surface water outflow accounted for greater than 99% of the outputs during the study period. Thus, groundwater recharge and ET, the other potential sources of output, were insignificant on an annual basis. This is typical for wetlands that have a low wetland-to-watershed area ratio (1.1 and 1.7% for B3I and PC12, respectively) and for wetlands that lie in a groundwater discharge area. For wetlands with low wetland-to-watershed ratios, inputs from the larger watershed (i.e., surface water flows) often dwarf the contributions from "in-wetland" components, such as groundwater and ET, because of the relatively small wetland area. Also, if groundwater exhibits mostly a discharge pattern as a result of topography and wetland location, then groundwater recharge is likely a minimal source of water output.

Surface water inflows accounted for 86 and 77% of the inputs for B3I and PC12, respectively, on an annual basis. Groundwater discharge to the wetlands accounted for most of the remaining input (13 and 21% for B3I and PC12, respectively). Direct precipitation inputs were quite small in the overall balance. During individual months or groups of months, however, groundwater and precipitation contributed substantially more to the wetland water inputs, particularly at PC12.

Differences also existed in the magnitudes of inputs and outputs for the wet and dry seasons. This was particularly true for precipitation, with 75 to 80% occurring during the wet season and ET, and with approximately 90% occurring during the dry season. At B3I, 60% of annual surface water flow occurred during the wet season, whereas at PC12 this component totaled approximately 80%. At PC12, ET accounted for greater than 50% of the output from July to September, 1989 when baseflows were minimal. During the same period, ET at B3I was less than 5% of the output, because of the stable and relatively high baseflow. The direct groundwater input to B3I was fairly steady throughout the year. However, at PC12, nearly 83% of the groundwater contribution to the wetland occurred during the wet season.

Urbanization and Other Factors Affecting Wetland Hydrology

The dynamics of wetland hydrology are governed by factors that may change seasonally or slowly over time. Seasonal changes result from variation in climate (e.g., precipitation, solar radiation), plant growth, and groundwater recharge. Longer-term changes result from human activities, including watershed development, groundwater withdrawal, wetland outlet modification, and drainage activities. Although this study was not designed to investigate change over time, some general conclusions can be drawn from comparisons between urbanized and nonurbanized catchments.

The runoff-to-precipitation ratios were 0.80 and 0.53 for B3I and PC12 wetland watersheds, respectively. Thus, more water is captured in the nonurbanized catchment, resulting in less runoff to the wetland. Potential pathways for the difference in water reflected in these numbers are ET, regional groundwater recharge, and withdrawal in the watershed itself. The ET in the forested nonurbanized catchment of PC12 is undoubtedly greater than in the developed urbanized catchment of B3I. Regional or deep groundwater recharge within the PC12 watershed is also likely greater than in the case of B3I, because of milder topography and less impervious surface. Finally, groundwater withdrawal to meet local water needs is likely more significant in the PC12 watershed.

Water level fluctuation is perhaps the best single indicator of wetland hydrology, because it integrates nearly all hydrologic factors. The mean WLFs were 0.15 and 0.49 m for the PC12 and B3I wetlands, respectively. The higher mean occasion WLF at B3I reflects the effect of many factors, including its urbanized catchment, piped storm-drain system, and constricted outlet. The maximum study period WLFs were quite similar. This apparent discrepancy occurred because of the evaporation and lowered water level in PC12 during the summer. In summary, both wetlands experienced similar long-term fluctuations; however, the urban wetland was exposed to much more frequent and greater WLFs.

Hydrologic Components Error Analysis

By measuring all components of the water balance shown in Equation 1-1 (Chapter 1), it was possible to determine error estimates for the seasonal balances. The seasonal errors (Table 8-4) ranged from –6.4 to +23.5% of the total hydrologic outputs. For the entire study period, the errors were 6.0 and 5.2% for B3I and PC12, respectively.

This reduction reflects the cancellation effect of positive and negative errors when summed over a longer time period. Generally, the larger percentage errors occurred during the dry seasons, reflecting the increased importance of groundwater inputs and ET in the overall balance at that time.

The type and magnitude of the errors associated with hydrologic or water balances may be characterized in several ways. These include errors associated with: (1) the equipment (e.g., inaccurate calibration), (2) the measurements (e.g., representability of measurement), (3) the calculations (e.g., weak stage-discharge correlations, groundwater calculations), and (4) the summation of balance components. It is important to note that these errors can improve or degrade the apparent accuracy of a water balance depending on the interaction between errors.

If precautions are taken to minimize the errors associated with the equipment, measurements and calculations and if all components are included in a water balance, it is possible to reduce potential errors greatly. An assessment of the importance of the different components of a balance is a critical task in this process. Because of the above-noted errors, it is recommended that no components of a balance be estimated by difference. Using this technique simply masks the errors in the unknown or unmeasured component (usually ET, groundwater, or both).

CONCLUSIONS

The quantity of stormwater entering many palustrine wetlands in the Puget Sound region has changed as a result of rapid development in urbanizing areas. The purpose of this chapter has been to characterize the hydrology of wetlands affected by urban stormwater, in comparison to unaffected or forested systems. This information, then, may help to explain observed changes in wetland soils, plants, and animals over time. Additionally, if observed effects of stormwater on wetlands can be documented, it may be possible to mitigate these effects through watershed controls and stormwater management efforts.

The hydrology of wetlands as measured by water level fluctuation was highly variable. Differences in water level fluctuation were attributed to level of watershed imperviousness, forested cover, and wetland outlet constriction. A multivariate model using these three parameters, calibrated to the study sites, was found to predict water level fluctuations accurately although the model should be verified and tested further using similar data sets from all years of collection in future research efforts.

For the two study wetlands, surface water inflow and outflow were the dominant components in the water balance on an annual basis. It was concluded that this is typical for wetlands with low wetland-to-watershed ratios. The ET was insignificant in the overall water budget on an annual basis; however, it was the major source of water output from the PC12 wetland from July to September, when outflows were minimal. Both wetlands were identified as primarily groundwater discharge zones, with groundwater contributing significant inputs. Like ET, the influence of groundwater was greatest at PC12 during the summer months.

Differences were also identified in the hydrology of both wetlands because of the level of watershed urbanization. In the urbanized watershed, a greater proportion of the precipitation was realized as surface inflow to the wetland. Storm runoff was delivered more quickly and in greater short-term volumes to the urban wetland. The result of these conditions was greater and more rapid water level fluctuations in the urban wetland. This characteristic would probably be replicated in most wetlands where development occurs in the watershed.

ACKNOWLEDGMENTS

Many thanks to Mike Surowiec for his study of the hydrogeologic and groundwater components of two wetlands (Bellevue 3I and Patterson Creek 12).

REFERENCES

1. Mitsch, W. J. and J. G. Gosselink, *Wetlands*, 2nd ed., Van Nostrand Reinhold, New York, 1993.
2. Taylor, B. L., The Influence of Wetland and Watershed Morphological Characteristics on Wetland Hydrology and Relationships to Wetland Vegetation Communities, M.S.C.E. thesis, Department of Civil Engineering, University of Washington, Seattle, 1993.
3. Wilkinson, L., *SYSTAT: The System for Statistics*, SYSTAT, Inc., Evanston, IL, 1990.
4. Zar, J. H., *Biostatistical Analysis*, Prentice-Hall, Englewood Cliffs, NJ, 1984.
5. Reinelt, L. E., M.S. Surowiec, and R. R. Horner, Urbanization effects on palustrine wetland hydrology as determined by a comprehensive water balance, (Manuscript), 1993.
6. Reinelt, L. E. and R. R. Horner, Characterization of the Hydrology and Water Quality of Palustrine Wetlands Affected by Urban Stormwater, King County Resource Planning, King County, WA, 1990.
7. Cedergren, H. R., Seepage, Drainage and Flow Nets, John Wiley & Sons, New York, 1978.
8. Chapuis, R. P., Shape factors for permeability tests in boreholes and piezometers, *Ground Water,* 27(5), 647, 1989.
9. Surowiec, M. S., A Hydrogeologic and Chemical Characterization of an Urban and Nonurban Wetland, M.S.E. thesis, Department of Civil Engineering, University of Washington, Seattle, 1989.
10. Surface Water Management, King County Surface Water Design Manual, King County Department of Public Works, Seattle, 1990.
11. Chin, N. T., Watershed Urbanization Effects on Palustrine Wetlands: A Study of the Hydrologic, Vegetative and Amphibian Community Response During Eight Years, M.S.C.E. thesis, Department of Civil Engineering, University of Washington, Seattle, 1996.
12. Wilcox, D. A., R. J. Shedlock, and W. H. Hendrickson, Hydrology, water chemistry and ecological relations in the raised mound of Cowles Bog, *J. Ecol.,* 74, 1103, 1986.
13. Siegel, D. I. and P. H. Glaser, Groundwater flow in a bog-fen complex, Lost River Peatland, Northern Minnesota, *J. Ecol.,* 75, 743, 1987.

9 Effects of Watershed Development on Water Quality and Soils

Richard R. Horner, Sarah S. Cooke, Lorin E. Reinelt, Kennneth A. Ludwa, Nancy T. Chin, and Marion Valentine

CONTENTS

Introduction 237
Effects of Watershed Development on Water Quality 238
  Moderately Urbanized Wetlands 238
  Highly Urbanized Wetlands 239
Other Findings 240
  Treatment of Wetlands 242
  Observations for Individual Treatment Wetlands 243
  Profile of Treatment Wetlands 244
Effects of Watershed Development on Soils 247
  Urbanized Wetland Soil Profiles 247
  Other Findings 248
Treatment Wetlands 250
References 253

INTRODUCTION

This chapter emphasizes water and soil quality in wetlands with significant urbanization in their watersheds. Like other chapters in this section, its purpose is to characterize particular elements of Puget Sound Basin freshwater wetlands with urbanizing watersheds. The urbanized cases were divided into two major categories. The treatment group included the wetlands whose watersheds had a more than 10% rise in urbanization between 1989 and 1995. Conversely, the control category consisted of wetlands whose watersheds experienced a less than 10% increase in urban land cover between 1989 and 1995. The control wetlands were further subdivided into three classifications:

1. The most highly urbanized sites (H) which had watersheds that were both ≥20% impervious and ≤7% forest by area;

1-56670-386-7/00/$0.00+$.50

2. Moderately urbanized wetlands (M) which had watersheds that were 4 to 20% impervious and 7 to 40% forested by area; and
3. Nonurbanized wetlands (N) with both <4% impervious land cover and ≥40% forest.

This latter category is discussed in Chapter 2, but will be mentioned in this chapter at times for comparison. Table 1-1 of Section 1 gives characteristics of the individual wetlands and watersheds.

This chapter first describes water quality conditions in urban control wetlands and then discusses changes in these conditions in treatment wetlands. It then proceeds to cover soil characteristics in a similar way. Chapter 2 covers the methods with which the data were collected. Tables 2-1, 2-2, and 2-4 of Chapter 2 summarize water quality results for the urban control wetlands as well as the nonurbanized cases. Tables 2-5 and 2-6 in that chapter perform the same function for the soils data. These tables are not repeated in this chapter but are referenced here several times.

As has been stated elsewhere in this volume, the research program concentrated on palustrine wetlands of the general type most prevalent in the lower elevations of the Central Puget Sound Basin. The results and conclusions presented here are probably applicable to similar types of wetlands areas to the north and south of the Central Puget Sound Basin, but may not be representative of higher, drier, or more specialized systems, like true bogs and low nutrition fens.

EFFECTS OF WATERSHED DEVELOPMENT ON WATER QUALITY

This section first profiles the urban control wetlands, both moderately and highly urbanized, using the statistical summary data in Chapter 2, Tables 2-1, 2-2, and 2-4. Following the profiles is a more general summary of other applicable findings from the research.

MODERATELY URBANIZED WETLANDS

A water quality portrait of Puget Sound Basin lowland palustrine wetlands moderately affected by humans would show slightly acidic (median pH = 6.7) systems, with DO often well below saturation and, in fact, sometimes quite low (<4 mg/l). Dissolved substances are fairly high relative to nonurbanized wetlands (median conductivity about three times as high) but somewhat variable. Suspended solids are only marginally higher than N wetlands but, like them, quite variable. Median total dissolved nitrogen concentrations (the sum of ammonia, nitrate, and nitrite) are more than 20 times as high as dissolved phosphorus, a ratio very similar to the nonurbanized wetlands but with higher magnitudes in both cases. Again, plant and algal growth is generally limited by P, rather than N. TP at the median level is more than twice as high in the M compared to the N wetlands (70 μg/l). The median fecal coliform concentration is close to the 50 CFU/100 ml criterion applied as a geometric mean to lakes and the highest class of streams by Washington State water quality criteria. Fecal coliforms were highly variable in all wetlands, extremely so in wetlands of the M class. More than half of the individual FC values for moderately urbanized flow-through wetlands exceeded the maximum 200 CFU/100 ml criterion applied to the lowest-class streams (however, their geometric mean may not do so). As with N wetlands, both mean and median heavy metals concentrations in the moderately urbanized sites are in the low parts per billion range, with standard deviations just about identical to the means. The median lead concentration, however, is close to the chronic water quality criterion set for lakes and streams having hardness of 50 mg/l as $CaCO_3$.

In summary, the following general statements can be made to characterize the water quality of Puget Sound Basin lowland palustrine wetlands in a moderately urbanized state:

- These wetlands are highly likely (≥71% of cases observed) to have median conductivity >100 μS/cm but median TSS in the range 2–5 mg/L, NH_3-N <50 μg/l, and total Zn <10 μg/l.
- Moderately urbanized wetlands are highly likely (71% of cases) to have median fecal coliforms <50 CFU/100 ml but also to have many individual measurements above 200.
- They are highly likely (100% of cases) to have TP >20 μg/l and likely (57% of cases) to have TP >50 μg/l and NO_3 + NO_2-N <100 μg/l. The latter variable is highly likely (86% of case) to be <500 μg/l.
- The pH and DO in these wetlands are unpredictable from consideration of urbanization status alone, being dependant on other factors.

Highly Urbanized Wetlands

Highly urbanized wetlands are harder to profile because of the small set of only two in the control group. Also, both of these wetlands are the flow-through type, not giving any picture of how morphology might affect the conclusions. What can be said is the following:

- There is some tendency for these wetlands to be the closest to neutral in pH among the three urbanization status categories. They tend to fall in the same range as the other classes in dissolved oxygen.
- They most likely would have median conductivity >100 μS/cm.
- Like the other two classes, highly urbanized wetlands are very likely to have median NH_3-N <50 μg/l.
- Unlike the other two classes, they are very likely to have median NO_3 + NO_2-N >100 μg/l and TP >50 μg/l.
- These sites are likely, but somewhat less than the other categories, to have autotrophic growth limited by phosphorus.
- They are likely to exceed the 50 CFU/100 ml level of fecal coliforms.
- These wetlands have a higher tendency than the other categories to have Zn >10 μg/l but in most instances still not to exceed the chronic criterion of 59 μg/l for relatively soft waters.

OTHER FINDINGS

In a synoptic study of 43 urban and 27 nonurban wetlands during 1987, before routine sampling began, the program found that FC and enterococcus were significantly higher in urban wetlands.[4] Although the mean counts for both types of bacteria were within water quality standards, bacteria substantially exceeded standards in wetlands in high-density areas that showed evidence of human intrusion. The watersheds of most of the wetlands in which bacteria also exceeded standards were in watersheds characterized by low-density residential development, while some of the watersheds of the remainder of the wetlands had high-density residential development. None of the watersheds with bacteria in excess of standards was dominated by commercial development. Other than pH, FC count was the only water quality variable measured by the survey.

After four years of regular data accumulated, a major effort was undertaken to relate water quality conditions to watershed and wetland morphological circumstances. In this work it was found that certain water quality parameters varied in response to the changes in watershed wetland characteristics that can accompany urbanization.[1] The characteristics used as independent variables in this analysis included.

1. Percent forest cover,
2. Percent total impervious area,
3. Percent effective impervious area (the area actually linked to a storm drain system),
4. The ratio of wetland to watershed area,
5. The ratio of forest to wetland area,
6. Wetland morphology (open water or flow-through), and
7. Outlet constriction.

These measures may be expressed as either continuous (ranges) or categorical (binary or ternary) variables. Multivariate linear regressions were used to determine if there is an adequate relationship between these characteristics and water quality parameters. If there is such a relationship, the equations could be used to analyze probable changes in water quality following development.

Specific watershed land uses and wetland morphological values were significantly associated with most water quality values.[1] The dependent water quality variables exhibiting the best associations and most correctly predicted when verified with a portion of the data set held aside for verification were conductivity, pH, and TSS.[1] Pollutants that are often adsorbed to particulates, specifically TP, Zn, and FC, showed similar degradation across key levels of the independent variables. Conductivity, TSS, FC, and enterococcus were consistently degraded the most between more highly developed watersheds and those that were moderately urbanized or rural.[7] Conductivity, TSS, Zn, DO, and FC varied by the greatest amounts.[1]

Based on program data from 1988 to 1993, percent forest cover was the best predictor of water quality for Pacific Northwest palustrine wetlands, followed by percent total impervious area, forest-to-wetland areal ratio, and morphology.[1] All variables except $NO_3 + NO_2$-N were higher in wetlands with no forest in their watersheds compared to those in watersheds with at least 14.7% forest cover.[7] Conductivity, TP, and FC rose significantly when the percentage of impervious surface exceeded the values of 3.5 and 20%.[7] Forest-to-wetland areal ratio strongly influenced conductivity, TSS, $NO_3 + NO_2$-N, TP, SRP, and FC, where the ratio was less than 7.2.[7] Conductivity, TSS, $NO_3 + NO_2$-N, TP, SRP, and FC had significantly higher means in relatively channelized wetlands, although it should be noted that these results may have been influenced by extraneous factors.[7] Outlet constriction and wetland-to-watershed areal ratios had inconsistent roles in influencing water quality.[7] It should be noted that because the breakpoint values are expressed as fixed ranges and it is unknown if thresholds exist or what they are, it is entirely possible that other water quality constituents also vary significantly with characteristics of urbanization on a continuous basis.[7]

The analysis indicated that, for similar watersheds in the region, there is a definite degree of deforestation and development above which average wetland water quality will become degraded. However, if amounts of forestation remain above some minimum level, water quality will comply with criteria. It should be noted that because extremes in water quality often have greater impacts on biological resources than average conditions, attention should also be given to the relationships of conditions with minimum and maximum values. Minimums and maximums were found to vary widely across urbanization and morphological levels, so that entire ranges of water quality variables shift significantly to degraded conditions.

Previous research found that the strongest regression relationships were for mean, maximum and minimum conductivity, TSS, and DO.[1,8] In view of the correlation coefficients, urbanization was consistently related to all water quality values except NH_3-N and SRP. The strong regressions of TSS and conductivity with urbanization suggested that an increase in watershed imperviousness will facilitate the movement to wetlands of runoff containing inorganic particulate and dissolved matter. Total suspended solids and conductivity are directly and indirectly harmful to wetland biological communities. Wetland morphological factors had similar effects, although they were less consistent for outlet constriction.

Predictions were generally better for mean and maximum values, since these values exhibited more variability than minimum values from site to site. The choice of factors to be included in the regression equations was manipulated to improve predictive value, although the process was somewhat subjective. Wetland-to-watershed areal ratio was the most frequently used factor. Although little can be done to affect this ratio other than by changing wetlands physically or by diverting inflows, the importance of this ratio does not suggest that development or deforestation are unimportant. To the contrary, where a wetland covers a smaller portion of a watershed, the effect of deforestation may be magnified. Effective impervious area, which expresses how much land is actually drained by a storm drainage system, had more predictive power than total impervious area. However, there was no consistent relation between outlet constriction and water quality.

The crucial values for water quality lie between 4 and 12% for total impervious area and 0 to 15% for forested area. Theory and observations in other regional ecosystems have demonstrated that there is likely to be a continuous and relatively rapid decline in various measures of ecosystem condition as forests begins to decrease in favor of impervious surfaces. As conversion progresses the decline is likely to slow in rate but to continue. Therefore, with a continuous pattern of variation normally prevailing, numerical values should not be regarded as thresholds but as points where degradation becomes evident and demonstrable and where standards generally accepted as necessary to support biota probably will not be met, at least at times.

Previous related work recommended that the total impervious area in Pacific Northwest watersheds, with strong wetland protection goals, be not more than 10%, and that a forest cover of at least 15% be maintained.[1] Whether more effective implementation of urban runoff best management practices would permit these thresholds to be shifted toward more urbanization is a matter only for conjecture. However, development and deforestation will ultimately have to be limited if high quality and well functioning wetland systems are to be preserved. Channelized sites usually had lower water quality, hence a shift to more channelized conditions intentionally or by inadvertant changes in hydroperiod resulting from increased urbanization should be avoided.

Treatment of Wetlands

The treatment wetlands studied by the program were Big Bear Creek 24 (BBC24), East Lake Sammamish 61 (ELS61), Jenkins Creek 28 (JC28), North Fork Issaquah Creek 12 (NFIC12), and Patterson Creek 12 (PC12). Their watersheds experienced increases of urbanization in the range of 10.2 to 10.5% in three of the five cases (JC28, ELS61, and PC12), 42.2% for BBC24, and 100% in the case of NFIC12. The most common change in land use was from forest to single-family residential, a development pattern typical of the early stages of urbanization.[2] Table 9-1 shows land cover in 1995 and the changes since 1989.

This distribution of changes gave an opportunity to observe the relative effects, which were substantial compared to more limited watershed alterations. The timing of development in relation to the program's schedule also offered the chance to observe effects during the construction phase, when soils are often bare for long intervals, vs. the subsequent period when areas finished with construction are restabilized.

TABLE 9-1
Land Cover in 1995 and the Changes from 1989

	Forest		Impervious Surface	
Wetland	**1995 (%)**	**Change (%)**	**1995 (%)**	**Change (%)**
JC28	19.8	−14.6	20.6	+0.6
ELS61	3.7	+1.2	10.6	+5.5
PC12	64.7	−10.5	6.8	+1.7
BBC24	47.4	−42.1	10.6	+7.2
NFIC12	0.0	−100.0	40.0	+38.0

After the development that occurred through 1995 in the treatment watersheds NFIC12 would be categorized as H and all of the others as M, according to the urbanization groupings used to classify the control wetlands. Morphologically, JC28 is a flow-through type and the remainder are all open water.

Observations for Individual Treatment Wetlands

Wetlands in urbanizing watersheds are especially vulnerable to erosion during the construction phase of development. Total suspended solids concentrations often increase greatly during such periods but return to approximately predevelopment levels as bare land is covered by structures and vegetation. During periods of construction, mean TSS values increase more dramatically than median values because of the influence of especially high concentrations. For instance, the ELS61 wetland recorded a median TSS concentration of 10.4 mg/l in 1989 and had a maximum concentration of 59 mg/l in August, as a result of construction site runoff. An increase in TSS at JC28 in 1989 was also linked to land disturbances. At both of these sites, TSS declined in the following year.

Elevated sediment in runoff from construction sites also corresponds to increasing concentrations of other pollutants, especially phosphorus and nitrogen, that are contained in soils.[9] Subsequent to construction, application of fertilizers can further increase nutrient concentrations in runoff. In the JC28 wetland, land disturbances, including expansion of an adjacent golf course in 1989, marked the commencement of a regime of higher nutrient concentrations. Median $NO_3 + NO_2$-N and SRP values increased by 63 and 96%, respectively, between 1988 and 1989 and continued to climb steadily from 1990 to 1995. The initial increases probably resulted from land disturbance, with the subsequent rises attributable to fertilizer runoff from the golf course. Mean NH_3-N also rose sharply in 1989, with a maximum value of 619 μg/l, and median NH_3-N was higher in subsequent years. More than half of the $NO_3 + NO_2$-N readings exceeded 500 μg/l.

At the ELS61 wetland, NH_3-N and $NO_3 + NO_2$-N initially rose in 1989 but declined in 1993, although not to predevelopment levels. Concentrations of NH_3-N climbed again in 1993, while $NO_3 + NO_2$-N greatly increased in 1995. Many NH_3-N and $NO_3 + NO_2$-N concentrations exceeded 100 and 500 μg/l, respectively, during these years. Average SRP and TP concentrations were actually the highest in 1988, perhaps because of the operations at a small livestock farm next to the wetland. However, after declining from 1988 to 1990, SRP and TP concentrations were substantially higher in 1993 and 1995. One of the two highest chlorophyll *a* concentrations in the first two years of the program was recorded at ELS61.[3]

NFIC12, the wetland that had the greatest amount of development in its watershed between 1989 and 1995, increasing from 0 to 100%, displayed different water quality patterns than ELS61 and JC28. Average values for TSS rose modestly from 1989 to 1995, with a maximum peak value of 16 μg/l in 1993. Average concentrations of NH_3-N and $NO_3 + NO_2$-N did not appear to rise during this period but NH_3-N and $NO_3 + NO_2$-N did reach maximum concentrations of 120 and 1400 μg/l, respectively, in 1993. Concentrations of SRP and TP, however, rose steadily, reaching median concentrations of 148 and 202 μg/l, respectively, in 1995. For all years, mean and median TP concentrations exceeded 50 μg/l.

Results were less conclusive for the other two treatment wetlands, PC12 and BBC24, demonstrating that there is not necessarily a link between development and water quality degradation, even for wetlands whose watersheds have undergone similar amounts of development. A possible explanation for the difference in results may be that the watersheds of PC12 and BBC24 remained approximately half forested, retarding transport of pollutants to the wetlands in runoff. The watersheds of JC28, ELS61, and NFIC12, on the other hand, were only 0 to 19.8% forested by area. In addition, a large wet pond meeting current design standards was constructed adjacent to PC12 to treat storm runoff from the development. These observations are signs that concerted action to maintain forest cover and impose structural storm water management measures can avoid water quality degradation.

Increases in nutrient loadings can have serious consequences for normally nutrient-limited bogs and fens. In one of the bog-like wetlands covered by the program in a special study, East Lake Sammamish 34 (ELS34), also known as Queen's Bog, the *Sphagnum mat* was observed to be measurably decomposing and reducing in area. The likely cause was increased nitrogen from inflow as nitrogen exponentially increases decomposition rates.

Profile of Treatment Wetlands

Table 9-2 shows statistics for the five treatment wetlands in the baseline period, when little or no urbanization had occurred (1988 to 1990), and then the later years (1993 and 1995) after most of the changes in land use were either well underway or complete. Very little change in pH was evident. DO exhibited some fluctuation in three wetlands but only ELS61 registered a notable decline in the median level with time (≈2 mg/l).

Most direct comparisons for all of the other water quality variables and all wetlands indicated no change or reduction during the program but there were some exceptions to that generality that bear examination. NH_3-N appeared to rise in ELS61 from predominantly less to more than 50 μg/l values. $NO_3 + NO_2$-N showed increases in all but NFIC12. Median concentrations still stayed mostly <100 μg/l in PC12 and ELS61. The increase in BBC24 kept the median still below 500 μg/l. JC28 increased from an already relatively high median >500 to >1000 μg/l. In NFIC12, relatively high concentrations of SRP and TP increased further after development, reaching among the highest levels seen in the entire program. Increases in TP also occurred in JC28 but stayed in the 20–50 μg/l range, and also in ELS61, where the median increased from the range of 50 to 100 to >100 μg/l. Relatively small rises in fecal coliform statistics were registered in JC28 and ELS61 but medians remained below 50 CFU/100 ml. Although relatively high detection limits in the early years make comparisons more difficult for the metals, there was no sign that any of the three metals increased substantially anywhere or threatened a violation of the water quality criteria applied to other water bodies.

Wetlands in moderately and highly urbanized watersheds are generally profiled earlier in this chapter. Whether or not the treatment wetlands fit these profiles after

TABLE 9–2
Water Quality Statistics for Treatment Wetlands in Baseline and Postdevelopment Years

Site/ Years	Statistic	pH	DO (mg/l)	Cond. (µS/cm)	TSS (µg/l)	NH_3-N (µg/l)	NO_3 + NO_2-N (µg/l)	SRP (µg/l)	TP (µg/l)	FC (CFU/100 ml)	Cu (µg/l)	Pb (µg/l)	Zn (µg/l)
JC28	Mean	6.59	6.9	99 <	5.68 <	72	710	17	44 <	237 <	5.0 <	5.5 <	28.9
88–90	St. Dev.	0.24	1.4	28 >	9.30 >	159	414	29	45 >	578 >	0.0 >	1.1 >	37.2
	CV	4%	21%	28%	164%	220%	58%	174%	101%	244%	0%	19%	129%
	Median	6.67	7.1	94	2.9	13	653	4	29	20 <	5.0 <	5.0 <	20.0
	n	19.00	19.0	16	19.0	19	19	19	19	19	8	8	8
93–95	Mean	6.74	6.9	98 <	4.9 <	34	1002 <	27	84	83 <	0.7 <	1.3 <	8.7
	St. Dev.	0.20	1.4	9 >	3.8 >	37	448 >	37	90	102 >	0.3 >	1.0 >	8.5
	CV	3%	20%	10%	78%	109%	45%	136%	107%	123%	45%	75%	98%
	Median	6.77	7.0	95	3.6	20	1080	8	43	36	0.6 <	1.0 <	5.0
	n	6.00	14.0	14	13.0	14	14	12	14	14	6	14	14
PC12	Mean	6.72	7.0	68 <	3.0 <	75 <	456	11	52 <	63 <	5.0 <	5.0 <	15.2
88–90	St. Dev.	0.32	3.4	11 >	2.6 >	76 >	551	10	45 >	146 >	0.0 >	0.0 >	5.4
	CV	5%	48%	15%	88%	101%	121%	89%	87%	233%	0%	0%	36%
	Median	6.62	7.5	71	2.4	35	108	7	40	8 <	5.0 <	5.0	16.0
	n	23.00	22.0	20	23.0	23	22	23	23	23	9	9	9
93–95	Mean	6.55	6.5	73 <	2.5 <	33 <	786 <	11 <	88	46 <	0.7 <	0.8 <	3.0
	St. Dev.	0.12	3.1	15 >	2.1 >	26 >	980 >	9 >	218	124 >	0.3 >	0.4 >	1.7
	CV	2%	48%	20%	87%	79%	125%	79%	248%	271%	40%	47%	57%
	Median	6.57	6.3	75	2.0	20	430	8	24	2	0.7	0.8	2.5
	n	8.00	16.0	16	16.0	15	13	15	15	16	8	16	16
ELS61	Mean	6.59	5.3	101	13.9 <	43 <	725 <	76	166 <	100 <	5.8 <	5.1 <	17.2
88–90	St. Dev.	0.27	3.1	19	16.8 >	94 >	1086 >	85	125 >	188 >	1.9 >	0.3 >	6.6
	CV	4%	58%	19%	121%	218%	150%	112%	76%	188%	32%	6%	38%
	Median	6.61	4.9	103	8.0	25	109	58	149	20	5.0	5.0	20.0
	n	23.00	23.0	20	23.0	23	17	22	23	23	10	10	10
93–95	Mean	6.31 <	3.8	91 <	9.5 <	136 <	527	35	101	321 <	0.9 <	0.8 <	2.3
	St. Dev.	0.19 >	2.8	19 >	20.9 >	190 >	592	41	95	992 >	0.3 >	0.3 >	1.2

TABLE 9–2 *(continued)*
Water Quality Statistics for Treatment Wetlands in Baseline and Postdevelopment Years

Site/ Years	Statistic	pH	DO (mg/l)	Cond. (µS/cm)	TSS (mg/l)	NH_3-N (µg/l)	NO_3 + NO_2-N (µg/l)	SRP (µg/l)	TP (µg/l)	FC (CFU/100 ml)	Cu (µg/l)	Pb (µg/l)	Zn (µg/l)
	CV	3%	73%	21%	219%	140%	112%	116%	94%	309%	30%	34%	52%
	Median	6.28	3.4	90	3.0	74	344	21	62	39	< 0.9	< 0.8	< 2.5
	n	8.00	13.0	16	16.0	15	9	16	16	16	8	16	16
BBC24	Mean	6.76	6.1	83	< 2.0	< 44	210	5	23	< 186	< 5.0	< 5.0	< 15.1
88–90	St. Dev.	0.25	1.9	19	> 2.0	> 35	183	3	13	> 542	> 0.0	> 0.0	> 5.6
	CV	4%	31%	23%	97%	80%	87%	64%	55%	292%	0%	0%	37%
	Median	6.77	5.4	84	1.1	31	189	4	21	14	< 5.0	< 5.0	17.0
	n	22.00	23.0	20	23.0	23	23	23	23	23	10	10	17
93–95	Mean	6.84	6.7	90	< 3.5	< 34	< 396	< 6	< 27	< 411	< 0.5	< 0.8	< 1.8
	St. Dev.	0.22	2.0	38	> 6.6	> 20	> 347	> 3	> 12	> 1492	> 0.0	> 0.4	> 1.3
	CV	3%	29%	43%	187%	57%	88%	51%	44%	362%	4%	50%	72%
	Median	6.82	7.5	82	1.7	30	323	5	28	18	0.5	0.8	2.5
	n	8.00	16.0	16	16.0	16	15	13	15	16	8	16	15
NFIC12	Mean	5.08	3.4	50	< 2.3	< 41	< 54	75	119	< 2	< 5.0	< 5.0	< 19.0
88–90	St. Dev.	0.69	1.0	46	> 2.3	> 29	> 45	104	94	> 0	> 0	> 0	> 7
	CV	14%	29%	90%	98%	71%	83%	140%	79%	21%	0%	0%	37%
	Median	4.84	3.5	37	1.0	39	34	53	80	2	< 5.0	< 5.0	22.0
	n	12.00	12.0	10	12.0	12	10	12	12	12	5	5	5
93–95	Mean	4.72	< 3.6	43	< 4.0	< 40	< 477	126	253	8	2.9	< 2.2	19.6
	St. Dev.	0.18	> 2.3	13	> 5.1	> 41	> 799	95	303	16	1.1	> 1.7	9.6
	CV	4%	63%	31%	129%	102%	167%	76%	120%	207%	37%	78%	49%
	Median	4.74	3.2	39	2.0	20	20	115	177	2	2.9	1.9	18.9
	n	4.00	9.0	10	10.0	10	3	10	10	10	4	10	10

going through development will now be examined. The four M wetlands all fit the profile for that category in the cases of conductivity, NH_3-N, zinc and fecal coliforms. It should be noted that they almost always fit the same profile in the baseline years; thus, preexisting factors are most responsible for how these wetlands profile. ELS61 and JC28 failed to fit the profile for TP and $NO_3 + NO_2$-N, respectively, being higher in both cases. In consequence, ELS61 did not appear to be generally phosphorus-limited in photosynthetic production, in contrast to the profile. A lack of fit occurred in only one other instance, TSS in BBC24, but the median was lower than the profile value. The only highly urbanized treatment wetland, NFIC12, exhibited fewer fits to the general H profile but usually because it had lower values. This was the case for conductivity, $NO_3 + NO_2$-N, and fecal coliforms. It did fit for NH_3-N, TP, and Zn but fit in those cases before development, too. This wetland also appears to tend toward nitrogen rather than phosphorus limitation, unlike the profile. Finally, it demonstrated no tendency toward more neutral pH, as the profile states. The humic acid-producing vegetation and peat prominent in this wetland apparently were not affected by the extensive urbanization, at least not yet.

EFFECTS OF WATERSHED DEVELOPMENT ON SOILS

This section first profiles the soils of the urban control wetlands, both moderately and highly urbanized, using the statistical summary data in Chapter 2, Tables 2-5 and 2-6. Following the profiles is a more general summary of other applicable findings from the research.

URBANIZED WETLAND SOIL PROFILES

A soils portrait of Puget Sound Basin lowland palustrine wetlands moderately affected by urbanization shows a somewhat acidic condition, more so (by about one pH unit) in open water than flow-through wetlands. The range of median values can be expected to be approximately 5.1 to 6.1. These soils will be aerobic in many instances but their redox potentials not infrequently are below the levels where oxygen is depleted. TP is likely to be in the range 500 to 1000 mg/kg and TKN up to 10 times as high. Median levels of soil organic content are approximately 15%. No general statement is possible concerning particle size distribution. Metals appear to be less variable than in nonurban sites but still have coefficients of variation ranging from about 60 to 100%. It is most likely for Cu concentration to be in the vicinity of 15 mg/kg, for Pb and Zn to be very roughly twice as high, and for As to be about half as concentrated. The lowest effect threshold for copper and lead in freshwater sediment criteria would be violated in some samples.

Only two sites represent the highly urban control sites, which is a very small sample from which to construct a profile. This group appears to have much less acidic soils than the other two, with median pH of 6.5. Soils in this urbanization category are the most likely to be anaerobic, with median redox less than 100 mV. Nutrients are no higher than in the moderately urbanized wetlands and may even be a bit lower. Median organic content in the small sample suggests a level of about 20%. Again, PSD is a function of local factors. The available values show metals to be distinctly higher than in the soils of the other urbanization categories, about double the values given in the preceding paragraph. These concentrations would routinely exceed lowest effect thresholds for Cu and Pb but not for Zn. Severe effect thresholds would still not be approached often.

Other Findings

Before regular sampling began, the research program conducted a synoptic survey of 73 wetlands, about 60% urban and the balance nonurban. Samples were analyzed in the laboratory for 31 of the wetlands. In the data from this study, significant differences appeared in soil Pb concentrations between urban and nonurban wetlands in the inlet and emergent zones.[4] There were also significant differences at $\alpha = 0.10$ between the concentrations of both Cd and Zn in the emergent zones of urban and nonurban wetlands.

Metals accumulations may be linked to soil toxicity, as estimated by the Microtox method. The Microtox test assesses the potential toxicity of an environmental sample by measuring the reduction of the light output of bioluminescent bacteria when exposed to the sample for a period of time. The method yields the effective concentration (EC), which indicates the reduction of light output after a certain length of exposure. The lower the EC value, the more toxic the sample.[4] In the synoptic study, urban open water zone soils had significantly lower ECs in both 5- and 15-min tests and were, therefore, relatively more toxic. There was also a significant difference in the emergent zone in the 15-min test. However, there were no significant differences between the inlet and scrub-shrub zones of urban and nonurban wetlands.

Microtox analysis of wetland soils in 1993 failed to confirm the previous conclusion that urban wetland soils were more toxic.[4] It should be noted that only one 1993 sample from each wetland underwent Microtox analysis and there was no attempt to compare the toxicity of various wetland zones, as in the synoptic study. Nevertheless, the 1993 results generally indicated that urban wetland soils were certainly no more toxic than those of rural wetlands. In fact, the three soils with the most toxic compounds came from less urbanized wetlands. The extraction and concentration of naturally occurring organic soil compounds in the laboratory, and not the presence of anthropogenic toxic substances, probably explained the results for these wetlands.[5] The results suggested that the soils of FC1, an urban wetland, and AL3, a rural wetland, possibly contained anthropogenic toxicants, because the results indicated toxicity in the absence of visible organic material. There were no evident accumulations of toxicants in the AL3 soil in 1993. The FC1 wetland, on the other hand, had the highest result for Cu (59 mg/kg), a Pb concentration (60 mg/kg) second only to the highly urban B3I, and a total petroleum hydrocarbon concentration (TPH) (840 mg/kg) more than three times greater than for any wetland except B3I, which exhibited an equally high value. That metals and TPH should be high at FC1 and B3I is not surprising in view of the intensity of commercial and transportation land uses in their watersheds.

Working with 1993 data, the efficacy of using regression relationships between widely distributed crustal metals (aluminum, Al and lithium, Li) and toxic metals in relatively unimpacted wetlands, was studied to evaluate whether particular wet-

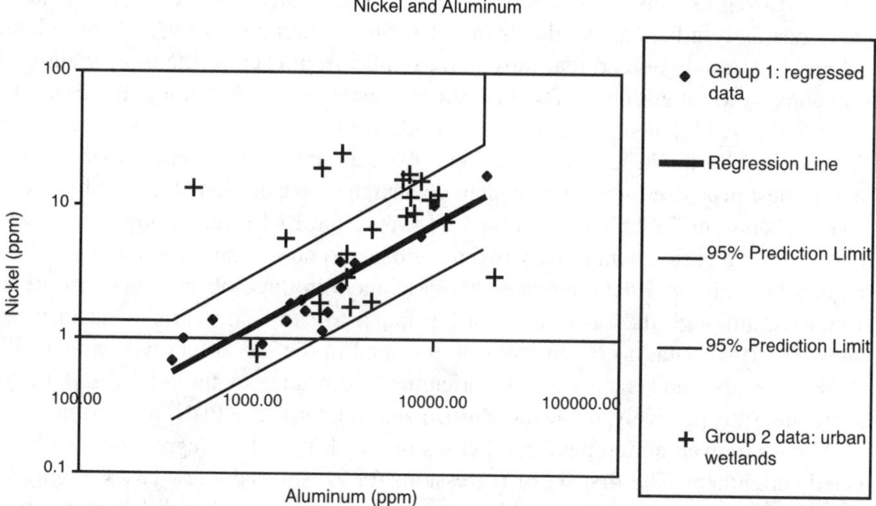

FIGURE 9-1 The assessment tool for Ni using Al as the reference element (1995 data).

lands have enriched concentrations of toxic metals in their soils.[6] The method was applied independently to the 1995 data. The regression analysis is based initially on relationships between crustal and heavy metals in relatively pristine wetlands that, it is assumed, have not received significant metal loadings of anthropogenic origin. These regressions must be developed for each region, since the natural background of metals varies with soils. If, in a given wetland, the concentration of a toxic metal is above a given confidence limit (95%) of the linear regression, it is probable that there has been anthropogenic toxic metal pollution of the wetland's soil.

The data collected in the program wetlands were analyzed using the same experimental groups outlined earlier in this chapter. The nomenclature is slightly different, Group 1 for nonurban (N) wetlands, Group 2 for moderately urbanized (M) ones, and Group 3 for highly urbanized (H) cases. In 1996, only two groups were employed, one less urban and the other more urban. Using the 1993 soil metals data, it was found that Li may be as good or better a reference metal than Al for As, Pb, and Zn. Nickel (Ni) bore a stronger relationship to Al, while Cu correlated equally well with both Al and Li.[6]

Figure 9-1 illustrates the assessment tool using a nickel-aluminum pairing with 1995 data. The regression line represents the best-fit line of the Group 1 (N) wetlands. The 95% confidence limits are the upper and lower bounds for one additional sample that is being assessed for Ni contamination. Sample contamination is gauged by considering the corresponding point's location on the graph. If the point lies on or above the upper 95% confidence limit, then the sample is judged to be enriched with the contaminating metal. Thus, in Figure 9-1, for example, seven samples above the line were judged to have Ni contamination of anthropogenic origin. The relationships from both the 1993 and 1995 data sets exhibited a close correspondence.

The study findings indicated that the most urbanized wetlands had a higher rate of soil metals enrichment than moderately urbanized wetlands, considering both the

Al and Li regressions.[6] There were far fewer indications of metals enrichment in the moderately urbanized wetlands in 1993. The regressions of 1993 Pb with both Al and Li strongly agreed that most soil samples from each of the most urbanized wetlands were Pb enriched. The first set of regressions of As with both Al and Li generally agreed that soil samples from each of the most urbanized wetlands were As enriched. Both Cu regressions using 1993 data indicated Cu enrichment in two of the most urbanized wetlands, which are also the wetlands listed as highly urbanized wetlands in Table 1 of Section I. However the Li-Cu regression using 1995 data indicated enrichment in only two nonurban and one urban wetlands. The Al-Ni regression with 1993 data indicated Ni enrichment in three of the four most urban wetlands, although the Li-Ni relationship failed to show any enrichment in these sites. For 1995 data, no Ni enrichment appeared in the less urban wetlands, while there were five and six cases of enrichment according to the Al-Ni and Li-Ni regressions, respectively, in the more urbanized wetlands. The Li-Ni regression using 1995 data showed enrichment in all cases in which the Al-Ni regression also indicated enrichment. The first set of regressions for Zn showed a few cases of enrichment in the most urban group, although fewer in number and with less agreement between the regressions than for the other metals. The Li-Zn relationship in the 1995 data indicated Zn enrichment in ten of the more highly urbanized wetlands, in comparison to only one of the less urbanized wetlands.

Although some of the wetlands were classified in different groups than they would be in according to the GIS analysis, the results of the study agree well with observations based on the soil data statistics. Therefore, anthropogenic sources clearly impact the sediments of palustrine wetlands in the Central Puget Sound Basin. While wetlands can remove metals from the water column, the accumulation of metals could still harm wetland functions. Also noted in the study is that long-term effects of atmospheric emissions from past operations of the ASARCO smelter in Tacoma on wetland soils are unknown. It is possible that such distant sources could play a role in the enrichment of toxic metals wetlands. Rainfall removes such suspended metals from the atmosphere and provides the runoff which transports the metals to wetlands, where they accumulate in the sediments.

TREATMENT WETLANDS

Table 9-3 shows soil statistics for the five treatment wetlands in the baseline period, when little or no urbanization had occurred (1988 to 1990) and then the later years (1993 and 1995), after most of the changes in land use were either well underway or complete. It appears from program data that soil pH may have increased over the years at BBC24 and, especially at NFIC12, both wetlands whose watersheds underwent the greatest amounts of development. For NFIC12, rises in pH were entirely expected because (1) this wetland was a late successional peat bog that had the lowest soil pH readings of any of the wetlands, with no place to go except up, and (2) its watershed went from 0 to 100% urbanization between 1989 and 1995.

The treatment wetlands exhibited the particle size distributions in 1989 and 1995 shown in Table 9-4. BBC24 and ELS61 both registered transition from relatively

TABLE 9-3
Soil Statistics for Treatment Wetlands in Baseline and Postdevelopment Years

Year	Statistic	pH	Redox Potential (mV)	TP (mg/kg)	TN (mg/kg)	Volatile Solids (%)	Cu (mg/kg)	Pb (mg/kg)	Zn (mg/kg)	As (mg/kg)
JC28	Mean	5.89	12	964	6661	49.6	19.5	33.2	27.1	6.4
88–90	St. Dev.	0.11	240	637	4578	26.5	3.2	15.1	23.7	2.0
	CV	2%	2003%	66%	69%	53%	16%	46%	88%	31%
	Median	5.87	–20	578	6056	59.9	20.4	32.6	18.0	7.1
	n	7.00	7	7	8	9.0	7.0	7.0	7.0	7.0
93–95	Mean	5.63	545	153	1392	13.1	5.0	8.1	3.4	1.8
	St. Dev.	0.32	102	150	1241	1.2	2.6	2.3	2.4	1.6
	CV	6%	19%	98%	89%	9%	51%	28%	71%	89%
	Median	5.53	608	176	1980	12.8	4.3	8.5	2.7	1.2
	n	5.00	5	5	5	5.0	5.0	5.0	5.0	5.0
ELS61	Mean	5.70	162	755	4409	33.6	31.8	60.3	47.8	9.0
88–90	St. Dev.	0.34	187	716	3821	22.6	20.3	20.3	24.8	5.4
	CV	6%	115%	95%	87%	67%	64%	34%	52%	60%
	Median	5.84	70	1024	3443	29.5	31.0	32.4	56.6	9.2
	n	11.00	11	13	14	14.0	13.0	13.0	13.0	13.0
93–95	Mean	5.70	164	748	4382	33.8	31.6	60.2	47.6	9.0
	St. Dev.	0.71	298	973	5092	29.3	158.4	77.9	64.1	16.8
	CV	13%	182%	130%	116%	87%	502%	129%	135%	187%
	Median	5.67	547	69	537	8.1	12.7	9.5	14.3	2.8
	n	191.00	188	194	204	208.0	195.0	195.0	195.0	195.0
PC12	Mean	6.08	107	1285	8431	52.5	26.1	170.7	52.7	14.3
88–90	St. Dev.	0.29	291	1122	6813	23.2	8.7	197.8	31.5	7.2
	CV	5%	273%	87%	81%	44%	33%	116%	60%	50%
	Median	6.04	114	1089	6347	59.2	26.9	105.1	39.7	12.5
	n	7.00	7	6	7	6.0	6.0	6.0	6.0	6.0

TABLE 9-3 *(continued)*
Soil Statistics for Treatment Wetlands in Baseline and Postdevelopment Years

Year	Statistic	pH	Redox Potential (mV)	TP (mg/kg)	TN (mg/kg)	Volatile Solids (%)	Cu (mg/kg)	Pb (mg/kg)	Zn (mg/kg)	As (mg/kg)
93–95	Mean	6.15	351	111	911	6.9	3.7	9.1	7.5	1.7
	St. Dev.	0.14	99	115	866	1.5	0.9	6.0	3.2	0.5
	CV	2%	28%	104%	95%	21%	24%	65%	43%	32%
	Median	6.10	310	109	1060	7.0	3.6	10.3	6.6	1.8
	n	5.00	5	5	5	5.0	5.0	5.0	5.0	5.0
BBC24	Mean	5.90	–197	469	4171	22.7	19.1	44.9	39.0	11.0
88–90	St. Dev.	0.35	15	434	4802	17.6	12.4	25.2	16.5	5.5
	CV	6%	–8%	93%	115%	77%	65%	56%	42%	50%
	Median	5.73	–200	297	1692	15.3	17.9	51.6	48.0	10.7
	n	6.00	6	6	7	7.0	7.0	6.0	6.0	7.0
93–95	Mean	5.97	326	43	441	7.4	5.0	4.4	6.0	1.4
	St. Dev.	0.24	49	54	596	3.7	2.9	0.7	0.5	0.3
	CV	4%	15%	126%	135%	50%	58%	16%	9%	26%
	Median	5.97	326	43	441	7.4	5.0	4.4	6.0	1.4
	n	2.00	2	2	2	2.0	2.0	2.0	2.0	2.0
NFIC12	Mean	3.97	144	2188	15984	76.0	19.5	48.6	10.4	9.2
88–90	St. Dev.	0.26	345	1138	7440	18.9	10.1	40.0	5.3	5.6
	CV	7%	240%	52%	47%	25%	52%	82%	51%	61%
	Median	3.94	140	2182	15698	81.0	16.3	44.4	10.9	9.9
	n	4.00	4	4	4	4.0	4.0	4.0	4.0	4.0
93–95	Mean	5.05	633	167	1950	12.5	2.5	10.3	3.0	2.2
	St. Dev.	1.05	2	67	113	0.7	0.3	0.8	2.2	0.3
	CV	21%	0%	40%	6%	6%	13%	8%	71%	13%
	Median	5.05	633	167	1950	12.5	2.5	10.3	3.0	2.2
	n	2.00	2	2	2	2.0	2.0	2.0	2.0	2.0

TABLE 9-4
Particle Size Distributions for Treatment Wetlands

	% Sand/Silt/Clay	
Wetland	**1989**	**1995**
BBC24	61/29/10	45/47/8
ELS61	35/49/16	15/69/16
JC28	52/35/13	46/36/18
NFIC12	4/59/37	3/32/65
PC12	58/37/5	63/30/7

sandy to silty soils, while clay stayed constant. NFIC12 exhibited a clay increase while sand was constant. The first result could be explained by sedimentation of finer particles over the predevelopment substrate, but a 30% increase in clay is not easily explained.

Otherwise, there was a strong trend for redox to rise but for nutrients, organic content, and metals all to fall from the predevelopment to the postdevelopment years. The reasons for these unexpected results can only be given speculation. What can be said is that, other than the pH increase at NFIC12, there is no obvious signal in the soils yet that negative changes may have accompanied recent urbanization.

REFERENCES

1. Ludwa, K. A., Urbanization Effects on Palustrine Wetlands: Empirical Water Quality Models and Development of a Macroinvertebrate Community-Based Biological Index, M.S.C.E. thesis, University of Washington, Department of Civil Engineering, Seattle, WA, 1994.
2. Chin, N. T., Watershed Urbanization Effects on Palustrine Wetlands: A Study of the Hydrologic, Vegetative and Amphibian Community Response During Eight Years, M.S.C.E. thesis, University of Washington, Department of Civil Engineering, Seattle, WA, 1996.
3. Reinelt, L. E. and R. R. Horner, Characterization of the Hydrology and Water Quality of Palustrine Wetlands Affected by Urban Stormwater, PSWSMRP, Seattle, WA, 1990.
4. Horner, R. R., F. B. Gutermuth, L. L. Conquest, and A. W. Johnson, Urban stormwater and Puget Trough wetlands, in *Proc. First Annu. Meet. Puget Sound Res.*, Seattle Center, Seattle, WA, March 18–19, 1988, Puget Sound Water Quality Authority, Seattle, WA, 1988, p. 723.
5. Houck, C. A., The Distribution and Abundance of Invasive Plant Species in Freshwater Wetlands of the Puget Sound Lowlands, King County, Washington, M.S. thesis, University of Washington, Seattle, WA, 1996.
6. Valentine, M., Assessing Trace Metal Enrichment in Puget Lowland Wetlands, Non-thesis paper for the M.S.E. degree, University of Washington, Department of Civil Engineering, Seattle, WA, 1994.

7. Taylor, B., K. Ludwa, and R. R. Horner, Urbanization effects on wetland hydrology and water quality, in *Puget Sound Research '95 Proc.*, Seattle, WA, January 12–14, 1995, Puget Sound Water Quality Authority, Olympia, WA, 1995, p. 146.
8. Azous, A., An Analysis of Urbanization Effects on Wetland Biological Communities, M.S. thesis. University of Washington, Department of Civil Engineering, Seattle, WA, 1991.
9. Novotny, V. and H. Olem, *Water Quality; Prevention, Identification and Management of Diffuse Pollution,* Van Nostrand Reinhold, New York, 1994.

10 Wetland Plant Communities in Relation to Watershed Development

Amanda L. Azous and Sarah S. Cooke

CONTENTS

Introduction 255
Methods 256
Results 258
  Watershed Urbanization and Vegetation Richness 258
  Water Level Fluctuation and Vegetation Richness 258
Discussion 259
Acknowledgments 261
References 262

INTRODUCTION

The vegetation communities associated with wetlands located in the Central Puget Sound Basin were examined between 1988 and 1995 to observe whether stormwater runoff from developing watersheds played a role in determining vegetation richness in receiving wetlands. The question we asked was whether plant richness would decrease in wetlands subjected to increasingly frequent stormwater runoff events that occur as a result of urbanization in the watershed. Plant richness in wetland habitats was examined over time and compared to conditions of watershed development. In addition, the study compared richness to a hydrologic measure, mean annual water level fluctuation (WLF), which increases in value the greater the disparity between minimum and maximum water depths in the approximately monthly sampling interval.

1-56670-386-7/00/$0.00+$.50

METHODS

The 26 wetlands evaluated in this study are inland palustrine wetlands ranging in elevation from 50 to 100 m above mean sea level and characterized by a mix of forested (PFO), scrub-shrub (PSS), emergent (PEM), and aquatic bed (PAB) wetland vegetation classes. In addition to the nineteen wetlands, surveyed for the long-term study three or more times between 1988 and 1995, the data set also includes 7 other wetlands that were surveyed one or more times during the years 1993, 1994, and 1995 as part of related studies.

Wetlands were selected so that approximately half would be affected by urbanization sometime after the baseline year. Sites that remained unaffected by urbanization were expected to be the controls for those wetlands receiving urbanization treatment. The wetlands were matched, wherever possible, as treatment (new urban disturbance) and control (no new urban disturbance) pairs on the basis of morphological characteristics and vegetation zones.[1-3]

These categories were later revised because, unfortunately, few of the watersheds developed as predicted. Only six watersheds developed beyond 10% of the developed area at the start of the study and not all of these were the wetlands we expected to be treated. The unexpected slowness of development in the study watersheds affected the ability to identify differences between control and treatment pairs due to stormwater and urbanization. Under the circumstances, the plan to compare control and treatment pairs of wetlands had to be abandoned. The six wetlands that underwent increases in development greater than 10% became the treatment wetlands. Of these six, only three wetlands had significant increases in watershed development of 100, 73, and 42%, with the remaining three having increases of only 10.5, 10.3, and 10.2% (see Table 1 in Section I).

Plant richness was selected to measure vegetation community function. Richness has been reported to be the component of plant community structure most sensitive to stress from suburbanization, and is measurable before there are quantifiable changes in species dominance.[4-6] Community physiognomy, the number and type of habitats, the selection of dominant species, and their contribution to coverage are resilient to calculable changes, whereas richness directly indicates the loss or gain of species occurring at low frequencies and low levels of cover.[6,7] Losses or gains of such species over a lengthy period could indicate larger changes affecting ecosystem functions.

The revised experimental categories compared richness among rural controls (RC) (wetlands with less than 12% impervious area and greater than 40% forest), urban controls (UC) (greater than 12% impervious area and less than 40% forest), and treatment (T) (wetlands with watersheds that increased developed area by at least 10% during the study period) (see Table 1 in Section I). In addition, vegetation richness was compared with annual and seasonal water level fluctuation for all wetlands over the study period. Richness in aquatic bed, emergent, scrub-shrub, and forested habitats was compared along with total wetland richness to wetland water level fluctuation.

Geographic Information System (GIS) analysis was used to delineate land use and impervious areas within the watersheds.[8] Land use classifications included agricultural lands, single- and multi-family residential housing, commercial and industrial development, transportation corridors, and any other development within a watershed that reduced forest cover.

Hydrologic measurements, including instantaneous (considered typical water levels) water levels from staff gauges and peak levels (representing storm events) from crest gauges, were recorded at least eight times annually (every four to six weeks) while water was present in the wetlands.[9] It was not possible to directly monitor each sample station, so hydroperiod was calculated based on the elevation of the station in relationship to the water levels measured at the wetland staff and crest gauges. This was expressed as:

$$D = G - E \tag{10-1}$$

where D = water depth at the vegetation plot, G = gauge reading, depth above zero water level, and E = plot survey elevation above zero water level (see Figure 10-1).

This method depended on water levels being evenly distributed throughout the wetland, varying only as elevation varied. In most cases this was a reasonable assumption, however, some wetland locations had more complex hydroperiods and were not well represented by the resulting data. Those sample stations where calculated hydroperiod was identified as inconsistent with observed conditions were eliminated from the data set.

Instantaneous and maximum water levels were calculated for each monitoring occasion. Water level fluctuation (WLF) was the difference between the maximum level and the average of the current and previous instantaneous water levels for each

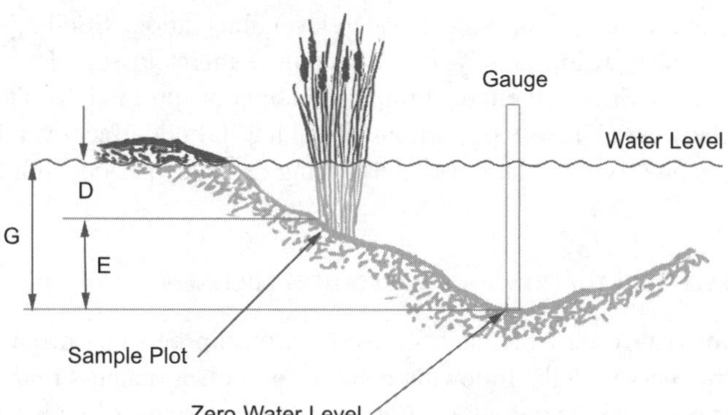

FIGURE 10-1 Relationship between water depth at vegetation sample station and water depth at the gauge. (From Taylor, B., Masters thesis, University of Washington, Seattle, 1993.)

four- to six-week monitoring period (see Equation 1-4, Chapter 1). Mean WLF was calculated by averaging all WLFs for the entire year. Seasonal water level fluctuation was the average of monthly water level fluctuations for each of four seasons: early growing season, defined to be from March 1 through May 15; intermediate growing season which lasts from May 16 to August 31; senescence, lasting from September 1 to November 15; and dormancy, November 16 to February 28.

RESULTS

WATERSHED URBANIZATION AND VEGETATION RICHNESS

Total wetland richness and richness within wetland habitats was examined for significant changes between years over the study period. Changes to all wetlands over time were evaluated as well as changes that potentially resulted from alterations in watershed land-use (treatment wetlands). There were no significant differences in wetland richness between years among all wetlands (Friedman test, $p = .32$, $\chi^2 = 4.6$) although the mean rank among the wetlands steadily diminished between 1989 and 1995. Among all wetlands, those with watersheds that underwent an increase in urbanization, treatment wetlands, showed the most significant drop in overall plant richness (Friedman test, $p = .11$, $\chi^2 = 6.06$). Urban and rural controls showed little change between study years ($p = .56$ and .48, respectively).

Among different wetland habitats, there were no significant differences in plant richness over the study period, nor any that could be statistically related to watershed land use. There were, however, losses of habitat in specific wetlands during the study period. For example, the watershed of ELS39 underwent almost complete urbanization during 1989. Development activities within that watershed eliminated a significant portion of forested wetland not accounted for in the statistical enumeration. In addition, a portion of the same wetland that was emergent when the study began, quickly converted to scrub-shrub after urbanization of the watershed. This change was attributed to alterations in hydroperiod that included an increased dry season as well as increased mean annual water level fluctuation. Similarly, three other wetlands, ELS61, JC28, and NFIC12, sustained direct losses of habitat through development activities that either eliminated habitat or substantially altered it during the study. However, the loss of portions of habitat did not affect overall richness of remaining habitats within these wetlands during the study period, with the exception of ELS39.

WATER LEVEL FLUCTUATION AND VEGETATION RICHNESS

All years of data were examined for the relationship between mean annual WLF and plant richness with the following results. Total plant richness found in wetlands was unrelated to the degree of WLF, however, plant richness within some habitat types was found to vary in relation to water level fluctuation. Within forested (PFO) wetland habitats, plant richness was unrelated to mean annual WLF. In contrast, both the emergent (PEM) and scrub-shrub (PSS) habitats showed decreasing plant richness with increasing mean annual WLF. Figure 10-2 shows plant richness in

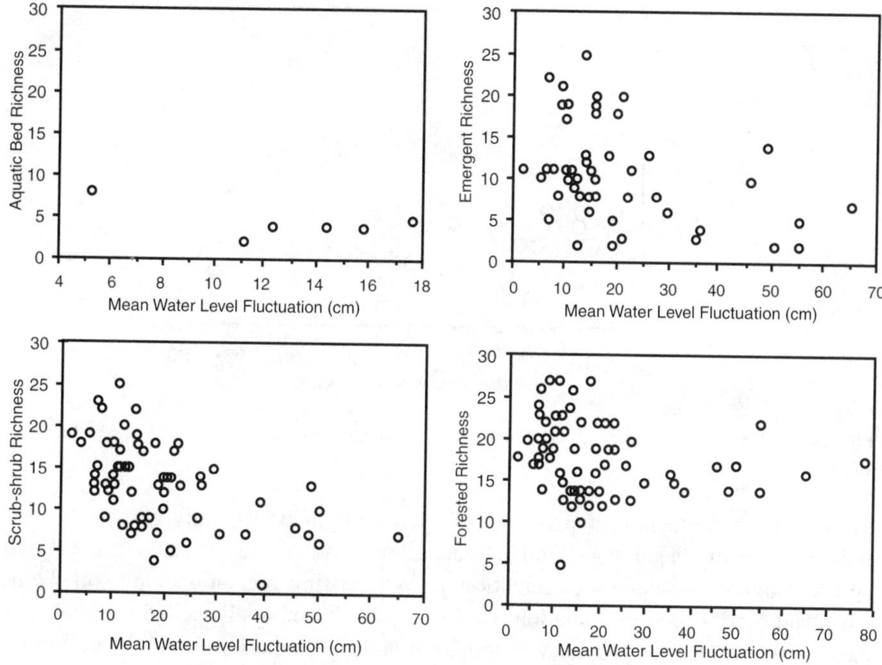

FIGURE 10-2 Plant richness in aquatic bed, emergent, scrub-shrub, and forested habitats compared to mean annual WLF.

wetland habitats related to mean annual WLF for all years of data and all wetlands, and illustrates the statistical relationship found in the PEM and PSS communities (Fisher's r to z test (Frz), PEM: r = –.38, p = .006; PSS: r = .5, p = .0001) as compared with forested habitats. There were too few sample stations in aquatic bed (PAB) habitats that remained over the study period for adequate comparison.

Seasonal differences in WLF within wetlands were also compared to plant community richness. As with mean annual WLF, there were no differences in total plant richness attributable to seasonal differences in WLF. Among different habitat types, however, plant richness in emergent zones showed a strong negative correlation with increasing water level fluctuation in the early growing season (Fisher's r to z test (Frz), r = –.54, p = .002) (Figure 10-3). No correlation with WLF was observed in the intermediate growing, senescence, or dormancy periods, and WLF in the early growing season correlated with richness in emergent habitats only. Again, there were too few samples of aquatic bed habitats for statistical comparison.

DISCUSSION

In general, many wetlands with more frequent flooding and longer durations of inundation have lower vegetation richness than do less-frequently flooded areas.[10,20] As a consequence, we expected to find that richness would decrease in wetlands and habitats subjected to runoff from urbanizing watersheds. Specifically, we thought

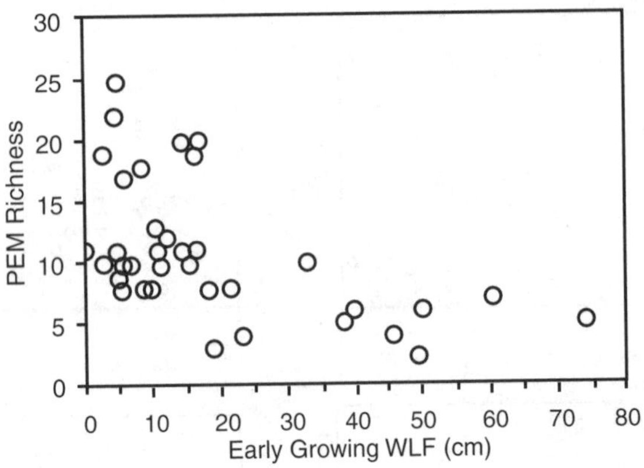

FIGURE 10-3 Plant richness in emergent habitats compared to mean early growing season WLF.

the increased frequency of peak runoff events might be partially responsible for reduced diversity in some wetland habitats. Large storm events often disturb wetland soils through erosion and sedimentation, uproot existing vegetation, alter soil chemistry, and hinder seed germination. Burial, especially of small seeded wetland species, could substantially reduce germination and both burial and inundation hinder oxygen availability and photosynthesis in living plants.[11-14] Moreover an increase in drying events could constitute another factor affecting species presence, particularly for aquatic species.[13,20] An increasing frequency of such events could conceivably affect which species were able to maintain populations in the affected areas.

In a related laboratory study, sediment deposition and hydroperiod alteration were examined.[14] These are two significant ecosystem perturbations that occur when watersheds urbanize. Sediment deposits negatively affected both sedges and sapling trees by reducing biomass and decreasing photosynthesis. The sedges, *Carex utriculata* and *Carex stipata,* were resilient to cycles of flooding and drying, but high water levels would further diminish photosynthesis when plants were partially buried. Saplings were differentially affected by flooding, with *Alnus rubra* (red alder) more sensitive than *Fraxinus latifolia* (Oregon ash), which was largely unaffected.[14] Another study examining flooding effects speculated that reduced growth at periodically flooded sites may result from frequent restructuring of root systems in response to alternately flooded and drained conditions.[16]

The slow pace of development during the study within most of the treatment wetland watersheds hindered our ability to clarify the relationship between watershed urbanization and vegetation richness. Although comparisons of richness between control and treatment pairs did not meet our criteria for statistical significance, declines in richness during the study were the most dramatic among the treatment wetlands.

More direct correlations with plant richness were measured between mean annual WLF and to a lesser extent early growing season WLF. These occurred in the emergent and scrub-shrub habitats but not forested habitats. This difference between habitat types is consistent with a review of flooding effects on different types of species. For example, large trees are known to show greater tolerances to flooding than seedlings, suggesting that woody species in forested wetlands would likely be impacted to a lesser extent than the herbaceous perennial and annual plants that dominate the emergent, and to a lesser extent, the scrub-shrub habitats.[16] On the other hand, one study found that weekly tree growth was impacted significantly by water level fluctuations when the mean water table was low. Reduced growth at intermittently flooded sites was thought to have resulted from frequent restructuring of root systems to accommodate alternately flooded and dry conditions.[15] A high water table will cause many herbaceous plants to root more superficially and in a smaller area.[15,17] As a result, herbaceous plants in areas subjected to high water level fluctuations may be exposed to subsequent drought and temperature fluctuations in the surface soil when water recedes. Frequent hydrologic disturbances are therefore more likely to affect emergent and scrub-shrub habitats where herbaceous species are more prevalent (see Table 3-3, Chapter 3, for a list of species categories within habitats).

Chapter 3 gave evidence of the hydrologic regimes where some common wetland species are found, suggesting that depth of inundation during the year may be a factor in the development of species distributions. If the hydrologic profile changes, such as through upstream controls, outlet design, or changing land use in the watershed, it is likely the plant community will shift towards the conditions produced by the new hydrologic conditions. This can have beneficial or negative consequences depending on the conditions created by management of the upstream watershed. In this chapter, we saw that increasing mean water level fluctuation is associated with lower plant richness. Possible implications are that continued urbanization will reduce biological diversity over time as well as extirpate locally uncommon wetland species if the impact of hydrologic changes is not adequately managed. Direct evidence of this possibility is supported by the results of a survey of 13 wetlands conducted in 1987, as part of this on-going study, where sedges were found to be missing from the most urbanized sites and present in undisturbed sites.[18,19]

The vegetation community changes observed in this study included direct losses of habitats through development activities, decreased plant diversity along a hydrologic gradient measured by mean water level fluctuation, increases in coverage of common and introduced species, and the loss of uncommon native species. Better management of changes in wetland hydroperiod will address three of these four concerns regarding wetlands. Controlling negative changes to wetland hydroperiod will help stay the loss of unique wetland plants and habitats and reduce available conditions for more tolerant and usually dominant species.

ACKNOWLEDGMENTS

Special thanks to Dr. Kern Ewing for his contribution to this study and his informative research on hydroperiod and water quality effects. We also would especially like to thank Dr. Fred Weinmann for his comments and suggestions on original drafts of this chapter.

REFERENCES

1. Cooke, S. S., R. R. Horner, C. Conolly, O. Edwards, M. Wilkinson, and M. Emers, Effects of Urban Stormwater Runoff on Palustrine Wetland Vegetation Communities -Baseline Investigation (1988), Report to U.S. Environmental Protection Agency, Region 10, King County Resource Planning Section, Seattle, WA, 1989.
2. Cooke, S. S., K. O. Richter, and R. R. Horner, Puget Sound Wetlands and Stormwater Management Research Program: Second Year of Comprehensive Research, Planning and Resources Department, Seattle, WA, 1989.
3. Cooke, S. S., R. Horner, and C. Conolly, Effects of Urban Stormwater Runoff on Palustrine Wetland Communities — Baseline Investigation, Natural Resources and Parks Division, King County, WA, 1989.
4. Ehrenfeld, J. G. and J. P. Schneider, The response of Atlantic white cedar wetlands to varying levels of disturbance from suburban development in the New Jersey Pinelands, in *Wetland Ecology and Management: Case Studies*, D. F. Whigham, R. E. Good, and J. Kvet, Eds., Junk, Dordrecht, 1990, p. 63.
5. Ehrenfeld, J. G. and J. P. Schneider, *Chamaecyparis thyoides* wetlands and suburbanization: Effects on hydrology, water quality and plant community composition, *J. Appl. Ecol.,* 28, 467, 1991.
6. Schindler, D. W., Detecting ecosystem response to anthropogenic stress, *Can. J. Fish. Aquatic Sci.,* 44, 6, 1987.
7. Ehrenfeld, J. G., The effects of changes in land-use on swamps of the New Jersey Pine Barrens, *Biol. Conserv.,* 25, 253, 1983.
8. Taylor, B., The Influence of Wetland and Watershed Morphological Characteristics on Wetland Hydrology and Relationships to Wetland Vegetation Communities, Masters thesis, University of Washington, Seattle, 1993.
9. Reinelt, L. E. and R. R. Horner, Characterization of the Hydrology and Water Quality of Palustrine Wetlands Affected by Urban Stormwater, King County Resource Planning, King County, WA, 1990.
10. Mitsch, W. J. and J. G. Gosselink, *Wetlands*, 2nd ed., Van Nostrand Reinhold, New York, 1993.
11. Galinato, M. I. and A. G. van der Valk, Seed germination traits of annuals and emergents recruited during drawdowns in the Delta Marsh, Manitoba, Canada, *Aquatic Bot*., 26, 89, 1986.
12. Zhang, J. and M. A. Maun, Effects of sand burial on seed germination, seedling emergence, survival, and growth of *Agryopyron psammophilum*, *Can. J. Bot*., 68, 304, 1990.
13. Leck, M. A., Germination of macrophytes from a Delaware River tidal freshwater wetland, *Bull. Torrey Bot. Club*, 123(1), 48, 1996.
14. Ewing, K., Tolerance of four wetland plant species to flooding and sediment deposition, *J. Exp. Bot.,* Vol. 36(2), 131, 1996.
15. Keeland, B. D. and R. R. Sharitz, The effects of water-level fluctuations on weekly tree growth in a Southeastern USA swamp, *Am. J. Bot.,* 84(1), 131, 1997.
16. Bedinger, M. W., Relation between forest species and flooding, in *Wetland Functions and Values: The State of Our Understanding*, P. E. Greeson et al., Eds., American Water Resources Association, Minneapolis, MN, 1979.
17. Jackson, M. B. and M. C. Drew, Effects of flooding on growth and metabolism of herbaceous plants, in *Flooding and Plant Growth*, T. T. Kozlowski, Ed., Academic Press, San Diego, 1984.

18. Horner, R. R., F. B. Gutermuth, L. L. Conquest, and A. W. Johnson, Urban stormwater and Puget Trough wetlands, in *Proc. First Annu. Meet. Puget Sound Res.*, Seattle Center, Seattle, March 18–19, 1988, Puget Sound Water Quality Authority, Seattle, 1988, p. 723.
19. Horner, R. R., Long-term effects of urban runoff on wetlands, in *Design of Urban Runoff Controls*, L. A. Roesner, B. Urbonas, and M. B. Sonnen, Eds., American Society of Civil Engineers, New York, 1989.
20. van der Valk, A. G., Response of wetland vegetation to a change in water level, in Wetland Management and Restoration, C. M. Finlayson and T. Larson, Eds., Rep. 3992, Swedish Environmental Protection Agency, Stockholm, 1991.

11 Emergent Macroinvertebrate Communities in Relation to Watershed Development

Kenneth A. Ludwa and Klaus O. Richter

CONTENTS

Introduction 265
Methods 266
Results 268
Summary 271
Acknowledgments 273
References 273

INTRODUCTION

Aquatic invertebrates play important roles in the food chain of freshwater wetlands. They are the pivotal link between the primary production, detrital trophic organisms, and higher consumer levels including fish, herpetofauna, birds, and mammals.[1] Moreover, aquatic macroinvertebrates have historically served as biological indicators of riverine and lacustrine environments.[2] For example, studies in King County and elsewhere have demonstrated the usefulness of macroinvertebrates as indicators of wetland health as well as in assessing the impacts associated with urbanization.[3-9]

Macroinvertebrate communities are noted for their response to the four major wetland stresses identified by EPA's Environmental Monitoring and Assessment Program (EMAP): (1) altered hydroperiod, (2) excess sediment, (3) changes in nutrients, and (4) environmental contaminants.[10] Unlike other terrestrial animal taxa with the exception of some amphibians, the larval forms of aquatic macroinvertebrates are completely confined to water or moist soil within a particular wetland, over entire growing seasons or years until emergence. Therefore, the aquatic macroinvertebrate community may be an excellent integrator of wetland impacts because it does not register impacts that may occur to other wetland taxa that migrate beyond wetlands.

1-56670-386-7/00/$0.00+$.50

The goal of this chapter is to establish the impacts of watershed development and, particularly, urban stormwater inputs on macroinvertebrate communities. Specific objectives include (1) developing a preliminary wetland macroinvertebrate community-based biotic index modeled on methods proven for streams, and (2) applying this index to examine the impacts of watershed urbanization on macroinvertebrate communities over a range of watersheds with different levels of existing development and within developing watersheds. The latter objective is based upon several hypotheses we developed regarding the response of the aquatic macroinvertebrate community to anthropogenic changes to wetlands and their watersheds. These hypotheses predict (1) changes in macroinvertebrate taxa richness and numbers of individual organisms will reflect changing land use, environmental pollution, direct habitat degradation, and general wetland health, (2) proportions of sensitive and pollution intolerant taxa will decrease with increasing watershed urbanization and wetland degradation, and (3) proportions of functional taxa groups will change with alterations to a wetland's nutrient cycle.

METHODS

The emergence traps, used in sampling macroinvertebrates and described in Chapter 4, collected mostly adult aquatic insects. Therefore, our term macroinvertebrates refers mostly to insects although some macroinvertebrates from other classes were also captured. Emergent aquatic macroinvertebrates were captured monthly in 19 palustrine wetlands in the Puget Sound Basin from 1988 to 1995. Wetlands surveyed were located in watersheds in various stages of urban and suburban development, as described in Section I.

Trapping protocols, their invertebrate captures, and their strengths and weaknesses are extensively described in Chapter 4. Briefly, we attempted to place traps in conditions of open still water, fine sediment, permanent water, and other conditions as similar as possible between wetlands. The location of traps was particularly important because the presence or absence of certain vegetation types, substrate conditions, and water depths can substantially influence the character of the aquatic macroinvertebrate community at sampling stations.[11-15] We deployed the traps in each wetland in 1989, 1993, and 1995 over the time periods listed in Table 11-1. Field staff collected trapped invertebrates and replaced the preservative on an approximately monthly basis from April to September, with a year-end collection made in October or November depending on the likelihood of additional captures. No collections were made from December through March because of low invertebrate activity during this period. Consequently, the traps provided a cumulative measure of macroinvertebrate emergence between each occasion that the traps are emptied.

We identified and enumerated the macroinvertebrates collected in 1989 to the lowest taxonomic level possible, in most cases to genus or species. We identified insects collected in 1993 and 1995 only to family for Diptera and to order for other taxa. We standardized identifications to a consistent level within each taxonomic grouping for all samples.

TABLE 11-1
Approximate Aquatic Invertebrate Emergence Trap Sampling Periods for Growing Seasons 1989, 1993, and 1995

	1989	1993	1995
Start collection	September 1, 1988	April 10, 1993	January 1, 1995
End collection	September 31, 1989	April 9, 1994	October 30, 1995

Note: Monitoring at 14 sites started in September, 1988; 5 more sites were added in April, 1989.

Using the 1989 data set, we developed a multimetric biological index based on principles of the *Benthic Index of Biotic Integrity.*[16] We proceeded by first testing metrics to determine whether they differentiated between the two best and two worst sites; we then confirmed these metrics by testing them over the whole range of 19 sites.[3] We then tested and adapted existing lotic macroinvertebrate community metrics to the wetland insect community, and tested and added new metrics unique to palustrine communities.

We designed and calculated the 1989 species/genus-level metrics using data split into distinct sampling periods: April to June, July to September, and October to November. The data split into these periods, especially the two summer periods, responded more strongly to urbanization parameters than did the year-long data set. In contrast, the 1989 order/family-level metrics were designed and calculated using the year-round data sets. Taxa richness values for the coarser-level data were too low within the individual sampling periods to differentiate between sites.

We assumed that the difference between the lengths of sampling periods between the three years (Table 11-1) did not significantly affect taxa richness values, but it may have affected total numbers of individuals collected. The metrics developed for the order/family-level data were taxa richness and proportion oriented, therefore we assumed that different sampling period lengths did not affect metric design or calculation.

The level of taxonomic effort was considerably coarser for the 1993 and 1995 collections, consequently we found it necessary to develop and test a new set of metrics suitable for that coarser level of information. We performed this step by elevating the 1989 taxonomic data to the same levels as the 1993 and 1995 collections. Most of these coarser-level metrics were based on original metrics for the 1989 collections.

We tested the overall index scores against land use and wetland morphology thresholds previously reported using the Mann-Whitney test, the nonparametric equivalent of the independent groups *t*-test.[3,17,18] We also tested index scores against parameters for wetland hydrology and water quality, and separately against wetland morphology and watershed land use, using multiple regressions. All statistical analyses were performed at a significance level of p >.05.[18]

RESULTS

The metrics recommended for further testing from emergent collections identified to genus-species level taxonomy are listed in Table 11-2. Although taxa belonging to orders Ephemeroptera, Plecoptera, and Trichoptera (including order Odonata, these orders are referred to as EPOT) are often the basis of stream biological metrics, we found a paucity of these taxa in our wetland insect collections using emergence traps (see Chapter 4). Although EPOT richness and abundance yielded two metrics, most of the metrics (numbers 7 through 22, including all new wetland-oriented metrics) related to order family Chironomidae of order Diptera (aquatic midges and true flies). Chironomids are a highly diverse family sparsely detailed in ecological literature, and generally considered to be negative indicators for running waters. Chironomids, however, are adapted to lentic environments, and therefore may be more appropriate indicators of wetland health.

Using an index composed of the metrics listed in Table 11-2, we calculated index scores and compared them to direct and indirect measures of wetland stress. We emphasize that further verification of this index and its component metrics is necessary before it can be used as an independent measure of wetland ecological health. Conclusions drawn from these analyses follow.

There were two primary periods of insect emergence: in the early summer and again in the late summer/early autumn, correlating with sampling periods in April-June and July-September. Consequently, these were most appropriate for calculation of biotic index scores. Collections made in October-November did not appear to be as effective for bioassessment.

Biotic index scores responded significantly to land use and wetland morphology parameters. A multiple regression revealed that scores responded negatively to total watershed impervious area, wetland channelization, and incidence of dryness. The regression explained 67% of the variance in index scores. Threshold analyses also revealed that index scores were significantly higher with increasing watershed forest coverage and lower with increasing impervious area. Highly channelized sites had significantly lower scores, consistent with the observation of degraded water quality for most parameters in highly channelized sites.

Multiple regressions indicated that water quality and hydrologic parameters explained a significant amount of variation of the index scores (as high as 73%). Index scores responded negatively to hydrogen ion concentration (antilog pH), conductivity, suspended solids, water level fluctuation, and incidence of wetland dryness. Suspended solids, conductivity, and water level fluctuation were demonstrated in other related studies to be the water quality and hydroperiod parameters most significantly degraded by increases in watershed impervious area and decreases in forest cover. Collectively, our work illustrates the interrelationship between a wetland's watershed, its physical and chemical parameters, and the health of its biological communities.[3,17,19]

The order/family-level metrics developed with the 1989 data are listed in Table 11-3. Table 11-4 lists the resulting index scores calculated with these metrics for 1988, 1993, and 1995. Although the order/family-level metrics responded to indi-

TABLE 11-2
Biotic Index Metrics Recommended for Use with Wetlands, Based on Emergent Macroinvertebrate Collections with Genus/Species-Level Identification

Metrics Included in Final Wetland Biotic Index Recommended for Further Research (Genus/Species-Level Taxonomy)

Adapted from Stream Metrics:

1. Taxa richness
2. Scraper and/or piercer taxa presence
3. Shredder taxa presence
4. Collector taxa richness
5. EPOT[a] taxa richness
6. Percent individuals as EPOT
7. Percent individuals as Tanytarsini tribe
8. Tanytarsini tribe richness

Unique Wetland Metrics:

9. Percent individuals as Chironomini tribe
10. Chironomini tribe taxa richness
11. Percent individuals as Tanypodinae subfamily
12. Tanypodinae subfamily taxa richness
13. Presence *Thienemanniella*
14. Presence *Endochironomus nigricans*
15. Presence *Parachironomus* sp. 2
16. Presence *Polypedilum* gr. 1 and 2
17. Presence *Alabesmyia*
18. Presence *Apsectrotanypus algens*
19. Presence *Paramerina smithae*
20. Presence *Psectrotanypus dyari*
21. Presence *Zavrelimyia thryptica*
22. Presence *Tanytarsus*

[a] EPOT = Ephemeroptera, Plecoptera, Odonata, and Trichoptera.

cators of urbanization, the overall index comprised of these metrics had much less power to discern between sites with different levels of urban impact than the previous one. For example, the multiple regressions of 1989 genus/species index scores vs. total impervious area, wetland channelization, and incidence of dryness, explained 67% of the index score variance. The same regression explained only 21% of the 1989 index score variance for the order/family data.

The next year, after 1989, in which land use data were available was 1995. The 1995 index scores, however, were not significantly related to total impervious area or forested area, nor did the scores respond significantly in the multiple regressions against total watershed impervious area, wetland channelization, and incidence of

TABLE 11-3
Biotic Index Metrics Recommended for Use with Wetlands, Based on Emergent Macroinvertebrate Collections with Genus/Species-Level Identification

Metrics Included in Final Wetland Biotic Index Recommended for Further Research (Order/Family-Level Taxonomy)
• Family/Order Richness
• Shredder Presence
• Collector Richness
• EPOT Order Richness
• % Individuals as EPOT
• % Individuals as Dixidae

TABLE 11-4
Order/Family Macroinvertebrate Index Scores

	Index Score		
	1989	**1993**	**1995**
AL3	16	10	20
B3I	12	8	6
BB24	26	10	16
ELS39	10	12	12
ELS61	18	10	18
ELW1	8	6	6
FC1	16	14	10
HC13	22	14	24
JC28	22	10	26
LCR93	28	16	6
LPS9	8	10	18
MGR36	20	12	16
NFIC12	10	10	24
PC12	18	10	10
RR5	10	6	18
SC4	16	10	12
SC84	14	14	12
SR24	18	10	14
TC13	10	12	10

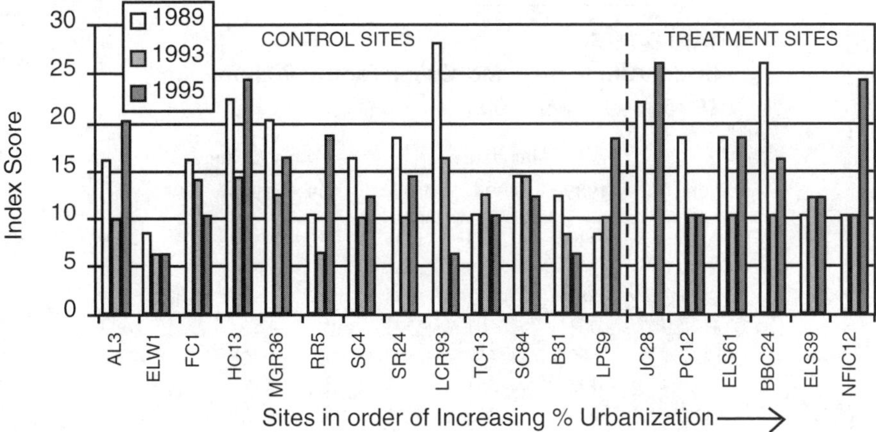

FIGURE 11-1 Wetland macroinvertebrate index scores for 1989, 1993, and 1995 change in watershed urbanization.

wetland dryness. Furthermore, the changes in index scores between 1989 and 1995 did not correspond to changes in land use. For example, NFIC12, which experienced an increase in impervious area from 2 to 40%, showed the highest percent increase in its index score, exactly opposite that which would be predicted (Figure 11-1).

In addition to relating index scores to changing watershed characteristics, we also examined changing taxa richness and abundance to describe the impact of urbanization on emergent macroinvertebrates. Table 11-5 lists abundance and taxa richness values for each site in each year. Multiple regressions and threshold tests revealed no significant patterns in order/family taxa richness related to impervious area between sites or years. In other wetland animal communities, taxa richness of sensitive species is often more responsive to wetland degradation than is overall taxa richness.[20] The index developed for the species/genus-level data incorporates this concept by including 16 metrics based on the presence of taxa that are assumed to be more sensitive to disturbance. The order/family data do not allow enough resolution to indicate sensitive taxa. Numbers of individuals decreased from 1989 to 1995 in 14 out of 19 sites. Although this could be a function of a somewhat different period in 1989, the major captures of species from late spring through early autumn suggest this may not be the case.

SUMMARY

Wetlands, in general, are considered dynamic aquatic environments, requiring organisms that occupy these ecosystems to be adapted to a wide range of environmental conditions. Wetlands within urban environments may exhibit even greater perturbations than those in undisturbed landscapes, consequently they may exhibit a different and more cosmopolitan invertebrate community than observed in more pristine conditions. Results from our 1989 comparisons of insect data across wetlands with different levels of watershed development suggest that urbanization affects emergent

TABLE 11-5
Insect Abundance and Order/Family Richness for 1988, 1993, and 1995

	Abundance			Taxa Richness		
	1989	**1993**	**1995**	**1989**	**1993**	**1995**
AL3	4408	3619	1946	12	11	13
B3I	3027	2219	988	14	10	8
BB24	8857	14742	5815	14	10	13
ELS39	7337	6267	3773	12	12	12
ELS61	20828	13457	2808	16	10	12
ELW1	1239	503	157	10	7	7
FC1	4736	13332	5751	14	9	9
HC13	8748	4436	2934	15	11	13
JC28	1133	5778	1251	13	8	13
LCR93	9689	12148	40464	15	12	7
LPS9	5127	1006	5490	12	10	12
MGR36	7365	13276	1918	14	10	10
NFIC12	8869	24866	2015	12	11	13
PC12	5893	10701	4350	15	11	11
RR5	8621	4748	2150	12	10	11
SC4	2952	2794	2962	12	10	12
SC84	3692	2159	1254	13	9	11
SR24	5598	4982	1140	14	8	12
TC13	4657	4204	4657	13	9	13

macroinvertebrate communities by (1) decreasing overall taxa richness, (2) eliminating or reducing taxa belonging to scraper and shredder functional feeding groups (leaving a dominance of collector taxa), (3) reducing EPOT taxa richness and relative abundance, and (4) eliminating or reducing specific Dipteran taxa, particularly those belonging to the Chironomidae family.

Our results also suggest that an index, similar to the one we developed, may be as useful as comparable indices established for running waters. Clearly, testing of the metrics proposed by this study is necessary before the index can be used as an independent wetland assessment tool. Although our genus and species-level analysis identifies correlations between macroinvertebrates and wetland condition, we strongly recommend further development of aquatic macroinvertebrate community-based biological indices for assessment of wetland biological health. Specifically, we recommend genus and species-level taxonomic identification of macroinvertebrates for use of taxa richness values and calculation of biological indices. Coarser-level identifications do not appear to adequately discern insect functional groups, tolerance levels, and specific sensitive genera or species. Also, additional refinement of insect tolerance to wetland hydrologic and water quality disturbances as well as an additional understanding of feeding group information may allow an index to be used as a diagnostic tool. Alternatively, a set of proposed guidelines for assessing wetland health advocate a broad multitaxa approach that not only includes invertebrates but plants and vertebrates as well.[21]

ACKNOWLEDGMENTS

We would like to thank Robert Wisseman, associated staff, and taxonomic specialists for identifying invertebrates collected during this study. (See additional acknowledgments in Chapter 4).

REFERENCES

1. Cummins, K. W. and R. W. Merritt, Ecology and distribution of aquatic insects, in *An Introduction to the Aquatic Insects of North America*, R. W. Merritt and K. W. Cummins, Eds., Kendall-Hunt, Dubuque, IA, 1996, p. 74.
2. Rosenberg, D. M. and V. H. Resh, Use of Aquatic Insects in Biomonitoring, in *An Introduction to the Aquatic Insects of North America*, R. W. Merritt and K. W. Cummins, Eds., Kendall-Hunt, Dubuque, IA, 1996, p. 87.
3. Ludwa, K. A., Urbanization Effects on Palustrine Wetlands: Empirical Water Quality Models and Development of Macroinvertebrate Community-Based Biological Index, M.S. thesis, College of Engineering, University of Washington, Seattle, WA, 1994.
4. Azous, A. L., An Analysis of Urbanization Effects on Wetland Biological Communities, University of Washington, Seattle, WA, 1991.
5. Murkin, H. R. and B. D. J. Batt, The interactions of vertebrates and invertebrates in peatlands and marshes, *Mem. Entomol. Soc. Can.,* 140, 15, 1987.
6. Rosenberg, D. M. and H. V. Danks, Aquatic insects of peatlands and marshes in Canada, *Mem. Entomol. Soc. Can.,* 140, 174, 1987.
7. Wrubleski, D. A., Chironomidae (Diptera) of peatlands and marshes in Canada, *Mem. Entomol. Soc. Can.,* 140, 141, 1987.
8. Hicks, A. L., Impervious Surface Area and Benthic Macroinvertebrate Response as an Index of Impact from Urbanization on Freshwater Wetlands, University of Massachusetts, Amherst, MA, 1995.
9. Hicks, A. L., Aquatic Invertebrates and Wetlands: Ecology, Biomonitoring and Assessment of Impact from Urbanization, University of Massachusetts, Amherst, MA, 1996.
10. Liebowitz, N. C. and M. T. Brown, Indicator strategy for wetlands, in *Environmental Monitoring and Assessment Program: Ecological Indicators*, U.S. Environmental Protection Agency, EPA/600/3-90/060, Office of Research and Development, Washington, D.C., 1990.
11. Voigts, D. K., Aquatic invertebrate abundance in relation to changing marsh vegetation, *Am. Midl. Nat.,* 95, 313, 1976.
12. Parsons, J. K. and R. A. Matthews, Analysis of the associations between macroinvertebrates and macrophytes in a freshwater pond, *Northwest Sci.,* 69, 265, 1995.
13. McGarrigle, M. L., The distribution of Chironomid communities and controlling sediment parameters, in *Chironomidae: Ecology, Systematic, Cytology and Physiology*, D.A. Murray, Ed., Permagon Press, New York, 1980, p. 275.
14. Minshall, G. W., Aquatic insect-substratum relationships, in *The Ecology of Aquatic Insects*, V. H. Resh and D. M. Rosenberg, Eds., Praeger, New York, 1984, p. 358.

15. Murkin, H. R. and J. A. Kadlec, Responses of benthic macroinvertebrates to prolonged flooding of marsh habitat, *Can. J. Zool.,* 64, 65, 1986.
16. Fore, L. S., J. R. Karr, and R. W. Wisseman, A Benthic Index of Biotic Integrity for Streams in the Pacific Northwest, North American Benthological Society, Lawrence, KS, 1995, p. 2.
17. Taylor, B. L., K. A. Ludwa, and R. R. Horner, Urbanization effects on wetland hydrology and water quality, in Puget Sound Research '95 Proc., E. Robichaud, Ed., Puget Sound Water Quality Authority, Olympia, WA, 1995, p. 146.
18. Zar, J. H., *Biostatistical Analysis*, Prentice-Hall, Englewood Cliffs, NJ, 1984.
19. Chin, N. T., Watershed Urbanization Effects on Palustrine Wetlands: A Study of the Hydrologic, Vegetative and Amphibian Community Response during Eight Years, M.S. thesis, University of Washington, Seattle, 1996.
20. Power, T., K. L. Clark, A. Harfenist, and D. B. Peakall, A Review and Evaluation of the Amphibian Toxicological Literature, Canadian Wildlife Service, Ottawa, Canada, 1989.
21. Brooks, R. P. and R. M. Hughes, Guidelines for assessing the biotic communities of freshwater wetlands, in *Proc. Natl. Wetlands Symp.: Mitigation of Impacts and Losses*, J. A. Kusler, M. L. Quamen, and G. Brooks, Eds., Association of State Wetland Managers, Berne, NY, 1988, p. 276.

12 Bird Communities in Relation to Watershed Development

Klaus O. Richter and Amanda L. Azous

CONTENTS

Introduction....275
Methods....276
Results....277
Discussion....280
References....283

INTRODUCTION

Wetlands are recognized because of the disproportionate value they provide for birds when compared with other habitats (see Chapter 6). Concomitantly, wetlands are under increasing threat from watershed development especially in urbanizing areas. As cities and suburbs expand, wetlands are directly destroyed by channelization, filling, draining, and impounding.[1] Annual losses are estimated at 400,000 ha of wetlands due to single and multiple residences, commercial businesses, roads, and other associated infrastructure for developments.[2] Additionally, wetlands are indirectly impacted by altered hydrology, water quality deterioration, and vegetation changes attributable to increased runoff from impervious surfaces associated with development.[3-7]

Dramatic population changes have been documented in birds along gradients of urbanization. A review of researchers' findings of bird distributions along gradients of urbanization summarized that the number of species almost always decreased with increasing urbanization and that bird density or abundance generally increased with urbanization.[8] More specifically, urbanization has been correlated with the increased density of a few dominant urban ground gleaners and decreased numbers of forest insectivores, canopy foliage gleaners, and bark drillers.[9]

Despite these well-documented changes to birds in terrestrial urbanizing environments, we are unaware of studies investigating riparian bird communities associated with palustrine wetlands along a gradient of urbanization or watershed development. Consequently, we describe the changing bird communities in wetlands with watershed development and also compare the pre- and post-development bird rich-

1-56670-386-7/00/$0.00+$.50

ness and abundances. We hypothesize that bird species richness and abundance should change with increasing wetland impact and watershed development. Specifically, total bird richness should reflect the intensity of development by first increasing during early stages of watershed impacts, attributable to the range expansion of disturbance-tolerant species, and secondly, remaining constant with moderate impacts as a result of species replacement between urban-intolerant and urban-tolerant species. Finally, total bird richness should decline with high urbanization from displacement of insectivorous urban-avoiders by a few competitive and predatory aggressive species, which are also more abundant. In summary, with increasing urbanization we predict that the richness and abundance of avoider species should decline, whereas the richness and abundance of urban adapters, exploiters, and exotics should increase. Finally, we predict the loss of intolerant wetland obligates as wetland watersheds become increasingly developed and buffers become the only remaining habitat.

In part, these predictions are based on the findings that bird distribution and abundances are widely accepted as dependent on vegetation structure, diversity, and anthropogenic impacts. Consequently, as watersheds and wetlands are altered, bird species richness, diversity, and relative abundance should reflect these changes, with those wetlands in watersheds exhibiting the greatest alterations also exhibiting the greatest avifaunal changes.

METHODS

Bird survey methods, distribution, abundance, and ecological groupings are described in Chapter 6. Here, we additionally identify nonnative species, native aggressive species, and urban exploiters within an ecological grouping whose respective bird richness and abundance may change with wetland and watershed development. Nonnatives are introduced species that spread without the direct assistance of people and include five species in King County — California Quail, Rock Dove, European Starling, Brown-headed Cowbird, and House Sparrow. Native aggressive species include the Northwest Crow. Both native and nonnative species may be further grouped according to their response to human encroachment and activity. Thus we identify urban exploiters, urban avoiders, and suburban adaptable species using criteria that are based on species sensitivity to human-induced changes.[8] Our listings for birds in respective categories are provided in Table 6-1 in Chapter 6.

Richness was calculated based on the number of species counted in each year's survey. Average richness was calculated based on the average of the number of species observed among wetland stations. We categorized wetlands into three experimental categories according to baseline watershed, wetland, and adjacent conditions (in 1988) and the additional disturbance wetlands received during the course of our study (up to 1995). Our three experimental categories included wetlands of rural areas in which watershed development did not change (Rural Controls), wetlands of urbanized areas in which no additional watershed development occurred (Urban Controls), and wetlands of either rural or urbanized areas in which watershed development increased by at least 10% or more regardless of initial base-line condition (Treatments).

We identified single-family houses with forests, open water, shorelines, undeveloped meadow, and shrub-land as suitable bird habitats adjacent to wetlands. We classified developed, cleared, and agricultural lands as generally unsuitable habitats for birds. The proportional changes in suitable and unsuitable habitats within watersheds were then investigated for correlations with our bird distributions and abundance to determine possible cause and effect relationships.

Statistical analyses of correlations utilized parametric tests when assumptions of normality were met and nonparametric tests when assumptions could not be met by data transformations or other methods. We chose $p \leq .05$, and $p \geq .5$ and $p \leq .10$ as significant and weakly significant, respectively, for reporting our results. Nevertheless, significance should be interpreted cautiously because of the variability in sampling populations of species and the low number of wetlands undergoing impacts that could be observed in changing bird sightings during the period of our study.

RESULTS

Total species richness decreased significantly among all wetlands between 1989 and 1995 (Friedman test (F), $\chi^2 = 26.8$, $p < .0001$). Total species richness also decreased between 1898 and 1995 among all wetlands when analyzed by experimental category. Richness in rural control wetlands declined by 34% from 1989 to 1991 and then another 4% from 1991 to 1995 (F, $\chi^2 = 12.25$, $p = .002$). Wetlands in both already developed (urban controls) and treatment watersheds also showed significant declines in total richness between 1989 and 1991. Richness was 26% lower among urban controls (F, $\chi^2 = 6.5$, $p < .04$) and 40% lower in treatment wetlands (F, $\chi^2 = 8.4$, $p = .02$) between 1989 and 1991. However, richness increased slightly among most wetlands of both groups in 1995 (Figure 12-1).

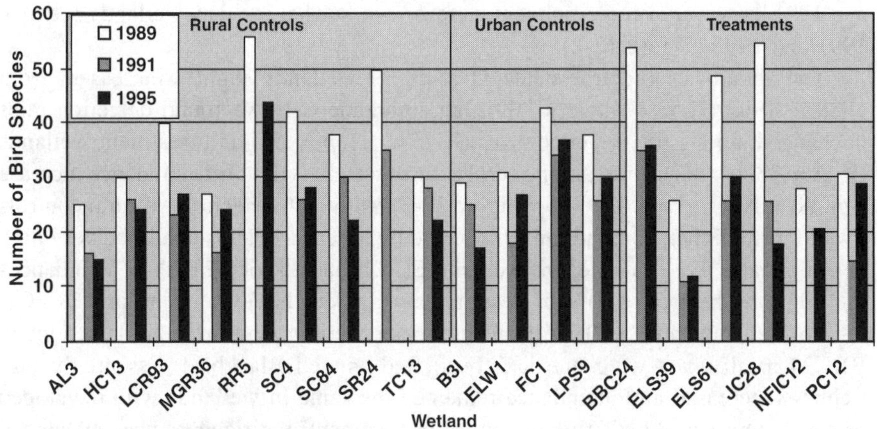

FIGURE 12-1 Bird richness in wetlands over the study period by experimental category.

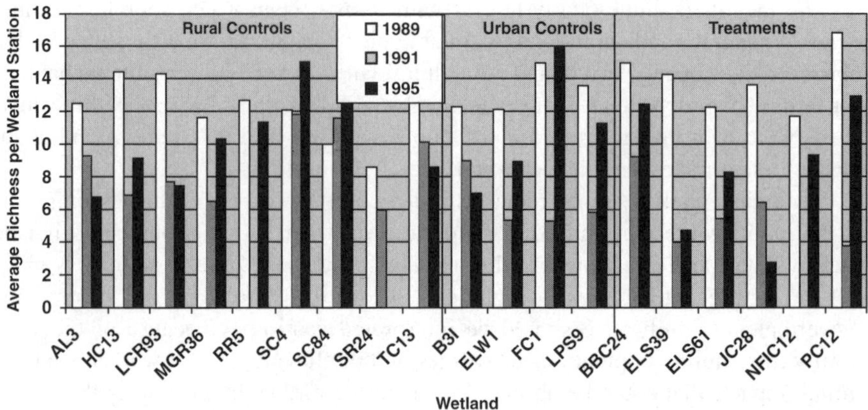

FIGURE 12-2 Average avian detection rate over the study period by wetland and experimental category.

Total richness in a single wetland and year ranged from 11 to 56 species over the study period and averaged 38 among all wetlands in 1989, the year of highest recorded richness. During that same year, we observed an average of 37 bird species in the rural control wetlands, 35 in the urban control, and an average of 40 in the treatment wetlands. By the last year of our surveys, 1995, total richness averaged 26 within rural control wetlands with undeveloped uplands, 28 in the urban control wetlands, and 24 in the treatment wetlands.

Average wetland richness, similar to total wetland richness, decreased significantly between 1989 and 1995 in all wetlands (F, $\chi^2 = 15.5$, $p = .0004$). However, average richness did not decrease among rural control wetlands (F, $\chi^2 = 3.71$, $p = .16$) only among treatment wetlands (F, $\chi^2 = 8.4$, $p = .02$), and to a lesser extent urban controls (F, $\chi^2 = 4.5$, $p = .1$). The most significant declines occurred between 1989 and 1991, however average richness increased somewhat in many wetlands between 1991 and 1995 (Figure 12-2).

The abundance of birds detected at all 19 wetlands slightly increased, from 1989 to 1995 (F, $\chi^2 = 4.8$, $p = .09$), but simultaneously we found detection rates unchanged among the urban controls (F, $\chi^2 = 1.2$, $p = .55$) and treatment wetlands (F, $\chi^2 = .33$, $p = .85$). Only among rural control wetlands did bird abundance increase significantly (F, $\chi^2 = 4.6$, $p = .1$) (Figure 12-3). Although detections in rural controls overall were higher, some urban control and treatment wetlands exhibited very high detection rates. For example, one wetland (ELW1) had one of the highest abundances in 1995, of which over 50% of detections were Canada Goose. Similarly in FC1, 28% of the abundance in 1995 were European Starling, and the high detections in BBC24 in all years were due largely to Red-winged Blackbird. Essentially bird richness decreased and abundance remained the same in wetlands with developed or developing watersheds (urban control or treatment) but richness was unchanged and abundance increased in wetlands with rural, relatively undeveloped watersheds (rural controls).

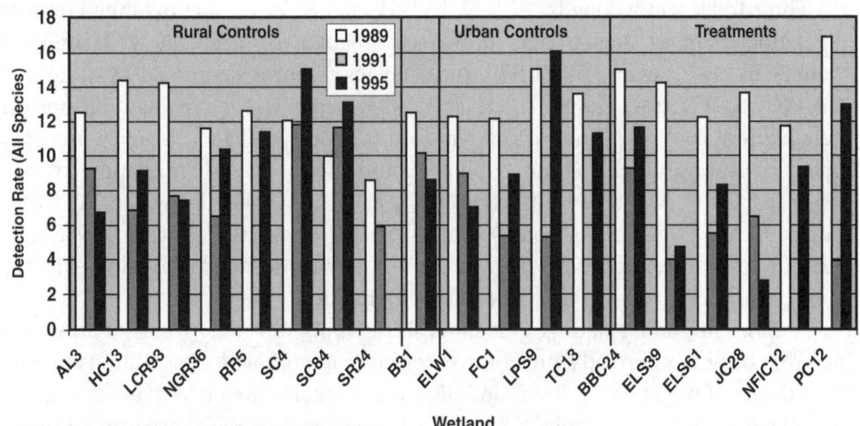

FIGURE 12-3 Bird abundance (detection rate) by wetland and experimental category.

Unexpectedly, we did not find overall richness to be related to urbanization within 100, 500, or 1000 m of the wetlands, but as expected we found increased richness of birds known to avoid human development (avoiders) in wetlands with high percentages of forested land within 500 m adjacent to the wetland (Mann-Whitney (MN), $p <.09$). Clearly, richness decreased among the already urban wetlands and in those where adjacent habitat declined due to land use changes.

Yearly richness and abundance of birds of each migratory status paralleled total species trends suggesting that most species were influenced by the same habitat conditions. Detections of long-distance migrants declined during the study among all wetlands combined (F, $\chi^2 = 31.6$, $p = .0001$) as did short-distance migrants (F, $\chi^2 = 6.4$, $p = .04$). Interestingly, detections of residents remained the same. Long-distance migrants also declined within all experimental categories (F, $\chi^2 \geq 7.1$, $p \leq .02$) but detections of short-distance migrants did not show any significant change within the experimental groups. Detections of permanent resident species did not change among the rural control and treatment wetlands but declined in the urban control wetlands (F, $\chi^2 = 5.1$, $p = .07$).

Across all wetlands, the number of detections of urban avoider species that avoid development and suburban adaptive species declined between 1989 and 1995 (F, $\chi^2 = 10.1$, $p = .007$), while densities of urban exploitive species stayed the same. Detections of avoiding species declined among the already urban and treatment wetlands but not the rural control wetlands (F, $\chi^2 = 9.1$, $p = .01$). The greatest declines of suburban adaptive species occurred in treatment wetlands (F, $\chi^2 = 7.5$, $p = .02$). While exploitive species detections were not significantly different between years in wetlands overall, among the rural control wetlands in nonurbanized areas densities of exploitive species increased significantly (F, $\chi^2 = 5.6$, $p = .06$) from 1989 to 1995. Density changes included increases in such invasive species as American Crow, European Starling, and House Sparrow. A complete list of detection rates for all species is available in Chapter 6, Appendix 6-1.

Three treatment wetlands, ELS39, ELS61, and NFIC12, that exhibited dramatic vegetation changes during our study (see Chapter 10) also showed significant changes in bird species. At ELS39, bird species richness decreased from 26 to 11 and then to 12, from 1989, 1991, and 1995, respectively. Species disappearing included Marsh Wren, Pine Siskin, and Red-breasted Nuthatch. Species increasing included, among others, urban habitat exploiters and suburban adapters such as American Crow, Mallard, California Quail, and Rufous-sided Towhee. At ELS61, species richness decreased from 49 to 30 species between 1989 and 1995, and at NFIC12 species decreased from 28 to 21. Within both wetlands, sightings of American Robin and Black-capped Chickadees increased.

Another important finding that suggests declining diversity trends with increasing urbanization is derived from comparisons of our work with observations of others in wetlands of urban areas. We found 38% more species in our May to June surveys of wetlands across our urbanization gradient than were found in April to June surveys of urban wetlands impacted by development as documented in another study.[10] Correspondingly, we found 30 to 62 species (excluding one-time sightings) with an average of 47 species (SE = 2.34) across all wetlands as opposed to only 13 to 29 species (including one-time sightings) with an average of 19 species (SE = 0.82) per wetland.[10] Furthermore, our average wetland diversity values (see Chapter 6) ranging from 2.83 to 3.54 with a mean of 3.15 (SE = 0.039), as compared to diversity measures ranging from 1.42 to 2.39 and averaging 1.89 (SE = 0.048), are considerably higher than the previous reported findings from wetlands in urban areas. Hence, urbanization at and adjacent to wetlands may exhibit the potential to significantly decrease wetland biodiversity.

Not only were species richness and diversity lower in urban wetlands, but also the relative abundances of some species differed. Song Sparrow and American Robin ranked within the top five most abundant species at all studied wetlands regardless of urbanization. Other abundant species jointly found at more pristine watersheds of our studies as well as those in urban settings include Willow Flycatcher, Black-capped Chickadee, Swainson's Thrush, American Crow, Spotted Towhee, and Red-winged Blackbird. However, abundant in our study wetlands and either absent or present in low numbers at urban wetlands were Pacific-slope Flycatchers, Winter Wren, Wilson's Warblers, and Black-headed Grosbeak. Missing from more urban wetlands are Virginia and Sora rail and scrub-shrub and forested wetland species including Hermit Thrush, MacGillivray's Warbler, Townsend's Warbler, and Western Wood-pewee. Conversely, species at urban wetlands and rarely found at more rural wetlands include Mute Swan, Peking Duck, domestic geese, Herring and Glaucous-winged Gull.

DISCUSSION

Wetland birds may be directly threatened by impacts to marshes, swamps, and bogs and secondarily by habitat changes attributable to urbanization surrounding wetlands. Foremost among wetland impacts include increased flooding frequency from urban stormwater, which may flood nest sites and disperse pollutants that bioaccumulate in birds through aquatic food chains. Runoff may alter the area of open water, existing hydroperiod, vegetation, and other wetland characteristics influencing cover, nesting habitat, and food distribution of birds.[11] At the same time, development may remove or alter bird habitat within wetland buffers and other areas immediately adjacent to wetlands. Finally, development usually fragments native habitat, further destroying access to water sources, breeding, nesting, or feeding habitat and changing competitive interactions among and between species.

Variations of densities among species should reflect specific habitat requirements for that species. Consequently, as vegetation changes within a wetland resulting from hydrologic changes, bird sightings presumably also change. We would expect changes to the richness, abundance, and distributions of wetland obligates to be impacted by different factors than what have been observed to impact forest and other terrestrial species. Because of the difficulty in controlling for all habitat factors besides urbanization, and because land-use changes during the study years were not as dramatic as expected for the study design, we were unable to demonstrate population changes clearly attributable to land uses. In addition, the broad-scale declines in species observed during the study affected all wetlands making differences due to land-use changes difficult to identify.

Our findings, generally, were that, average species richness decreased more in wetlands in watersheds affected by urbanization than in wetlands not affected by urbanization. In addition, we found that abundance of birds increased among the rural control wetlands and very dramatic increases were observed in two urban control wetlands. Finally, detections of many native species that avoid urbanization decreased in all but rural wetlands located in areas that remained undeveloped during the study.

Decreasing richness and increasing numbers of individuals of each species remaining in response to isolation have been observed by other researchers, who found that wetland size and isolation account for 75% of the variation in species richness observed within prairie marshes.[12] They also found that species richness was often greater in wetland complexes than in simple, larger isolated marshes. We expected to find that the proportion of suitable habitat within 100, 500, and 1000 m was correlated to avian richness or overall abundance, but we didn't. However, we did find that wetlands with significant forest land lying adjacent had increasing numbers of species that avoid urbanization, even though adaptable and exploitive species mostly declined during the same period.

For the most part, we found the wetland avifauna to be an extension of the upland avifauna. As expected, in wetlands of undisturbed landscapes such as SR24 and RR5 species diversity is dominated by native species whereas wetlands in more urban areas such as B3I and FC1 had bird diversity characterized by increasing numbers of nonnative species including American Crow, European Starling, House Sparrow, and some Brown-headed Cowbirds.

We have seen European Starlings displace cavity nesters including swallows and chickadees. European Starling increases have been shown to coincide with Purple Martin declines, suggesting they may eventually also cause local extinctions of wetland cavity nesters including swallows and chickadees.[13] Moreover, we have seen

American Crows raid passerine nests. The shift of bird communities from predominantly native species in undisturbed areas to invasive species in highly developed areas is well documented in terrestrial environments and we saw similar shifts among some, but not all, wetlands within this study.[8] Nevertheless, observations must be cautiously interpreted as recent literature suggests that determining bird richness and abundance is extremely difficult, and furthermore, may be driven by immigration from a few large regional source sites that produce surpluses rather than by local conditions.[14-16]

The inferences from these comparisons suggest that bird communities will become more cosmopolitan as urbanization and watershed development continues. Specifically, we would expect decreasing diversity and abundance of migrants and residents, and increasing nest predators including urban exploiters like the American Crow and European Starling, as well as nest parasites such as the Brown-headed Cowbird. Other factors contributing to declines in birds that avoid urbanization are the density of predators like domestic cats, introduced rodents such as Norway rats, brown rats, and other predators such as raccoons and opossums.[17-19] We especially expect significant reductions in ground nesting species as increasing numbers of predators are introduced with human development.

Many wetlands in our study still exhibit a wide variety of vegetation structure and microhabitats that enable a rich diversity of birds to be found. However, with increasing urbanization and habitat fragmentation that separates wetlands from larger upland habitats, and wetlands from each other, diversity of native species may be expected to decrease.[10] Watershed development brings alterations in the depth, duration, and frequency of flooding, which alters the availability of water and conditions for vegetation communities. Birds limited by need for open water and specific vegetation structure may disappear, whereas other generalist species may increase. Consequently, waterbirds (e.g., waterfowl, rail), Red-winged Blackbird, and other species use will change. To avoid these effects, we recommend that forest land with complex structure be retained to the greatest extent possible in areas adjacent to wetlands. Dense stands of herbs and shrubs should also be retained to provide cover to birds and restrict the movement of avian predators. Access via roads, trails, and footpaths that enable disturbance by humans and use by pets should be limited, and edge habitat minimized as edge-related problems of thermoregulation, predation, and nest parasitism increase along edges.

Our data support the increasingly accepted view that several measures of avian communities are needed to assess conditions because increasing diversity and abundance may be attributable to urban exploiters and urban adaptable species, and may in fact indicate deterioration in wetland function. To maintain regional biodiversity, it is critical to differentiate between native species with distinct habitat preferences, invasive species, and adaptable species associated with urbanization. We need to maintain habitat for native, specialized, species rather than for the increasingly common adaptable birds. Finally, wetlands must be viewed as dynamic ecosystems that must be managed for diversity over the entire landscape and not just as individual isolated wetland habitats.

This study intensively covers the wetlands of the Central Puget Sound Basin and represents a first comprehensive account of wetland bird diversity across a gradient of watershed urbanization over a seven-year period. Nevertheless, this work is a rough initial attempt to assess bird densities and population trends associated with watershed development. Our studies are a first step in establishing the causal relationship between avifauna and watershed development. However, because of the natural variability in bird populations from multiple sources and the limited amount of development that occurred, more years of data are needed to rigorously link changing bird demographics to watershed urbanization. Clearly, long-term investigative monitoring is essential for adequately demonstrating the influence of disturbance on species.

REFERENCES

1. Darnell, R. M., *Impacts of Construction Activities in Wetlands of the United States*, U.S. Environmental Protection Agency, Office of Research and Development, Corvallis, OR, 1976.
2. Adams, L. W., L. E. Dove, and D. L. Leedy, Public attitudes toward urban wetlands for stormwater control and wildlife enhancement, *Wildl. Soc. Bull.,* 12, 299, 1984.
3. Kusler, J. A. and G. Brooks, Eds., *Proc. Natl. Wetlands Symp. Wetland Hydrology*, Association of State Wetland Managers, Berne, NY, 1988.
4. Hopkinson, C. S. and J. W. J. Day, Modeling the relationship between development and stormwater nutrient runoff, *Environ. Manage.,* 4, 315, 1980.
5. Reinelt, L. E. and R. R. Horner, Pollutant removal from stormwater runoff by palustrine wetlands based on comprehensive budgets, *Ecol. Eng.,* 4, 77, 1993.
6. USEPA, *Natural Wetlands and Urban Stormwater: Potential Impacts and Management*, 843-R-001, U.S. Environmental Protection Agency, Office of Water, Washington, D.C., 1993.
7. Houck, C. A., The Distribution and Abundance of Invasive Plant Species in Freshwater Wetlands of the Puget Sound Lowlands, King County, Washington, M.S. thesis, University of Washington, Seattle, 1996.
8. Blair, R. B., Land use and avian species diversity along an urban gradient, *Ecol. Appl.,* 6, 506, 1996.
9. Beissinger, S. R. and D. R. Osborne, Effects of urbanization on avian community organization, *Condor,* 84, 75, 1982.
10. Milligan, D. A., The Ecology of Avian Use of Urban Freshwater Wetlands in King County, Washington, University of Washington, Seattle, 1985.
11. Weller, M. W., The influence of hydrologic maxima and minima on wildlife habitat and production values of wetlands, in *Proc. Natl. Wetlands Symp. Wetland Hydrology,* J. A. Kusler and G. Brooks, Eds., Association of State Wetland Managers, Berne, NY, 1988.
12. Brown, M. and J. J. Dinsmore, Implications of marsh size and isolation for marsh bird management, *J. Wildl. Manage.,* 50, 392, 1986.
13. Lewis, M R. and F. A. Sharpe, *Birding in the San Juan Islands*, The Mountaineers, Seattle, 1987.
14. James, F. C., C. E. McCulloch, and D. E. Wiedenfeld, New approaches to the analysis of population trends in land birds, *Ecology,* 77, 13, 1996.

15. Thomas, L. and K. Martin,. The importance of analysis method for breeding bird survey population trend estimates, *Biol. Conserv.,* 10, 479, 1996.
16. Brawn, J. D. and S. K. Robinson, Source-sink population dynamics may complicate the interpretation of long-term census data, *Ecology,* 77, 3, 1996.
17. Wilcove, D. S., Nest predation in forest tracts and the decline of migratory songbirds, *Ecology,* 66, 1211, 1985.
18. Yahner, R. H. and D. P. Scott, Effects of forest fragmentation on depredation of artificial nests, *J. Wildl. Manage.*, 52, 158, 1988.
19. Terborgh, J., Why American songbirds are vanishing, *Ci. Am.,* 266, 104, 1992.

Section IV

Management of Freshwater Wetlands in the Central Puget Sound Basin

13 Managing Wetland Hydroperiod: Issues and Concerns

Amanda L. Azous, Lorin E. Reinelt, and Jeff Burkey

CONTENTS

Introduction 287
Methods 288
Results 290
  Plant Richness and Inundation Depth, Frequency, and Duration 290
Discussion 293
  Basin Planning 294
  Wetland Management Areas 295
  Master Drainage Planning and Guidelines 295
Conclusions 297
References 297

INTRODUCTION

Land use changes and stormwater management practices typically alter hydrology within a watershed. A major finding of our studies as they progressed was that hydrologic changes were having more immediate and measurable effects on the composition of vegetation and amphibian communities than other environmental conditions we monitored, such as water quality. Early study results also showed wetland hydroperiod, which refers to the depth, duration, frequency, and pattern of wetland flooding, to be a key factor in determining biological responses. Consistently, we observed reduced richness of plant and amphibian species in wetlands with more urbanized watersheds, which in most cases had highly fluctuating water levels (see Figure 1-2 in Chapter 1 for hydroperiod categories).[1-5] In addition, we observed increases in the dry period at some wetlands, which effectively eliminated types of habitat in those wetlands thus reducing biotic diversity. For example, ELS39 was dominated by an emergent habitat class at the start of the study and became dominated by a scrub-shrub habitat class over the study period, eliminating open water for amphibian species and reducing emergent species and herbs that contributed to overall diversity.

1-56670-386-7/00/$0.00+$.50

As a result, substantial attention was given to understanding hydroperiod impacts and developing management guidelines for protecting wetland plants and animals. The local county stormwater management utility, King County Surface Water Management (KCSWM), expressed an interest in developing wetland management guidelines that could be used in continuous-flow simulation computer models. Because only a few of the wetlands in the original 19 study wetlands were at the extreme of water level fluctuations, we wanted to measure more plant communities with significant hydroperiod alterations. As a consequence, we augmented the data collected on the 19 wetlands by more intensively monitoring the hydroperiods of 6 wetlands, of which 4 were known to have high water level fluctuations. In these wetlands, water level data was collected with continuous recording gages and plant communities were surveyed. The purpose was to better understand the relationship between biological diversity and the regime of water depth, duration, and frequency of inundation in wetlands. This chapter will discuss the methods and results of this study and some of the management techniques local jurisdictions can use for protecting wetland diversity from hydrological changes attributable to urbanization.

METHODS

Continuous recording gages were installed in six wetlands in late 1994 and early 1995. The gages were programmed to record water surface elevations at 15-min increments. Three of the wetlands we monitored were already experimental controls in our ongoing study of 19 wetlands. Of these, two were located in relatively undisturbed watersheds and one was in a highly urbanized watershed. The remaining three wetlands included in this study were selected based on field observations of large water level fluctuations throughout the year and were located in watersheds dominated by suburban-density housing (one dwelling unit per 0.5 acre).

Water depths in all six wetlands were monitored over one year; however, due to unexpected seasonal lows in rainfall and some losses of data due to malfunctioning equipment, data were available for only a partial water year for some of the wetlands. The available hydroperiod data were used to calibrate the computer model Hydrologic Simulation Program — FORTRAN (HSPF), a continuous event model with the ability to simulate hydrologic processes in a watershed.[7] The model was used to predict rainfall runoff from different watershed conditions. Field measurements were used to adjust runoff from simulated rainfall events, with the outflows and stages resulting from actual events.

Four of the six wetlands we modeled had well-defined outlets, hydraulics, and bathymetry, which allowed reasonably accurate stage, storage, and discharge relationships to be developed. Two of the wetlands were stepped beaver dam systems, with stable base flows and low event fluctuations (see Chapter 1, Figure 1-2 for an illustration), which proved too complex to accurately model within the project scope. Since both beaver dam wetlands had been previously monitored as part of our ongoing study, we used data collected in these wetlands during 1988, 1989, 1991, and 1993 to supplement the hydrologic data obtained in 1995 from continuous recording gauges and simplified the modeling of the wetlands. In general, the margin of error in the spatial distribution of precipitation represented by nearby gages, and the length of the field record, limited the accuracy of the simulations of wetland water levels to plus or minus 0.5 ft (15 cm) of recorded water levels.

Emergent (PEM), scrub-shrub (PSS), and forested (PFO) wetland habitat classes were identified and surveyed for plant species using the protocols for vegetation field work documented in Chapter 3 and in previous work.[5] Plant richness within habitat classes and the presence and dominance of exotic species were calculated from the plant species presence and coverage data collected. PEM and PSS habitat classes dominated the wetlands in this study, with only two wetlands having areas of PFO.

Water depths recorded from gages and water depths predicted from HSPF computer simulations were used to compute the duration and the frequency of storm events. We used these data to analyze the differences in plant species composition in emergent, scrub-shrub, and forested habitat classes to determine if there were significant differences in plant species richness within each habitat class related to the frequency, duration, and depth of flooding in a wetland.

The data were analyzed using the *StatView*® statistical applications program.[8] The independent variables were analyzed in categories to provide more statistical rigor given the small data set and the 0.5 ft. (15 cm) margin of error, but were also evaluated in a series of continuous variable stepwise regressions. Categories were based on frequency distributions of the data. The plant richness data were not normally distributed; therefore the nonparametric Kruskal-Wallace (KW) and Mann-Whitney (MW) tests were used to compare the distributions among categories, depending on the number and type of variables being compared. Both statistical tests help establish whether the underlying distributions for different groups are similar. Both methods use ranked data and consequently are resistant to outliers. We chose $p \leq .05$, and $p > 0.05$, and $\leq .10$ as significant and weakly significant, respectively. Significance should be interpreted cautiously because of the high margin of error and the wide confidence intervals for the predictive level of significance attributable to analysis of only six wetlands. Discontinuities in wetland habitat characteristics also limit the data set.

We measured frequency of storm events in a hydroperiod by defining a hydrologic event as a water level increase greater than 0.5 ft (15 cm) above the monthly average depth. Duration was defined as the time period (e.g., number of days) of such a flooding event (Table 13-1).

TABLE 13-1
Category Definitions for Water Depth, Frequency, and Duration of Events

Frequency of Events	Water Depth	Duration of Events
Less than 6 per year	Greater than –2.0 ft depth (> –60 cm)	Less than 3 days
More than 6 per year	–2.0 to 0 ft (–60 to 0 cm)	3 to 6 days
	0 to +2.0 ft above average water depth (0 to 60 cm)	More than 6 days

RESULTS

Plant Richness and Inundation Depth, Frequency, and Duration

Plant richness in the sample stations ranged from 3 to 31 species in the PEM habitat class, 3 to 22 in the PSS class, and 14 to 25 in the forested class. Very few invasive species were found. When observed, they were dominant in only a few localized areas.

A categorical nonparametric analysis of plant richness in relation to the depth, frequency, and duration of inundation showed that plant richness was significantly lower at water depths exceeding 2 ft (60 cm) and highest in areas with less than 2 ft of flooding on average (KW, p <.0001). Consequently, to control for plant richness and water depth correlations, we evaluated frequency and duration of flooding events separately for correlations with plant richness at three ranges of average water depths, greater than –2 ft, zero to –2 ft, and zero to +2 ft above average water level, representing deep water, shallow water, and seasonally flooded habitats, respectively.

Our test for differences in richness for the three different ranges of water depth showed that, in general, plant communities in areas subjected to more than three hydrologic events per month tended to have lower richness regardless of average water depths. The range of plant richness and the average plant richness is higher in wetlands with fewer fluctuation events. The difference is statistically significant at all water depths (MW, p ≤.05) (Figure 13-1).

The duration of events was compared to plant richness and water depth. Duration alone was a significant factor only in the deepest zones of greater than 2 ft (KW, p <.001) (Figure 13-2). From –2 ft to +2 ft above average water levels, increased duration did not contribute to the variability of plant richness as much as the frequency of flood events.

When the effects of excursion frequency and duration were combined, the relationship with plant richness was statistically stronger. Plant richness was found

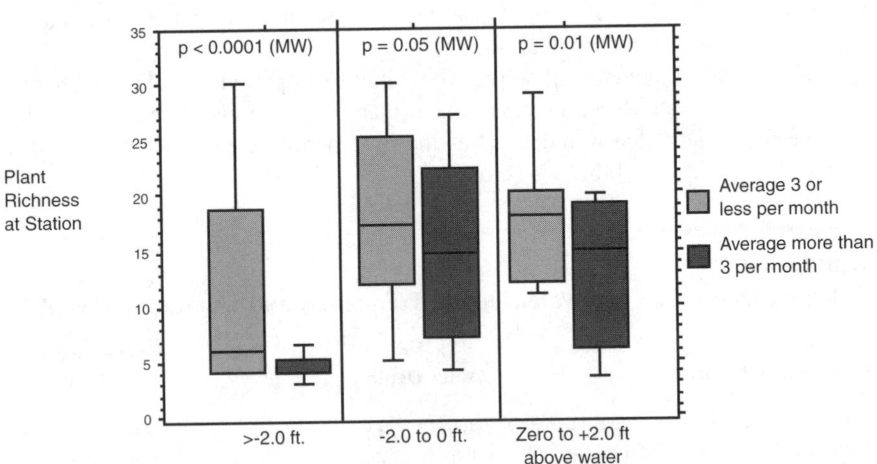

FIGURE 13-1 Plant richness, water depth, and frequency of events.

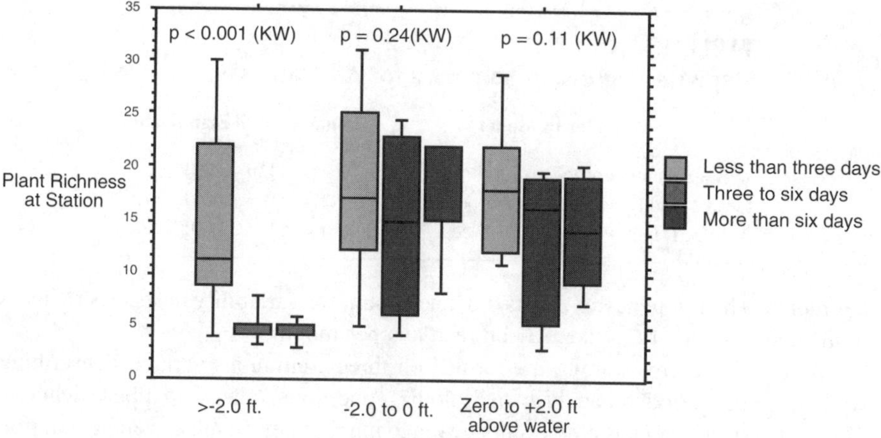

FIGURE 13-2 Plant richness, water depth, and duration of events.

to decrease significantly with events longer than six days duration, even with frequencies of less than three per month (KW, $p < .0001$). For excursion frequencies greater than three per month, richness was significantly lower in wetlands where durations exceeded three days per month (KW, $p < .0001$) (Figure 13-3).

These results were significant for both emergent and scrub-shrub classes; however, there were too few forested classes represented to be included. The results suggest that the average monthly duration of inundation can be a factor in determining plant species richness, when the frequency of inundation is greater than three times per month on average or when the length of inundation exceeds three days

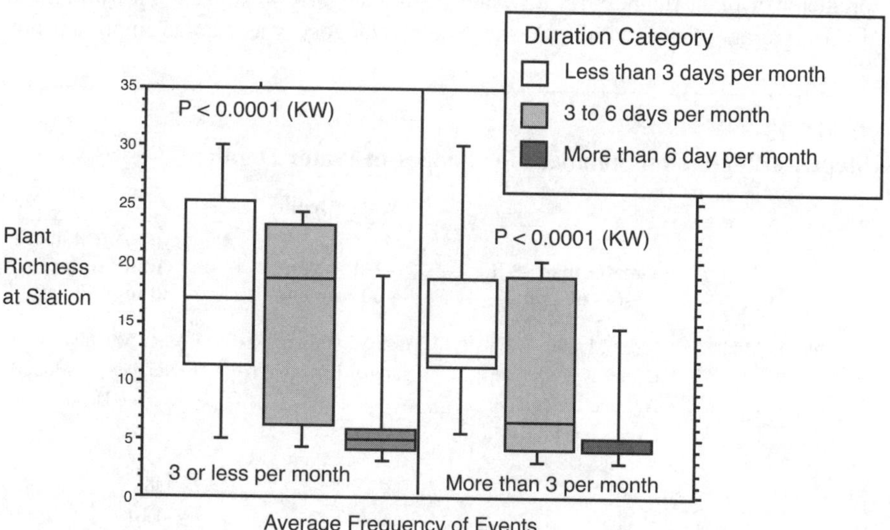

FIGURE 13-3 Plant richness, frequency, and duration of events.

TABLE 13-2
Stepwise Regression Summary for All Data

Variables in Model	Incremental Significance
Average duration of Excursion (hr)	• r = .44, p <.0001
Mean Monthly Events (Number per Month)	• r = .51, p <.0001
Average Depth	• r = .53, p <.0001

per month. The frequency of events did not account for variability in species richness until excursion durations exceeded three days per month.

A stepwise regression was performed on three continuous variables describing wetland hydroperiod to analyze hydrologic conditions related to plant richness. These included mean monthly events (average number per month), average duration of event (hr), and average depth of flooding. When the plant data for all water depths were analyzed collectively, the regression showed that duration of events was the biggest predictor of plant richness, followed by the mean number of monthly events and then average depth (see Table 13-2).

Plots of actual plant richness vs. predicted values suggested that there should be separate models for plant communities at different ranges of water depth. As a consequence, the stepwise regression was performed for the same three categories of average water depth used in the categorical analysis; greater than –2 ft, zero to –2 ft and zero to +2 ft. The same three variables of average number of events per month, average duration of event (days), and average depth (within the water depth category) were evaluated.

We found that the average number of monthly events was the most significant predictor of plant richness for all water depth categories analyzed (shown in Table 13-3). Average depth, within the water depth category, was also an important pre-

TABLE 13-3
Stepwise Regression Summary for Ranges of Water Depth

	Water Depth		
	Greater than –2 ft (> –60 cm)	**–2 ft to zero (–60 to 0 cm)**	**zero to +2 ft above water surface (0 to +60 cm)**
Variables in Model	• Mean Monthly Events (Number per Month)[a] • Average Duration of Excursion (hr) • Average Depth	• Mean Monthly Events (Number per Month) • Average Depth	• Mean Monthly Events (Number per Month) • Average Depth
Significance	r = .66 p = .005	r = .25 p = .0009	r = .46 p <.0001

[a] An excursion is a water level increase of more than 0.5 ft (15 cm) above the monthly average depth.

dictor of habitat richness although duration of inundation was not. Average number of monthly flood events and flood duration were the most critical variables in the greater than –2 ft water depth category, followed by average depth within the category. These predictive variables were most robust for the deeper water and seasonal wetland habitats and statistically weakest between them, from zero to –2 ft. Plots of plant richness values within this water depth category showed less spread between the highest and lowest richness as compared with the deeper water wetland habitats (greater than –2 ft) and drier seasonal wetland habitats (zero to +2 ft above average water level).

DISCUSSION

Our results suggest that the frequency of storm-caused flooding events can affect wetland plant diversity in both deep and shallow water and in seasonally flooded wetland habitats. Storm event frequency affected each of the three hydrologic regimes we studied; greater than –2 ft, zero to –2 ft, and zero to +2 ft above average water level. The duration of storm events is also a factor reducing plant richness, and compounds the impact of event frequency in deeper water communities. In fact, when all water depths greater than –2 ft, zero to –2 ft, and zero to +2 ft above average water level, were analyzed as a whole, the cumulative duration of the flooding event was the greatest predictor of plant richness.

The highest species richness was found in wetlands with an average of less than three flooding events per month in all water depth categories. Wetlands with cumulative durations of events exceeding three days of flooding per month were also lower in richness. The highest richness was found in wetlands with both an average of less than three events per month and a cumulative duration of flooding (above 0.5 ft of average) that was less than six days per month.

Current stormwater protection measures primarily rely on stormwater detention for protecting wetlands. Although detention reduces peak depth, it increases the duration of a stormwater runoff-flood event. Water is captured, stored, and released after the storm is over, thereby maintaining a longer flood condition at wetlands downstream. Stormwater ponds are a management tool designed, in large part, primarily for controlling floods, erosion, and water quality in streams, but may, in some cases, operate counter to wetland protection.

Our findings suggest that when urbanization affects wetland hydroperiod, limits on the frequency and the durations of flooding events may help mitigate losses to wetland plant diversity. Based on this study we recommend that the frequency of water levels greater than 0.5 ft (15 cm) above predevelopment levels be limited to an annual average of three or less per month; and that the cumulative durations of events (water levels greater than 0.5 ft (15 cm) above predevelopment average) be limited to less than six days per month.

Our data were limited, and many questions remain to be answered regarding the interrelationships between flood frequency and flood duration. For example, it might be possible to extend the durations of events in wetlands if the frequency of those events is further reduced, without any decrease in vegetation richness. Similarly, it may also be possible to reduce durations of flood events but increase frequency without effect. These issues should be explored further.

It will be difficult for urbanizing areas to meet these suggested standards in all areas, and is also not likely to happen if detention is the primary management tool. Achieving real resource protection of high-value wetlands will require a more comprehensive approach that recognizes wetlands are part of a system in a larger landscape and should be managed accordingly. This view has a number of implications for management:

- It is necessary to consider incidental effects on wetlands of activities in their watersheds, along with any engineering activities performed on the wetland itself for stormwater management purposes.
- Wetland response and management depend on a host of landscape factors, including retention of forest and other natural cover, maintenance of natural storage reservoirs and drainage corridors, separation of human activities from wetlands, and public awareness of wetland functions.
- Wetland protection means finding root cause solutions, e.g., source control practices that prevent or minimize quantities of runoff and release of pollutants. For example, utilize downstream retention and detention for quantity control and treatment for pollutant capture where source controls, such as retaining forest land and clearing limits, are inadequate to ensure wetland protection.

The experience of King County in its attempts to apply recommendations resulting from our research affords a view of some additional management strategies besides detention. Three comprehensive methods of providing wetland protection from runoff and urbanization impacts, used in King County and available to jurisdictions, are the Basin Planning process, the establishment of Wetland Management Areas, and Master Drainage Planning. The guidelines, outlined in Chapter 14, have been used in both basin and master drainage planning processes. Most applications have focused on minimizing changes in water level fluctuation, as it was identified as having the most direct effect on wetland biotic communities, particularly vegetation and habitat for breeding amphibians. Regulations governing factors that affect water level fluctuation have been targeted at new development on the urban side of an urban growth boundary, where the most significant impacts are predicted to occur. The general information on construction impacts generated by the research program has also led to the application of seasonal clearing limits in the drainage areas of high-valued wetlands.

Basin Planning

The basin planning process was developed to address the significant and rapid land use changes that have an impact on water resources, including flooding, water quality, and habitat protection. The basin planning process is a way to create and utilize a comprehensive set of management recommendations involving development regulations, capital improvement projects, education programs, improved maintenance practices, and monitoring. Basin planning can be utilized to identify wetlands of local significance that are unique and outstanding (Class 1 rating), which can then receive significant protection based on management goals. Basin planning is a tool for implementing long-term strategies for wetland and watershed protection.

Wetland Management Areas

Prior to adoption of the basin planning process, wetland protection in King County was achieved primarily through an ordinance to protect sensitive areas (SAO). The wetland protection in the SAO provides for discrete buffer widths as a function of assigned rating (e.g., 100 ft for highest class wetlands). Although these buffers confer some protection to wetlands, they are often not adequate to protect other functions influenced by the broader watershed and surrounding landscape. To address these issues, wetland management areas (WMA) were developed to focus on watershed-based controls to protect high-quality wetlands. The intent of these controls was to minimize the stormwater-related impacts on wetlands by minimizing impervious surfaces, retaining forests, clustering, and providing constructed infiltration systems, where feasible.

A major component of a wetland management strategy is the limitation of total impervious area in the catchments to 8%, where allowed by zoning. From the research program data, it was clear that there are significant increases in WLF between wetlands with watersheds less than 4% and those with watersheds greater than 12% impervious surface.[9,10] We were unable to define this more precisely because of the absence of watersheds in our study with impervious surfaces between 4 and 12%. Nevertheless, data sets have been summarized and show a loss of aquatic system function with impervious surface areas above about 10% as measured by changes in channel morphology, fish and amphibian populations, habitat, and water chemistry.[11] Forest retention has also been correlated with reduced WLF although no specific threshold was identified in this work.[6] While the precise threshold will vary by watershed and the effectiveness of mitigation strategies, 8–10% impervious surface is probably an appropriate threshold.

A requirement for 50% forest retention has been imposed in the catchments of some highly valued wetlands. Clustering of development away from hydrologic source areas (landscape features transmitting water to wetlands during the wet season) is recommended. An additional requirement in one wetland watershed is the use of constructed infiltration systems to reduce increases in stormwater volumes. This is feasible when soils in the watershed are amenable to substantial infiltration and water quality issues are not present. Finally, seasonal clearing limits for construction activities may be imposed to prevent clearing and grading during the wet season (October to April), when up to 88% of erosion occurs.[8]

Master Drainage Planning and Guidelines

The master drainage planning (MDP) process can be used for large or complex development sites to assess the potential impacts of development on aquatic resources.[12] The MDP process is typically required for urban plan developments

(UPD), for subdivisions with more than 100 single-family residences, and for projects that clear 500 acres or more within a subbasin. In addition, lower thresholds for development may be established in the drainage areas of Class 1 wetlands, regionally significant resource streams, or over sole-source aquifers. For Class 1 wetlands, an MDP is required if a project seeks to convert more than 10% of the wetland's total watershed area to impervious surface.

The updated guidelines for MDP monitoring and studies, supported in part by results of this research, require monitoring for purposes of:

1. Assessing wetland functions in storing and releasing stormwater
2. Determining hydroperiods in relation to vegetation and amphibian communities
3. Establishing baseline conditions from which to measure potential post-development changes[9]

Specific urbanization concerns are

1. Loss of live storage and infiltration functions of wetlands
2. Stability of outlet control conditions
3. The effects of increases in flow rates and volumes
4. Changes in hydroperiod and resultant habitat changes
5. Changes in groundwater and interflow

For purposes of assessing wetland impacts, the MDP guidelines require determination of the bathymetry (morphometry) of the wetland, outlet control description and measurement, stage-discharge volume relationships, surface area of open water including ordinary high water levels and the dead and live storage maximum elevation and volume. Specific monitoring requirements are:

1. Monthly instantaneous and crest water levels to determine water level fluctuation in the permanent pool area of the wetland
2. Inflow and outflow rates of the wetland
3. The duration of summer drying, if applicable

As an example, for the North Fork Issaquah Creek Wetland 7 Management Area and Grand Ridge MDP, the East Sammamish Community Plan limited development in the drainage area tributary to North Fork Issaquah Creek Wetland 7, a Class 1 wetland, to no more than 8% impervious surfaces and 65% forest retention. This condition applies to all development proposals submitted prior to adoption of the Issaquah Basin Plan and for all developments not going through the MDP process.[12] In a different case, the proposed Grand Ridge development in the North and East Fork Issaquah Creek basins involved two development options: rural estates at a density of 1 unit per 5 acres and an urban proposal consisting of 580 acres of urban development and 1400 acres of permanent open space. After studying the potential development scenarios carried out under the recommended stormwater management guidelines, a proposal was agreed upon that maintained greater forested area than in the original proposals and utilized infiltration to reduce stormwater volumes.

CONCLUSION

Managing stormwater to protect wetland ecosystems has no easy solutions under the pressures of increasing urbanization. Hydrologic changes occur affecting vegetation communities and, in turn, alter habitat conditions for wildlife. Controls focused on minimizing impervious surfaces and maximizing forest retention are the most effective strategies for minimizing stormwater impacts in the long term. However, additional mitigations that reduce stormwater volumes through engineered solutions will be needed to retain wetland biodiversity in urbanizing areas.

REFERENCES

1. Azous, A. L., An Analysis of Urbanization Effects on Wetland Biological Communities, M.S. thesis, Department of Civil Engineering, University of Washington, Seattle, 1991.
2. Cooke, S. S. and A. Azous, Effects of Urban Stormwater Runoff on Palustrine Wetland Vegetation, Final Rep. to U.S. Environmental Protection Agency, Region 10, and the Puget Sound Wetlands and Stormwater Research Program, King County Resource Planning Section, Seattle, March, 1993.
3. Richter, K. O. and A. L. Azous, Amphibian Occurrence and Wetland Characteristics in the Puget Sound Basin, *Wetlands*, 15(3), 305, 1995.
4. Platin, T. J. and K. O. Richter, Effects of Changing Wetland Hydrology and Water Quality on Amphibian Breeding Success, Proc. Third Puget Sound Res. Meet., Puget Sound Water Quality Authority, Olympia, WA, 1995.
5. Cooke, S. S. and A. Azous, Vegetation Species Responses to Changing Hydrology in Urban Wetlands, Proc. Third Puget Sound Res. Meet., Puget Sound Water Quality Authority, Olympia, WA, 1995.
6. Cooke, S. S., R. R. Horner, C. Conolly, O. Edwards, M. Wilkinson, and M. Emers, Effects of Urban Stormwater Runoff on Palustrine Wetland Vegetation Communities — Baseline Investigation (1988), Rep. to U.S. Environmental Protection Agency, Region 10, King County Resource Planning Section, Seattle, 1989.
7. Hydrocomp, Inc., Hydrologic Simulation Program — FORTRAN (HSPF), public domain software program distributed by the Center for Exposure Assessment Modeling (CEAM), U.S. Environmental Protection Agency, Washington, D.C., 1976.
8. Feldman, D. S., Jr., R. Hoffman, J. Gagnon, and J. Simpson, *StatView SE+,* Abacus Concepts Inc., Berkley, CA, 1988.
9. Taylor, B. L., The Influence of Wetland and Watershed Morphological Characteristics on Wetland Hydrology and Relationships to Wetland Vegetation Communities, M.S.C.E. thesis, Department of Civil Engineering, University of Washington, Seattle, 1993.
10. Taylor B. L., K. Ludwa, and R. R. Horner, Urbanization Effects on Wetland Hydrology and Water Quality, Proc. Third Puget Sound Res. Meet., Puget Sound Water Quality Authority, Olympia, WA, 1995.

11. Booth, D. B. and L. E. Reinelt, Consequences of Urbanization on Aquatic Systems — Measured Effects, Degradation Thresholds, and Corrective Strategies, Proc. Watersheds '93 Conf., Alexandria, VA, March 21–24, 1993.
12. Master Drainage Planning for Large Site Developments — Proposed Process and Requirement Guidelines, King County Public Works, Seattle, WA, 1993.

14 Wetlands and Stormwater Management Guidelines

Richard R. Horner, Amanda L. Azous,
Klaus O. Richter, Sarah S. Cooke,
Lorin E. Reinelt, and Kern Ewing

CONTENTS

Approach and Organization of the Management Guidelines 300
Introduction 300
Scope and Underlying Principles of the Guidelines 300
Support Material 302
Guide Sheet 1: Comprehensive Landscape Planning for Wetlands and Stormwater Management 302
Guide Sheet 1A: Comprehensive Planning Steps 303
Guide Sheet 1B: Stormwater Wetland Assessment Criteria 307
Guide Sheet 2: Wetland Protection Guidelines 309
Guide Sheet 2A: General Wetland Protection Guidelines 309
Guide Sheet 2B: Guidelines for Protection from Adverse Impacts of Modified Runoff Quantity Discharged to Wetlands 310
Guide Sheet 2C: Guidelines for Protection from Adverse Impacts of Modified Runoff Quality Discharged to Wetlands 313
Guide Sheet 2D: Guidelines for the Protection of Specific Biological Communities 315
Appendix A: Information Needed to Apply Guidelines 317
Appendix A: Guide Sheet 1 317
Appendix A: Guide Sheet 2 318
Appendix B: Definitions 318
Appendix C: Native and Recommended Noninvasive Plant Species for Wetlands in the Puget Sound Basin 321
Appendix D: Table Comparing Water Chemistry Characteristics in *Sphagnum* Bog and Fen vs. More Typical Wetlands 323
References 323

1-56670-386-7/00/$0.00+$.50

If you are unfamiliar with these guidelines, read the description of the approach and organization that follows. If you are familiar, proceed directly to the appropriate guide sheet(s) for guidelines covering your issue(s) or objective(s):

Guide Sheet 1: Comprehensive Landscape Planning for Wetlands and Stormwater Management

Guide Sheet 2: Wetlands Protection Guidelines

APPROACH AND ORGANIZATION OF THE MANAGEMENT GUIDELINES

INTRODUCTION

The Puget Sound Wetlands and Stormwater Management Research Program performed comprehensive research with the goal of deriving strategies that protect wetland resources in urban and urbanizing areas, while also benefiting the management of urban stormwater runoff that can affect those resources. The research primarily involved long-term comparisons of wetland ecosystem characteristics before and after their watersheds urbanized, and between a set of wetlands that became affected by urbanization (treatment sites) and a set whose watersheds did not change (control sites). Shorter-term and more intensive studies of pollutant transport and fate in wetlands, several laboratory experiments, and ongoing reviews of relevant work being performed elsewhere supplemented this endeavor. These research efforts were aimed at defining the types of impacts that urbanization can cause and the degree to which they develop under different conditions, in order to identify means of avoiding or minimizing impacts that impair wetland structure and functioning. The program's scope embraced both situations where urban drainage incidentally affects wetlands in its path, as well as those in which direct stormwater management actions change wetlands' hydrology, water quality, or both.

Presented here are preliminary management guidelines for urban wetlands and their stormwater discharges based on the research results. The set of guidelines is the principal vehicle to implement the research findings in environmental planning and management practice.

SCOPE AND UNDERLYING PRINCIPLES OF THE GUIDELINES

Note: For terms in **boldface** type see Item 1 under Support Materials in the next subsection.

1. These provisions currently have the status of guidelines rather than requirements. Application of these guidelines does not fulfill assessment and permitting requirements that may be associated with a project. It is, in general, necessary to follow the stipulations of the State Environmental Policy Act and to contact such agencies as the local planning agency; the state departments of ecology, fisheries, and wildlife; the U.S. Environmental Protection Agency; and the U.S. Army Corps of Engineers.

2. Using the guidelines should be approached from a problem-solving viewpoint. The "problem" is regarded to be accomplishing one or more particular planning or management objectives involving a **wetland** potentially or presently affected by stormwater drainage from an urban or urbanizing area. The objectives can be broad, specific, or both. Broad objectives involve comprehensive planning and subsequent management of a drainage catchment or other **landscape unit** containing one or more wetlands. Specific objectives pertain to managing a wetland having particular attributes to be sustained. Of course, the prospect for success is greater with the ability to manage the whole landscape influencing the wetland, rather than just the wetland itself.
3. The guidelines are framed from the standpoint that some change in the landscape has the potential to modify the physical and chemical **structure** of the wetland environment, which in turn could alter biological communities and the wetland's ecological **functions.** The general objective in this framework would be to avoid or minimize negative ecological change. This view is in contrast to one in which a wetland has at some time in the past experienced negative change, and consequent ecological degradation, and where the general objective would be to recover some or all of the lost structure and functioning through **enhancement** or **restoration** actions. Direct attention to this problem was outside the scope of the Puget Sound Wetlands and Stormwater Management Research Program. However, the guidelines do give information that applies to enhancement and restoration. For example, attempted restoration of a diverse amphibian community would not be successful if the water level fluctuation limits consistent with high amphibian species richness are not observed.
4. The guidelines can be applied with whatever information concerning the problem is available. Of course, the comprehensiveness and certainty of the outcome will vary with the amount and quality of information employed. The guidelines can be applied in an iterative fashion to improve management understanding as the information improves. Appendix A lists the information needed to perform basic analyses, followed by other information that can improve the understanding and analysis.
5. These guidelines emphasize avoiding structural, hydrologic, and water quality **modifications** of existing wetlands to the extent possible in the process of urbanization and the management of urban stormwater runoff.
6. In pursuit of this goal, the guidelines take a systematic approach to management problems that potentially involve both urban stormwater (quantity, quality, or both) and wetlands. The consideration of wetlands involves their area, **values,** and functions. This approach emphasizes a comprehensive analysis of alternatives to solve the identified problem. The guidelines encourage conducting the analysis on a landscape scale and considering all of the possible stormwater management alternatives, which may or may not involve a wetland. They favor **source control best management practices** (BMPs) and **pretreatment** of stormwater runoff prior to release to wetlands.

7. Finally, the guidelines take a holistic view of managing wetland resources in an urban setting. Thus, they recognize that urban wetlands have the potential to be affected structurally and functionally whether or not they are formally designated for stormwater management purposes. Even if an urban wetland is not structurally or hydrologically engineered for such purposes, it may experience altered hydrology (more or less water), reduced water quality, and a host of other impacts related to urban conditions. It is the objective of the guidelines to avoid or reduce the negative effects on wetland resources from both specific stormwater management actions and incidental urban impacts.

Support Material

1. The guidelines use certain terms that require definition to ensure that the intended meaning is conveyed to all users. Such terms are printed in **boldface** the first time that they appear in each guide sheet, and are defined in Appendix B.
2. The guideline provisions were drawn principally from the available results of the Puget Sound Wetlands and Stormwater Management Research Program, as set forth in Sections II and III of the program's summary publication, *Wetlands and Urbanization, Implications for the Future* (Horner et al. 1996).[1] Where the results in this publication are the basis for a numerical provision, a separate reference is not given. Numerical provisions based on other sources are referenced.
3. Appendix C presents a list of plant species native to wetlands in the Puget Sound Region. This appendix is intended for reference by guideline users who are not specialists in wetland botany. However, nonspecialists should obtain expert advice when making decisions involving vegetation.
4. Appendix D compares the water chemistry characteristics of *Sphagnum* bog and fen wetlands (termed **priority peat wetlands** in these guidelines) with more common wetland communities. These bogs and fens appear to be the most sensitive among the Puget Sound lowland wetlands to alteration of water chemistry, and require special water quality management to avoid losses of their relatively rare communities.

GUIDE SHEET 1: COMPREHENSIVE LANDSCAPE PLANNING FOR WETLANDS AND STORMWATER MANAGEMENT

Wetlands in newly developing areas will receive urban effects even if not specifically "used" in stormwater management. Therefore, the task is proper overall management of the resources and protection of their general **functioning,** including their role in storm drainage systems. Stormwater management in newly developing areas is distinguished from management in already-developed locations by the existence of many more feasible stormwater control options prior to development. The guidelines emphasize appropriate selection among the options to achieve optimum overall resource protection benefits, extending to downstream receiving waters and groundwater aquifers, as well as to wetlands.

The comprehensive planning guidelines are based on two principles that are recognized to create the most effective environmental management:

1. The best management policies for the protection of wetlands and other natural resources are those that prevent or minimize the development of impacts at potential sources; and
2. The best management strategies are self-perpetuating, that is they do not require periodic infusions of capital and labor.

To apply these principles in managing wetlands in a newly developing area, carry out the following steps.

Guide Sheet 1A: Comprehensive Planning Steps

1. Define the **landscape unit** subject to comprehensive planning. Refer to the definition of a landscape unit in Appendix B for assistance in defining it.
2. Begin the development of a plan for the landscape unit with attention to the following general principles:
   - Formulate the plan on the basis of clearly articulated community goals. Carefully identify conflicts and choices between retaining and protecting desired resources and community growth.
   - Map and assess land suitability for urban uses. Include the following landscape features in the assessment: forested land, open unforested land, steep slopes, erosion-prone soils, foundation suitability, soil suitability for waste disposal, aquifers, aquifer recharge areas, wetlands, floodplains, surface waters, agricultural lands, and various categories of urban land use. When appropriate, the assessment can highlight outstanding local or regional resources that the community determines should be protected (e.g., a fish run, scenic area, recreational area, threatened species habitat, farmland). Mapping and assessment should recognize not only these resources but also additional areas needed for their sustenance.
3. Maximize natural water storage and infiltration opportunities within the landscape unit and outside of existing wetlands, especially:
   - Promote the conservation of forest cover. Building on land that is already deforested affects basin hydrology to a lesser extent than converting forested land. Loss of forest cover reduces interception storage, detention in the organic forest floor layer, and water losses by evapotranspiration, resulting in large peak runoff increases and their negative effects or the expense of countering them with structural solutions.
   - Maintain natural storage reservoirs and drainage corridors, including depressions, areas of permeable soils, swales, and intermittent streams.

Develop and implement policies and regulations to discourage the clearing, filling, and channelization of these features. Utilize them in drainage networks in preference to pipes, culverts, and engineered ditches.

- In evaluating infiltration opportunities refer to the stormwater management manual for the jurisdiction and pay particular attention to the selection criteria for avoiding groundwater contamination and poor soils and hydrogeological conditions that cause these facilities to fail. If necessary, locate developments with large amounts of impervious surfaces or a potential to produce relatively contaminated runoff away from groundwater recharge areas. Relatively dense developments on glacial outwash soils may require additional runoff treatment to protect groundwater quality.

4. Establish and maintain **buffers** surrounding wetlands and in riparian zones as required by local regulations or recommended by the Puget Sound Water Quality Authority's wetland guidelines. Also, maintain interconnections among wetlands and other natural habitats to allow for wildlife movements.
5. Take specific management measures to avoid general urban impacts on wetlands and other water bodies (e.g., littering, vegetation destruction, human and pet intrusion harmful to wildlife).
6. To support management of runoff water quantity, perform a hydrologic analysis of the contributing drainage catchment to define the type and extent of flooding and stream channel erosion problems associated with existing development, redevelopment, or new development that require control to protect the beneficial uses of receiving waters, including wetlands. This analysis should include assembly of existing flow data and hydrologic modeling as necessary to establish conditions limiting to attainment of beneficial uses. Modeling should be performed as directed by the stormwater management manual in effect in the jurisdiction.
7. In wetlands previously relatively unaffected by human activities, manage stormwater quantity to attempt to match the **predevelopment hydroperiod** and **hydrodynamics.** In wetlands whose hydrology has been disturbed, consider ways of reducing hydrologic impacts. This provision involves not only management of high runoff volumes and rates of flow during the wet season, but also prevention of water supply depletion during the dry season. The latter guideline may require flow augmentation if urbanization reduces existing surface or groundwater inflows. Refer to *Guide Sheet 2, Wetland Protection Guidelines,* for detail on implementing these guidelines.
8. Assess alternatives for the control of runoff water quantities as follows:
   a. Define the runoff quantity problem subject to management by analyzing the proposed land development action.
   b. For existing development or **redevelopment,** assess possible alternative solutions that are applicable at the site of the problem occurrence, including:
      - Protect health, safety, and property from flooding by removing habitation from the flood plain.

- Prevent stream channel erosion by stabilizing the eroding bed and/or bank area with **bioengineering** techniques, preferably, or by structurally reinforcing it, if this solution would be consistent with the protection of aquatic habitats and beneficial uses of the stream (refer to local administrative code for the definition of beneficial uses).

b. For new development or redevelopment, assess possible regulatory and incentive land use control alternatives, such as density controls, clearing limits, impervious surface limits, transfer of development rights, purchase of conservation areas, etc.

c. If the alternatives considered in Steps 8a or 8b cannot solve an existing or potential problem, perform an analysis of the contributing drainage catchment to assess possible alternative solutions that can be applied **on-site** or on a **regional** scale. The most appropriate solution or combination of alternatives should be selected with regard to the specific opportunities and constraints existing in the drainage catchment. For new development or redevelopment, on-site facilities that should be assessed include, in approximate order of preference:
   - Infiltration basins or trenches;
   - Retention/detention ponds;
   - Below-ground vault or tank storage;
   - Parking lot detention.

   Regional facilities that should be assessed for solving problems associated with new development, redevelopment, or existing development include:
   - Infiltration basins or trenches;
   - Detention ponds;
   - **Constructed wetlands;**
   - Bypassing a portion of the flow to an acceptable receiving water body, with treatment as required to protect water quality and other special precautions as necessary to prevent downstream impacts.

d. Consider structurally or hydrologically engineering an existing wetland for water quantity control only if upland alternatives are inadequate to solve the existing or potential problem. To evaluate the possibility, refer to the *Stormwater Wetland Assessment Criteria* in *Guide Sheet 1B*.

9. Place strong emphasis on water resource protection during construction of new development. Establish effective erosion control programs to reduce the sediment loadings to receiving waters to the maximum extent possible. No preexisting wetland or other water body should ever be used for the sedimentation of solids in construction-phase runoff.
10. In wetlands previously relatively unaffected by human activities, manage stormwater quality to attempt to match predevelopment water quality conditions. To support management of runoff water quality, perform an analysis of the contributing drainage catchment to define the type and extent of runoff water quality problems associated with existing development, redevelopment, or new development that require control to protect

the beneficial uses of receiving waters, including wetlands. This analysis should incorporate the hydrologic assessment performed under Step 6 and include identification of key water pollutants, which may include solids, oxygen-demanding substances, nutrients, metals, oils, trace organics, and bacteria, and evaluation of the potential effects of water pollutants throughout the drainage system.

11. Assess alternatives for the control of runoff water quality as follows:
    a. Perform an analysis of the contributing drainage catchment to assess possible alternative solutions that can be applied on-site or on a regional scale. The most appropriate solution or combination of alternatives should be selected with regard to the specific opportunities and constraints existing in the drainage catchment. Consider both **source control BMPs** and **treatment BMPs** as alternative solutions before considering use of existing wetlands for quality improvement according to the following considerations:
        - Implementation of source control BMPs prevent the generation or release of water pollutants at potential sources. These alternatives are generally both more effective and less expensive than treatment controls. They should be applied to the maximum extent possible to new development, redevelopment, and existing development.
        - Treatment BMPs capture water pollutants after their release. This alternative often has limited application in existing developments because of space limitations, although it can be employed in new development and when redevelopment occurs in already developed areas. Following is a list of treatment BMPs that should be considered. Each has appropriate and inappropriate applications and advantages and disadvantages and must be carefully selected, designed, constructed, and operated according to the specifications of the stormwater management manual in use in the jurisdiction.
          — Infiltration basins or trenches;
          — Constructed wetlands;
          — Wet or extended-detention ponds;
          — Biofiltration facilities (vegetated swales or filter strips);
          — Filters with sand, compost, or other media;
          — Water quality vaults;
          — Oil/water separators.
    b. Consider structurally or hydrologically engineering an existing wetland for water quality control only if upland alternatives are inadequate to solve the existing or potential problem. Use of Waters of the State and Waters of the United States, including wetlands, for the treatment or conveyance of wastewater, including stormwater, is prohibited under state and federal law. Discussions with federal and state regulators during the research program led to development of a statement concerning the use of existing wetlands for improving stormwater quality (**polishing**), as follows. Such use is subject to analysis on a case-by-case basis and may be allowed only if the following conditions are met:

- If **restoration** or **enhancement** of a previously **degraded** wetland is required, and if the upgrading of other wetland functions can be accomplished along with benefiting runoff quality control, and
- If appropriate source control and treatment BMPs are applied in the contributing catchment on the basis of the analysis in Step 10a and any legally adopted water quality standards for wetlands are observed.

If these circumstances apply, refer to the *Stormwater Wetland Assessment Criteria* in *Guide Sheet 1B* to evaluate further.

12. Stimulate public awareness of and interest in wetlands and other water resources in order to establish protective attitudes in the community. This program should include:
    - Education regarding the use of fertilizers and pesticides, automobile maintenance, the care of animals to prevent water pollution, and the importance of retaining buffers.
    - Descriptive signboards adjacent to wetlands informing residents of the wetland type, its functions, the protective measures being taken, etc.
    - If beavers are present in a wetland, educate residents about their ecological role and value and take steps to avoid human interference with beavers.

Guide Sheet 1B:
Stormwater Wetland Assessment Criteria

This guide sheet gives criteria that disqualify a natural wetland from being structurally or hydrologically engineered for control of stormwater quantity, quality, or both. These criteria should be applied only after performing the alternatives analysis outlined in *Guide Sheet 1A*.

1. A wetland should not be structurally or hydrologically engineered for runoff quantity or quality control and should be given maximum protection from overall urban impacts (see *Guide Sheet 2, Wetland Protection Guidelines*) under any of the following circumstances:
    - In its present state it is primarily an **estuarine** or **forested wetland** or a **priority peat system.**
    - It is a rare or irreplaceable wetland type, as identified by the Washington Natural Heritage Program, the Puget Sound Water Quality Preservation Program, or local government.
    - It provides **rare, threatened, or endangered species** habitat that could be impaired by the proposed action. Determining whether or not the conserved species will be affected by the proposed project requires a careful analysis of its requirements in relation to the anticipated habitat changes.

    In general, the wetlands in these groups are classified in Categories I and II in the Puget Sound Water Quality Authority's draft wetland guidelines.

2. A wetland can be considered for structural or hydrological modification for runoff quantity or quality control if most of the following circumstances exist:
   - It is classified as a low value wetland which has monotypic vegetation of similar age and class, lacks special habitat features, and is isolated from other aquatic systems.
   - The wetland has been previously **disturbed** by human activity, as evidenced by agriculture, fill, ditching, and/or introduced or **invasive weedy plant species.**
   - The wetland has been deprived of a significant amount of its water supply by draining or previous urbanization (e.g., by loss of groundwater supply), and stormwater runoff is sufficient to augment the water supply. A particular candidate is a wetland that has experienced an increased summer dry period, especially if the drought has been extended by more than two weeks.
   - Construction for structural or hydrologic modification in order to provide runoff quantity or quality control will disturb relatively little of the wetland.
   - The wetland can provide the required storage capacity for quantity or quality control through an outlet orifice modification to increase storage of water, rather than through raising the existing overflow. Orifice modification is likely to require less construction activity and consequent negative impacts.
   - Under existing conditions the wetlands experience a relatively high degree of water level fluctuation and a range of velocities (i.e., a wetland associated with substantially flowing water, rather than one in the headwaters or entirely isolated from flowing water).
   - The wetland does not exhibit any of the following features:

     — Significant priority peat system or forested zones that will experience substantially altered hydroperiod as a result of the proposed action;

     — Regionally **unusual biological community types;**

     — Animal habitat features of relatively high value in the region (e.g., a protected, undisturbed area connected through undisturbed corridors to other valuable habitats, an important breeding site for protected species);

     — The presence of protected commercial or sport fish;

     — Configuration and topography that will require significant modification that may threaten fish stranding;

     — A relatively high degree of public interest as a result of, for example, offering valued local open space or educational, scientific, or recreational opportunities, unless the proposed action would enhance these opportunities;

- The wetland is threatened by potential impacts exclusive of stormwater management, and could receive greater protection if acquired for a stormwater management project rather than left in existing ownership.
- There is good evidence that the wetland actually can be restored or enhanced to perform other functions in addition to runoff quantity or quality control.
- There is good evidence that the wetland lends itself to the effective application of the *Wetland Protection Guidelines* in *Guide Sheet 2.*
- The wetland lies in the natural routing of the runoff. Local regulations often prohibit drainage diversion from one basin to another.
- The wetland allows runoff discharge at the natural location.

GUIDE SHEET 2: WETLAND PROTECTION GUIDELINES

This guide sheet provides information about likely changes to the ecological **structure** and **functioning** of **wetlands** that are incidentally subject to the effects of an urban or urbanizing watershed or are **modified** to supply runoff water quantity or quality control benefits. The guide sheet also recommends management actions that can avoid or minimize deleterious changes in these wetlands.

GUIDE SHEET 2A: GENERAL WETLAND PROTECTION GUIDELINES

1. Consult regulations issued under federal and state laws that govern the discharge of pollutants. Wetlands are classified as "Waters of the United States" and "Waters of the State" in Washington.
2. Maintain the wetland **buffer** required by local regulations or recommended by the Puget Sound Water Quality Authority's draft wetland guidelines.
3. Retain areas of native vegetation connecting the wetland and its buffer with nearby wetlands and other contiguous areas of native vegetation.
4. Avoid compaction of soil and introduction of exotic plant species during any work in a wetland.
5. Take specific site design and maintenance measures to avoid general urban impacts (e.g., littering and vegetation destruction). Examples are protecting existing buffer zones; discouraging access, especially by vehicles, by plantings outside the wetland; and encouragement of stewardship by homeowners' associations. Fences can be useful to restrict dogs and pedestrian access, but they also interfere with wildlife movements. Their use should be very carefully evaluated on the basis of the relative importance of intrusive impacts vs. wildlife presence. Fences should generally not be installed when wildlife would be restricted and intrusion is relatively minor. They generally should be used when wildlife passage is not

a major issue and the potential for intrusive impacts is high. When wildlife movements and intrusion are both issues, the circumstances will have to be weighed to make a decision about fencing.

6. If the wetland inlet will be modified for the stormwater management project, use a diffuse flow method, such as a spreader swale, to discharge water into the wetland in order to prevent flow channelization.

Guide Sheet 2B: Guidelines for Protection from Adverse Impacts of Modified Runoff Quantity Discharged to Wetlands

1. Protection of wetland plant and animal communities depends on controlling the wetland's **hydroperiod,** meaning the pattern of fluctuation of water depth and the frequency and duration of exceeding certain levels, including the length and onset of drying in the summer. A hydrologic assessment is useful to measure or estimate elements of the hydroperiod under existing **predevelopment** and anticipated **postdevelopment** conditions. This assessment should be performed with the aid of a qualified hydrologist. Postdevelopment estimates of watershed hydrology and wetland hydroperiod must include the cumulative effect of all anticipated watershed and wetland modifications. Provisions in these guidelines pertain to an anticipated full build-out of the wetland's watershed. This analysis hypothesizes a fluctuating water stage over time before development that could fluctuate more, both higher and lower after development; these greater fluctuations are termed **stage excursions.** The guidelines set limits on the frequency and duration of excursions, as well as on overall water level fluctuation, after development.

   To determine existing hydroperiod use one of the following methods, listed in order of preference:

   - Estimation by a continuous simulation computer model. The model should be calibrated with at least one year of data taken using a continuously recording level gauge under existing conditions and should be run for the historical rainfall period. The resulting data can be used to express the magnitudes of depth fluctuation, as well as the frequencies and durations of surpassing given depths. [*Note:* Modeling that yields high quality information of the type needed for wetland hydroperiod analysis is a complex subject. Providing guidance on selecting and applying modeling options is beyond the scope of these guidelines, but they are being developed by King County Surface Water Management Division and other local jurisdictions. An alternative possibility to modeling depths, frequencies, and durations within the wetland is to model durations above given discharge levels entering the wetland over various time periods (e.g., seasonal, monthly, weekly). This option requires further development.]
   - Measurement during a series of time intervals (no longer than one month in length) over a period of at least one year of the maximum

water stage using a crest stage gauge, and instantaneous water stage using a staff gauge. The resulting data can be used to express water level fluctuation (WLF) during the interval as follows:

WLF = Crest stage – average base stage;
Average base stage = (Instantaneous stage at beginning of interval + instantaneous stage at end of interval)/2

Compute mean annual and mean monthly WLF as the arithmetic averages for each year and month for which data are available.

To forecast future hydroperiod use one of the following methods, listed in order of preference:

- Estimation by the continuous simulation computer model calibrated during predevelopment analysis and run for the historical rainfall period. The resulting data can be used to express the magnitudes of depth fluctuation, as well as the frequencies and durations of events surpassing given depths. [*Note:* Postdevelopment modeling results should generally be compared with predevelopment modeling results, rather than directly with field measurements, because different sets of assumptions underlie modeling and monitoring. Making pre- and post-development comparisons on the basis of common assumptions allows cancellation of errors inherent in the assumptions.]
- Estimation according to general relationships developed from the Puget Sound Wetlands and Stormwater Management Program Research Program, as follows (in part adapted from Chin, 1996)[2]:

  — Mean annual WLF is very likely (100% of cases measured) to be <20 cm (8 in. or 0.7 ft) if total impervious area (TIA) cover in the watershed is <6% (roughly corresponding to no more than 15% of the watershed converted to urban land use).

  — Mean annual WLF is very likely (89% of cases measured) to be >20 cm if TIA in the watershed is >21% (roughly corresponding to more than 30% of the watershed converted to urban land use).

  — Mean annual WLF is somewhat likely (50% of cases measured) to be >30 cm (1.0 ft) if TIA in the watershed is >21% (roughly corresponding to more than 30% of the watershed converted to urban land use).

  — Mean annual WLF is likely (75% of cases measured) to be >30 cm, and somewhat likely (50% of cases measured) to be 50 cm (20 in. or 1.6 ft) or higher, if TIA in the watershed is >40% (roughly corresponding to more than 70% of the watershed converted to urban land use).

  — The frequency of stage excursions greater than 15 cm (6 in. or 0.5 ft) above or below predevelopment levels is somewhat likely (54% of cases measured) to be more than six per year if the mean annual WLF increases to >24 cm (9.5 in. or 0.8 ft).

— The average duration of stage excursions greater than 15 cm above or below predevelopment levels is likely (69% of cases measured) to be more than 72 hr if the mean annual WLF increases to >20 cm.

2. The following hydroperiod limits characterize wetlands with relatively high vegetation species richness and apply to all zones within all wetlands over the entire year. If these limits are exceeded, then species richness is likely to decline. If the analysis described above forecasts exceedences, one or more of the management strategies listed in Step 5 should be employed to attempt to stay within the limits.
   - Mean annual WLF (and mean monthly WLF for every month of the year) does not exceed 20 cm. Vegetation species richness decrease is likely with: (1) a mean annual (and mean monthly) WLF increase of more than 5 cm (2 in. or 0.16 ft) if predevelopment mean annual (and mean monthly) WLF is greater than 15 cm, or (2) a mean annual (and mean monthly) WLF increase to 20 cm or more if predevelopment mean annual (and mean monthly) WLF is 15 cm or less.
   - The frequency of stage excursions of 15 cm above or below predevelopment stage does not exceed an annual average of six.
   - The duration of stage excursions of 15 cm above or below predevelopment stage does not exceed 72 hr per excursion.
   - The total dry period (when pools dry down to the soil surface everywhere in the wetland) does not increase or decrease by more than two weeks in any year.
   - Alterations to watershed and wetland hydrology that may cause perennial wetlands to become **vernal** are avoided.
3. The following hydroperiod limit characterizes **priority peat wetlands** (bogs and fens, as more specifically defined by the Washington Department of Ecology) and applies to all zones over the entire year. If this limit is exceeded, then characteristic bog or fen wetland vegetation is likely to decline. If the analysis described above forecasts exceedence, one or more of the management strategies listed in Step 5 should be employed to attempt to stay within the limit.
   - The duration of stage excursions above the pre-development stage does not exceed 24 hours in any year.
   - *Note:* To apply this guideline a continuous simulation computer model needs to be employed. The model should be calibrated with data taken under existing conditions at the wetland being analyzed and then used to forecast post-development duration of excursions.
4. The following hydroperiod limits characterize wetlands inhabited by breeding native amphibians and apply to breeding zones during the period 1 February through 31 May. If these limits are exceeded, then amphibian breeding success is likely to decline. If the analysis described above forecasts exceedences, one or more of the management strategies listed in Step 5 should be employed to attempt to stay within the limits.

- The magnitude of stage excursions above or below the pre-development stage does not exceed 8 cm, and the total duration of these excursions does not exceed 24 hours in any 30 day period.
- *Note:* To apply this guideline a continuous simulation computer model needs to be employed. The model should be calibrated with data taken under existing conditions at the wetland being analyzed and then used to forecast post-development magnitude and duration of excursions.

5. If it is expected that the hydroperiod limits stated above could be exceeded, consider other strategies such as:
   - Reduction of the level of development;
   - Increasing runoff infiltration [*Note:* Infiltration is prone to failure in many Puget Sound Basin locations with glacial till soils and generally requires **pretreatment** to avoid clogging. In other situations, infiltrating urban runoff may contaminate groundwater. Consult the stormwater management manual adopted by the jurisdiction and carefully analyze infiltration according to its prescriptions];
   - Increasing runoff storage capacity; and
   - Selective runoff bypass.
6. After development, monitor hydroperiod with a continuously recording level gauge or staff and crest stage gauges. If the applicable limits are exceeded, consider additional applications of the strategies in step 5 that may still be available. It is also recommended that goals be established to maintain key vegetation species, amphibians, or both, and that these species be monitored to determine if the goals are being met.

Guide Sheet 2C: Guidelines for Protection from Adverse Impacts of Modified Runoff Quality Discharged to Wetlands

1. Require effective erosion control at any construction sites in the wetland's drainage catchment.
2. Institute a program of **source control BMPs** to minimize the generation of pollutants that will enter storm runoff that drains to the wetland.
3. Provide a water quality control facility consisting of one or more **treatment BMPs** to treat all urban runoff entering the wetland and designed according to the following criteria:
   - The facility should be designed to remove at least 80% of the total suspended solids in the runoff.
   - If the catchment could generate a relatively large amount of oil (e.g., certain industrial sites, bases handling large vehicles, areas where oil may be spilled or improperly disposed of), the facility should include an appropriate oil control device.
   - If the wetland is a **priority peat wetland** (also called a bog or fen), the facility should include a BMP with the most advanced ability to

control nutrients (e.g., an infiltration device, a wet pond, or constructed wetland with residence time in the pooled storage of at least two weeks). [*Note:* Infiltration is prone to failure in many Puget Sound Basin locations with glacial till soils and generally requires **pretreatment** to avoid clogging. In other situations infiltrating urban runoff may contaminate groundwater. Consult the stormwater management manual adopted by the jurisdiction and carefully analyze infiltration according to its prescriptions.] Refer to Appendix D for a comparison of water chemistry conditions in priority peat vs. more typical wetlands.

Refer to the stormwater management manual to select and design the facility. Generally, the facility should be located outside and upstream of the wetland and its buffer.

4. Design and perform a water quality monitoring program for priority peat wetlands and for other wetlands subject to relatively high water pollutant loadings. The research results (Horner, 1989)[3] identified such wetlands as having contributing catchments exhibiting either of the following characteristics.
   - More than 20% of the catchment area is committed to commercial, industrial, and/or multiple-family residential land uses; or
   - The combination of all urban land uses (including single-family residential) exceeds 30% of the catchment area.

   A recommended monitoring program, consistent with monitoring during the research program, is:
   - Perform predevelopment **baseline sampling** by collecting water quality grab samples in an open water pool of the wetland for at least one year, allocated through the year as follows: November 1–March 31: four samples, April 1–May 31: one sample, June 1–August 31: two samples, and September 1–October 31: one sample (if the wetland is dry during any period, reallocate the sample(s) scheduled then to another time). Analyze samples for pH; dissolved oxygen (DO); conductivity (Cond); total suspended solids (TSS); total phosphorus (TP); nitrate + nitrite-nitrogen (N); fecal coliforms (FC); and total copper (Cu), lead (Pb), and zinc (Zn). Find the median and range of each water quality variable.
   - Considering the baseline results, set water quality goals to be maintained in the postdevelopment period. Example goals are (1) pH — no more than "x" percent (e.g., 10%) increase (relative to baseline) in annual median and maximum or decrease in annual minimum; (2) DO — no more than "x" percent decrease in annual median and minimum concentrations; (3) other variables — no more than "x" percent increase in annual median and maximum concentrations; (4) no increase in violations of state and local water quality criteria.
   - Repeat the sampling on the same schedule for at least one year after all development is complete. Compare the results to the set goals.

If the water quality goals are not met, consider additional applications of the source and treatment controls described in Steps 2 and 3. Continue monitoring until the goals are met at least two years in succession.

Note: Wetland water quality was found to be highly variable during the research, a fact that should be reflected in goals. Using the maximum (or minimum), as well as a measure of central tendency like the median, and allowing some change from predevelopment levels are ways of incorporating an allowance for variability. Table 14-1 presents data from the wetlands studied during the research program to give an approximate idea of magnitudes and degree of variability to be expected. Non-urbanized watersheds (N) are those that have both <15% urbanization and <6% impervious cover. Highly urbanized watersheds (H) are those that have both lost all forest cover and have >20% impervious cover. Moderately urbanized watersheds (M) are those that fit neither the N nor H category.

Guide Sheet 2D: Guidelines for the Protection of Specific Biological Communities

1. For wetlands inhabited by breeding native amphibians:
   - Refer to Step 4 of *Guide Sheet 2B* for hydroperiod limit.
   - Avoid decreasing the sizes of the open water and aquatic bed zones.
   - Avoid increasing the channelization of flow. Do not form channels where none exist, and take care that inflows to the wetland do not become more concentrated and do not enter at higher velocities than accustomed. If necessary, concentrated flows can be uniformly distributed with a flow-spreading device such as a shallow weir, stilling basin, or perforated pipe. Velocity dissipation can be accomplished with a stilling basin or riprap pad.
   - Limit the postdevelopment flow velocity to <5 cm/s (0.16 ft/s) in any location that had a velocity in the range 0–5 cm/s in the predevelopment condition.
   - Avoid increasing the gradient of wetland side slopes.
2. For wetlands inhabited by forest bird species:
   - Retain areas of coniferous forest in and around the wetland as habitat for forest species.
   - Retain shrub or woody debris as nesting sites for ground-nesting birds and downed logs and stumps for winter wren habitat.
   - Retain snags as habitat for cavity-nesting species, such as woodpeckers.
   - Retain shrubs in and around the wetland for protective cover. If cover is insufficient to protect against domestic pet predation, consider planting native bushes such as rose species in the buffer.
3. For wetlands inhabited by **wetland obligate** bird species:
   - Retain **forested zones,** sedge, and rush meadows, and deep open water zones, both without vegetation and with submerged and floating plants.

TABLE 14-1
Water Quality Ranges Found in Study Wetlands

	N			M			H		
Metric	Median	Mean	Std.Dev./n[a]	Median	Mean	Std.Dev./n[a]	Median	Mean	Dev./n[a]
pH[b]	6.4	6.4	0.5/162	6.7	6.5	0.8/132	6.9	6.7	0.6/52
DO (mg/l)	5.9	5.7	2.6/205	5.1	5.53.6/173	6.3	5.4	2.9/67	
Cond. (μS/cm)	46	73	64/190	160	142	73/161	132	151	86/61
TSS (μg/l)	2.0	4.6	8.5/204	2.8	9.2	22/175	4.0	9.2	15/66
TP (μg/l)	29	52	87/206	70	93	92/177	69	110	234/67
N (μg/l)	112	368	485/206	304	598	847/177	376	395	239/67
FC (no./100 ml)	9.0	271	1000/206	46	2665	27,342/173	61	969	4753/66
Cu (μg/l)	<5.0	<3.3	>2.7/93	<5.0	<3.7	>1.9/78	<5.0	<4.1	<2.5/29
Pb (μg/l)	1.0	<2.7	>2.8/136	3.0	<3.4	>2.7/122	5.0	<4.5	>4.0/44
Zn (μg/l)	5.0	8.4	8.3/136	8.0	9.8	7.2/122	20	20	17/44

[a] Std. Dev. = standard deviation; n = number of observations.

[b] Values do not apply to priority peat wetlands. The program did not specifically study these wetlands but measured pH in three wetlands with "bog-like" characteristics. The minimum value measured in these wetlands was 4.5, and the lowest median was 4.8; but pH can be approximately 1 unit lower in wetlands of this type. Refer to Appendix D for a comparison of water chemistry conditions in priority peat vs. more typical wetlands.

- Retain shrubs in and around the wetland for protective cover. If cover is insufficient to protect against domestic pet predation, consider planting native bushes such as rose species in the buffer.
- Avoid introducing **invasive weedy plant species,** such as purple loosestrife and reed canarygrass.
- Retain the buffer zone. If it has lost width or forest cover, consider reestablishing forested buffer area at least 30 m (100 ft) wide.
- If human entry is desired, establish paths that permit people to observe the wetland with minimum disturbance to the birds.

4. For wetlands inhabited by fish:
   - Protect fish habitats by avoiding water velocities above tolerated levels (selected with the aid of a qualified fishery biologist to protect fish in each life stage when they are present), siltation of spawning beds, etc. Habitat requirements vary substantially among fish species. If the wetland is associated with a larger water body, contact the Department of Fisheries and Wildlife to determine the species of concern and the acceptable ranges of habitat variables.
   - If stranding of protected commercial or sport fish could result from a structural or hydrologic modification for runoff quantity or quality control, develop a strategy to avoid stranding that minimizes disturbance in the wetland (e.g., by making provisions for fish to return to the stream as the wetland drains, or avoiding use of the facility for quantity or quality control during fish presence).

APPENDIX A: INFORMATION NEEDED TO APPLY GUIDELINES

The following information listed for each guide sheet is most essential for applying the Wetlands and Stormwater Management Guidelines. As a start, obtain the relevant soil survey, the National Wetland Inventory, topographic and land use maps, and the results of any local wetland inventory.

GUIDE SHEET 1

1. Boundary and area of the contributing watershed of the wetland or other landscape unit
2. A complete definition of goals for the wetland and landscape unit subject to planning and management
3. Existing management and monitoring plans
4. Existing and projected land use in the landscape unit in the categories commercial, industrial, multifamily residential, single-family residential, agricultural, various categories of undeveloped, and areas subject to active logging or construction (expressed as percentages of the total watershed area)
5. Drainage network throughout the landscape unit

6. Soil conditions, including soil types, infiltration rates, and positions of seasonal water table (seasonally) and restrictive layers
7. Groundwater recharge and discharge points
8. Wetland category designation if the needed information is not available, a biological assessment will be necessary.
9. Watershed hydrologic assessment
10. Watershed water quality assessment
11. Wetland type and zones present, with special note of estuarine, priority peat, forested, sensitive scrub-shrub zone, sensitive emergent zone, and other sensitive or critical habitats designated by state or local government (with dominant plant species)
12. Rare, threatened, or endangered species inhabiting the wetland
13. History of wetland changes
14. Relationship of wetland to other water bodies in the landscape unit and the drainage network
15. Flow pattern through the wetland
16. Fish and wildlife inhabiting the wetland
17. Relationship of wetland to other wildlife habitats in the landscape unit and the corridors between them

Guide Sheet 2

1. Existing and potential stormwater pollution sources
2. Existing and projected landscape unit land use (see Number 4 under *Guide Sheet 1*)
3. Existing and projected wetland hydroperiod characteristics
4. Wetland bathymetry
5. Inlet and outlet locations and hydraulics
6. Landscape unit soils, geologic and hydrogeologic conditions
7. Wetland type and zones present (see Number 11 under *Guide Sheet 1*)
8. Presence of breeding populations of native amphibian species
9. Presence of forest and wetland obligate bird species
10. Presence of fish species

APPENDIX B: DEFINITIONS

Baseline sampling: Sampling performed to define an existing state before any modification occurs that could change the state.

Bioengineering: Restoration or reinforcement of slopes and stream banks with living plant materials.

Buffer: The area that surrounds a wetland and that reduces adverse impacts to it from adjacent development.

Constructed wetland: A wetland intentionally created from a non-wetland site for the sole purpose of wastewater or stormwater treatment. These wetlands are not normally considered Waters of the United States or Waters of the State.

Degraded (disturbed) wetland (community): A wetland (community) in which the vegetation, soils, and/or hydrology have been adversely altered, resulting in lost or reduced functions and values; generally, implies topographic isolation; hydrologic alterations such as hydroperiod alteration (increased or decreased quantity of water), diking, channelization, and/or outlet modification; soils alterations such as presence of fill, soil removal, and/or compaction; accumulation of toxicants in the biotic or abiotic components of the wetland; and/or low plant species richness with dominance by invasive weedy species.

Enhancement: Actions performed to improve the condition of an existing degraded wetland, so that functions it provides are of a higher quality.

Estuarine wetland: Generally, an eelgrass bed; salt marsh; or rocky, sand flat, or mudflat intertidal area where fresh and salt water mix. (Specifically, a tidal wetland with salinity greater than 0.5 parts per thousand, usually semienclosed by land but with partly obstructed or sporadic access to the open ocean).

Forested communities (wetlands): In general terms, communities (wetlands) characterized by woody vegetation that is greater than or equal to 6 m in height; in these guidelines the term applies to such communities (wetlands) that represent a significant amount of tree cover consisting of species that offer wildlife habitat and other values and advance the performance of wetland functions overall.

Functions: The ecological (physical, chemical, and biological) processes or attributes of a wetland without regard for their importance to society (see also **Values**). Wetland functions include food chain support, provision of ecosystem diversity and fish and wildlife habitat, flood flow alteration, groundwater recharge and discharge, water quality improvement, and soil stabilization.

Hydrodynamics: The science involving the energy and forces acting on water and its resulting motion.

Hydroperiod: The seasonal occurrence of flooding and/or soil saturation; encompasses the depth, frequency, duration, and seasonal pattern of inundation.

Invasive weedy plant species: Opportunistic species of inferior biological value that tend to out-compete more desirable forms and become dominant; applied to nonnative species in these guidelines.

Landscape unit: An area of land that has a specified boundary and is the locus of interrelated physical, chemical, and biological processes.

Modification, Modified (wetland): A wetland whose physical, hydrological, or water quality characteristics have been purposefully altered for a management purpose, such as by dredging, filling, forebay construction, and inlet or outlet control.

On-site: An action (here, for stormwater management purposes) taken within the property boundaries of the site to which the action applies.

Polishing: Advanced treatment of a waste stream that has already received one or more stages of treatment by other means.

Predevelopment, postdevelopment: Respectively, the situation before and after a specific stormwater management project (e.g., raising the outlet, building an outlet control structure) will be placed in the wetland, or a land use change occurs in the landscape unit that will potentially affect the wetland.

Pretreatment: An action taken to remove pollutants from runoff before it is discharged into another system for additional treatment.

Priority peat systems: Unique, irreplaceable fens that can exhibit water pH in a wide range from highly acidic to alkaline, including fens typified by *Sphagnum* species, *Rhododendron groenlandicum* (Labrador tea), *Drosera rotundifolia* (sundew), and *Vaccinium oxycoccos* (bog cranberry); marl fens; estuarine peat deposits; and other moss peat systems with relatively diverse, undisturbed flora and fauna. Bog is the common name for peat systems having the *Sphagnum* association described, but this term applies strictly only to systems that receive water income from precipitation exclusively.

Rare, threatened, or endangered species: Plant or animal species that are regional relatively uncommon, are nearing endangered status, or whose existence is in immediate jeopardy and is usually restricted to highly specific habitats. Federal and state authorities officially list threatened and endangered species, whereas rare species are unofficial species of concern that fit the above definitions.

Redevelopment: Conversion of an existing development to another land use, or addition of a material improvement to an existing development.

Regional: An action (here, for stormwater management purposes) that involves more than one discrete property.

Restoration: Actions performed to reestablish wetland functional characteristics and processes that have been lost by alterations, activities, or catastrophic events in an area that no longer meets the definition of a wetland.

Source control best management practices (BMPs): Actions that are taken to prevent the development of a problem (e.g., increase in runoff quantity, release of pollutants) at the point of origin.

Stage excursion: A postdevelopment departure, either higher or lower, from the water depth existing under a given set of conditions in the predevelopment state.

Structure: The components of an ecosystem, both the abiotic (physical and chemical) and biotic (living).

Treatment best management practices (BMPs): Actions that remove pollutants from runoff through one or more physical, chemical, and biological mechanisms.

Unusual biological community types: Assemblages of interacting organisms that are relatively uncommon regionally.

Values: Wetland processes or attributes that are valuable or beneficial to society (also see **Functions**). Wetland values include support of commercial and sport fish and wildlife species, protection of life and property from flooding, recreation, education, and aesthetic enhancement of human communities.

Vernal wetland: A wetland that has water above the soil surface for a period of time during and/or after the wettest season but always dries to or below the soil surface in warmer, drier weather.

Wetland obligate: A biological organism that absolutely requires a wetland habitat for at least some stage of its life cycle.

Wetlands: Lands transitional between terrestrial and aquatic systems that have a water table usually at or near the surface or a shallow covering of water, hydric soils, and a prevalence of hydrophytic vegetation.

APPENDIX C: NATIVE AND RECOMMENDED NONINVASIVE PLANT SPECIES FOR WETLANDS IN THE CENTRAL PUGET SOUND BASIN

Caution: Extracting plants from an existing wetland donor site can cause a significant negative effect on that site. It is recommended that plants be obtained from native plant nursery stocks whenever possible. Collections from existing wetlands should be limited in scale and undertaken with care to avoid disturbing the wetland outside of the actual point of collection. Plant selection is a complex task, involving matching plant requirements with environmental conditions. A qualified wetlands botanist should perform it. Refer to *Restoring Wetlands in Washington* by the Washington Department of Ecology for more information.

Plants Preferred in Puget Sound Basin Freshwater Wetlands

Open Water Zone:

Potamogeton species (pondweeds)
Nymphaea odorata (pond lily)
Nuphar luteum (yellow pond lily)
Polygonum hydropiper (smartweed)
Alisma plantago-aquatica (broadleaf water plantain)
Ludwigia palustris (water purslane)
Menyanthes trifoliata (bogbean)
Utricularia minor, U. vulgaris (bladderwort)

Emergent Zone:

Carex obnupta, C. utriculata, C. arcta, C. stipata, C. vesicaria C. aquatilis, C. comosa, C. lenticularis (sedge)
Scirpus atricinctus (woolly bulrush)
Scirpus microcarpus (small-fruited bulrush)
Eleocharis palustris, E. ovata (spike rush)
Epilobium watsonii (Watson's willow herb)
Typha latifolia (common cattail) (*Note:* This native plant can be aggressive but has been found to offer certain wildlife habitat and water quality improvement benefits; use with care.)
Veronica americana, V. scutellata (American brookline, marsh speedwell)
Mentha arvensis (field mint)
Lycopus americanus, L. uniflora (bugleweed or horehound)
Angelica sp. (angelica)
Oenanthe sarmentosa (water parsley)
Heracleum lanatum (cow parsnip)
Glyceria grandis, G. elata (manna grass)
Juncus acuminatus (tapertip rush)
Juncus ensifolius (daggerleaf rush)
Juncus bufonius (toad rush)
Mimulus guttatus (common monkey flower)

Scrub-Shrub Zone:

Salix lucida, S. rigida, S. sitchensis, S. scouleriana, S. pedicellaris (willow)
Ribes species (gooseberry)
Rhamnus purshiana (cascara)

Plants Preferred in Puget Sound Basin Freshwater Wetlands *(continued)*

Scrub-Shrub Zone:

Lysichitum americanus (skunk cabbage)
Athyrium filix-femina (lady fern)
Cornus sericea (redstem dogwood)
Rubus spectabilis (salmonberry)
Physocarpus capitatus (ninebark)
Sambucus racemosa (red elderberry) (occurs in wetland-upland transition)
Lonicera involucrata (black twinberry)
Oemleria cerasiformis (Indian plum)
Stachys cooleyae (Stachy's horsemint)
Prunus emarginata (bitter cherry)

Forested Zone:

Populus balsamifera, spp. *trichocarpa* (black cottonwood)
Fraxinus latifolia (Oregon ash)
Thuja plicata (western red cedar)
Picea sitchensis (Sitka spruce)
Alnus rubra (red alder)
Tsuga heterophylla (hemlock)*Acer circinatum* (vine maple)
Maianthemum dilatatum (wild lily-of-the-valley)
Luzula parviflora (small-flower wood rush)
Torreyochloa pauciflora (weak alkaligrass)
Ribes spp. (currants)

Bog:

Sphagnum sp. (sphagnum mosses)
Rhododendron groenlandicum (Labrador tea)
Vaccinium oxycoccos (bog cranberry)
Kalmia microphylla, spp. *occidentalis* (bog laurel)

Exotic Plants that Should Not Be Introduced to Existing, Created, or Constructed Puget Sound Basin Freshwater Wetlands

Hedera helix (English ivy)
Phalaris arundinacea (reed canarygrass)
Lythrum salicaria (purple loosestrife)
Iris pseudacorus (yellow iris)
Ilex aquifolia (holly)
Impatiens glandulifera (policeman's helmet)
Lotus corniculatus (birdsfoot trefoil)
Lysimachia thyrsiflora (tufted loosestrife)
Myriophyllum sp. (water milfoil, parrot's feather)
Polygonum cuspidatum (Japanese knotweed)
Polygonum sachalinense (giant knotweed)
Rubus discolor (Himalayan blackberry)
Tanacetum vulgare (common tansy)

Native Plants that Should Not Be Introduced to Existing, Created, or Constructed Puget Sound Basin Freshwater Wetlands

Potentilla palustris (Pacific silverweed)
Solanum dulcamara (bittersweet nightshade)
Juncus effusus (soft rush)
Conium maculatum (poison hemlock)
Ranunculus repens (creeping buttercup)

APPENDIX D: TABLE COMPARING WATER CHEMISTRY CHARACTERISTICS IN *SPHAGNUM* BOG AND FEN VS. MORE TYPICAL WETLANDS[4-10]

Water Quality Variable	Typical Wetlands	*Sphagnum* Bogs and Fens
pH	6–7	3.5–4.5
Dissolved oxygen (mg/l)	4–8	Shallow surface layer oxygenated, anoxic below
Cations	Divalent Ca, Mg common	Divalent Ca, Mg uncommon; univalent Na, K predominant
Anions	HCO_3^-, CO_3^{2-} predominant	Cl^-, SO_4^{2-} predominant; almost no HCO_3^-, CO_3^{2-} (organic acids form buffering system)
Hardness	Moderate	Very low
Total phosphorus (μg/l)	50–500	5–50
Total Kjeldahl nitrogen (μg/l)	500–1000	~50

REFERENCES

1. Horner, R. R., S. S. Cooke, K. O. Richter, A. L. Azous, L. E. Reinelt, B. L. Taylor, K. A. Ludwa, and M. Valentine, Wetlands and Urbanization, Implications for the Future, Puget Sound Wetlands and Stormwater Management Research Program, Final Rep., Engineering Professional Programs, University of Washington, Seattle, 1996.
2. Chin, N. T., Watershed Urbanization Effects on Palustrine Wetlands: A Study of the Hydrologic, Vegetative and Amphibian Community Response over Eight Years, M.S.C.E. thesis, University of Washington, Seattle, 1996.
3. Horner, R. R., Long-term effects of urban runoff on wetlands, in *Design of Urban Runoff Controls*, L. A. Roesner, B. Urbonas, and M. B. Sonnen, Eds., American Society of Civil Engineers, New York, 1989.
4. Horner, R. R., J. J. Skupien, E. H. Livingston, and H. E. Shaver, *Fundamentals of Urban Runoff Management: Technical and Institutional Issues,* Terrene Institute, Washington, D.C., 1994.
5. Clymo, R. S., Ion exchange in *Sphagnum* and its relation to bog ecology, *Ann. Bot.,* 27(106), 310, 1963.
6. Cooke, S. S., Puget Sound Wetlands and Stormwater Management Research Program, unpublished Queen's Bog data, 2000.
7. Meyer, J., L. Vogel, and T. Duebendorfer, East Lake Sammamish Wetland No. 21, unpublished data, submitted to L. Kulzer, King County Surface Water Management Division, King County, Washington, 2000.
8. Moore, P. D. and D. J. Bellamy, Peatlands, *The Geochemical template*, Elek Science, London, 1974, chap. 3.
9. Thurman, E. M., *Organic Geochemistry of Natural Waters*, Martinus Nijhoff/Dr W. Junk Publishers, Dordrecht, The Netherlands, 1985.
10. Vitt, D. H., D. G. Horton, N. G. Slack, and N. Malmer, *Sphagnum*-dominated peatlands of the hyperoceanic British Columbia coast: Patterns in surface water chemistry and vegetation, *Can. J. For. Res.,* 20, 696, 1990.

Index

A

Acer
  circinatum, 76, 82, 90
  macrophyllum, 76, 85, 90
Actea rubra, 93
Adenocaulon bicolor, 93
Adiantum pedatum, 94
Aeshna palmata, 135
Agrostis
  capillaris, 82, 91
  gigantea, 92
  oregonensis, 92
  scabra, 94
Aira caryophyllea, 94
Alderwood series, 38
Alisma plantago-aquatica, 93, 321
Alnus, 101
  rubra, 72, 73, 76, 79, 83, 87, 90, 260, 322
  sinuata, 94
Alopecurus pratensis, 94
Ambystoma
  gracile, 148, 149, 162, 163, 164, 165
  macrodactylum, 162, 163, 164, 165
Amelanchier alnifolia, 92
American coot, 176, 184, 194, 196
American crow, 173, 178, 179, 183, 194, 196
American goldfinch, 174, 183, 194, 196
American robin, 173, 178, 179, 183, 196
Amphibian(s), 12
  breeding, 20, 147, 158, 312
  populations, distribution of, 158
  richness, 153, 155
    lentic-breeding, 154
    of Puget Lowlands, 156
  surveys, 145
Amphibian distribution, abundance, and habitat use, 143–165
  1988 capture rates of amphibians in various wetlands, 162
  1989 capture rates of amphibians in various wetlands, 163
  1993 capture rates of amphibians in various wetlands, 164
  1995 capture rates of amphibians in various wetlands, 165
  discussion, 155–158
  methods, 145–147
    amphibian surveys, 145–146
    wetland characteristics, 146–147
  results, 147–155
    abundance, 151–152
    distributions, 147–151
    water level fluctuations, 154
    water permanence, 154–155
    watershed urbanization, 152–154
Anaphalis margaritacea, 82, 92
Anax junius, 135
Anisopodidae, 123, 124, 127, 128
Anna's hummingbird, 176, 184, 194, 196
Anthoxanthum odoratum, 94
Aphididae, 105, 106
Apsectrotanypus algens, 117
Aquatic food webs, 97
Aquatic macroinvertebrates, 99, 265
Arachnids, 97, 133
Arbutus menziesii, 94
Arthropods, 105, 106, 134
Asaurum caudatum, 94
Ascaphus truei, 148
Athyrium filix-femina, 72, 73, 76, 83, 90, 322
Avian detection rate, 278
Avian richness, distribution of among wetlands, 187
Avifauna, 169
Azolla
  filiculoides, 93
  mexicana, 73, 94

B

Bald eagle, 176, 181, 184, 194, 196
Band-tailed pigeon, 176, 181, 184, 194, 196
Banksiola crotchi, 109, 111
Bark drillers, 275
Barn swallow, 175, 183, 194, 196
Baseline sampling, 314
Basin planning, 294

Bathymetry, 6, 296
Belted kingfisher, 175, 184, 194, 196
Berberis
  aquifolium, 94
  nervosa, 91
Best management practices (BMPs), 301, 306, 313
Betula papyrifera, 92
Bewick's wren, 173, 178, 179, 183, 194, 196
Bibionidae, 123, 124, 127, 128
Bidens cernua, 81, 91
Bioengineering techniques, 305
Biofiltration facilities, 306
Biological communities, protection of, 315
Bird(s)
  abundance, 279
  classification of, 169
  diversity statistics, 177
  ground-nesting, 315
  of prairie marshes, 168
  richness, in wetlands, 277
  sightings, changing, 277
  of social significance, 180
  species
    diversity, 190
    within freshwater wetlands, 186
    richness, 168, 187
    sighted within study wetlands, 181
    uncommon, exotic, and aggressive, 182
  survey methods, 276
  wetland use, 189
Bird communities, relationships of to watershed development, 275–284
  discussion, 280–283
  methods, 276–277
  results, 277–280
Bird distribution, abundance, and habitat use, 167–199
  discussion, 186–190
  methods, 169–170
  results, 170–186
    annual variation, 170–171
    birds of social significance, 180–182
    community assemblages, 185–186
    migrants and residents, 182–185
    species diversity, 171–179
    uncommon, exotic, and aggressive species, 182
    versatility ratings, 185
    wetland characteristics and species richness, 179–180
Black rat, 205
Black-capped chickadee, 173, 178, 179, 183, 194, 196
Black-headed grosbeak, 174, 179, 183, 194, 196
Black-throated gray warbler, 173, 183, 194, 196
Blechnum spicant, 83, 90
Blue-green algae, 92
Blue-winged teal, 176, 184, 194, 196
BMPs, see Best management practices
Boreochlus, 115
Brachycera, 129
Brasenia schribneri, 73, 94
Brewer's blackbird, 175, 184, 194, 196
Brillia, 115, 119
Bromus ciliatus, 94
Brown creeper, 175, 184, 194, 196
Brown-headed cowbird, 173, 183, 194, 196
Buffers, surrounding wetlands, 304
Bufflehead, 181
Bufo boreas, 148, 162, 163, 164, 165
Bullfrogs, 154
Bullock's oriole, 175, 184, 194, 196
Bushtit, 173, 183, 194, 196
Bushy-tailed woodrat, 205

C

California quail, 175, 184, 194, 196
Callitriche heterophylla, 85, 91
Canada goose, 175, 184, 194, 196
Canopy foliage gleaners, 275
Capture rates, of terrestrial small mammals, 207
Carex, 101
  arcta, 83, 91
  athrostachya, 94
  deweyana, 83, 90
  exsiccata, 80, 81, 93
  hendersonii, 85, 91
  lenticularis, 81, 92
  obnupta, 80, 83, 90, 321
  stipata, 94, 260
  urticulata, 81, 91, 260
Carnivora, 204
Caspian tern, 176, 184, 194, 196
Cassin's vireo, 175, 184, 194, 197
Castor canadensis, 201
Cecidomyiidae, 123, 124, 127, 128
Cedar waxwing, 173, 179, 183, 194, 197
Ceratopogonidae, 123, 124, 127, 128
Chaetocladius, 115, 119
Chaoboridae, 123, 124, 127, 128
Chenopodium alba, 94
Chestnut-backed chickadee, 173, 183, 194, 197
Chipping sparrow, 175, 184, 194, 197
Chironomid
  distribution, 132
  -only communities, 132
  species richness, 129

Chironomidae, 272
Chloropidae, 130, 131
Chrinomus riparius, 116, 120
Circaea alpina, 92
Circium
arvense, 91
vulgare, 93
Cladina rangiferina, 94
Cladopelma viridula, 116, 120
Claytonia lanceolata, 94
Cliff swallow, 176, 184, 194, 197
Clostoeca disjuncta, 109, 111
Cloudbursts, dry-season, 34
Coefficients of variation (CV), 51
Coleoptera, 105, 106, 134
Collembola, 105, 106
Common goldeneye, 181
Common yellowthroat, 174, 179, 183, 194, 197
Community
assemblages, 185
physiognomy, 256
types, unusual biological, 308
Conceptual model, 40, 41
Conchapelopia dusena, 117, 121
Conductivity, 49, 58
Control wetlands, 15
Convolvulus
arvensis, 92
sepium, 94
Cooper's hawk, 176, 184, 194, 197
Cornus
canadensis, 95
nuttallii, 94
sericea, 76, 78, 83, 90, 322
Corophium sp., 12
Corylus cornuta, 85, 91
Corynoneura, 115, 119
Cowardin categories, 72, 77
Crataegus monogyna, 85, 92
Creeping vole, 205, 215
Crest stage, 39
Cricotopus, 115, 119
Cricotopus bifurcatus, 116
Crustaceans, 97
Culicidae, 123, 124, 127, 128
CV, see Coefficients of variation
Cytisus scoparius, 93

D

Dactylis glomerata, 92
Darcy's Law, 224
Dark-eyed junco, 173, 183, 194, 197
Deer mouse, 205, 215, 216, 217, 218
Dicamptodon tenebrosus, 148, 162, 163, 164, 165
Dicentra formosa, 83, 91
Dicrotendipes, 116, 120
Digitalis purpurea, 92
Diptera, 102, 108
Dissolved oxygen (DO), 48, 49
Diversity index, 132
Dixidae, 123, 124, 127, 128
DO, see Dissolved oxygen
Doithrix, 115, 119
Dolochipodidae, 130, 131
Douglas-fir forests, 202
Douglas squirrel, 205, 215
Downy woodpecker, 174, 183, 194, 197
Drainage systems, 221, 317
Drosera rotundifolia, 92
Dryopteris expansa, 83, 90
Dry period, 230
Dulichium arundinaceum, 93
Dusky shrew, 205, 216

E

EC, see Effective concentration
Echinochloa crusgalii, 95
Effective concentration (EC), 248
Effective Impervious Area (EIA), 21
EIA, see Effective Impervious Area
Eleocharis
ovata, 93
palustris, 93
Elodea
canadensis, 93
spp., 11
Elytrigia repens, 94
EMAP, see EPA Environmental Monitoring and Assessment Program
Emergence trap(s)
aquatic macroinvertebrate, 99
total and proportional abundance of aquatic and semiaquatic diptera taxa collected from, 113–114
Emergent macroinvertebrate communities, relationship of to watershed development, 265–274
methods, 266–267
results, 268–271
Empididae, 130, 131
Enallagma boreale, 104, 109, 111, 135
Endangered species, 156, 307
Endochironomus
nigricans, 116, 120
subtendens, 117, 120
Energy facilities, 37

Ensatina eschscholtzii, 148, 162, 163, 164, 165
Enterococcus, 240
EPA Environmental Monitoring and Assessment Program (EMAP), 265
Ephemeroptera, 102, 268
Ephydridae, 130, 131
Epilobium, 101
  angustifolium, 82, 91
  ciliatum, 83, 90
  watsonii, 321
Equisetum, 101
  arvense, 83, 90
  hyemale, 85, 91
  telmateia, 85, 91
Eriophorum chamissonis, 73, 94
Ermine, 215, 216
Erosion, 295, 303
Error analysis, hydrologic components, 233
Estuarine wetland, 307
ET, see Evapotranspiration
European starling, 175, 183, 194, 197
Evapotranspiration (ET), 35, 48, 221, 227
Evening grosbeak, 175, 184, 194, 197
Evenness value, 132
Exotic rats, at wetlands, 212

F

Facultative wetland, 70
Fauna, impacts to wetland, 11–12
  hydrologic impacts, 11–12
  water quality impacts, 12
FC, see Fecal coliforms
Fecal coliforms (FC), 49, 57, 240
  standard, 53
  stream median, 54
Fertilizers, 243
Festuca
  pratensis, 95
  rubra, 92
Fish, wetlands inhabited by, 317
Flood
  frequency, 293
  storage, 4
Flow-through (FT) wetlands, 57, 58
Food chains
  biomagnification of toxics in, 4
  shift to simpler, 4
  support, 98
Food webs, aquatic, 97
Forest cover, percent, 240
Forest(s)
  cover
    relationship between event WLF and, 229
    relationship between small mammal richness and, 209
  Douglas-fir, 202
  land, importance of adjacent to wetlands, 212
  with low-density single family housing, 170
  upland, 211
Forest deer mouse, 205, 216
Forested wetland, 188, 307
Formicidae, 105, 106
Fox sparrow, 176, 184, 194, 197
Fragaria virginiana, 95
F-ratio, 224
Fraxinus laitfolia, 79, 83, 87, 92, 260, 322
Friedman test, 277
FT, see Flow-through wetland

G

Gadwall, 175, 184, 194, 197
Galium
  aparine, 83, 92
  cymosum, 83, 92
  trifidum, 81, 91
Gauge reading, 257
Gaultheria shallon, 76, 81, 90
Geographical information system (GIS), 20, 36, 256
Geranium robertianum, 81, 92
Germination requirements, plant species having specific, 10
Geum macrophyllum, 83, 90
GIS, see Geographical information system
Glaucous-winged gull, 176, 184, 194, 197
Glecoma hederacea, 94
Glyceria, 101
  borealis, 83, 92
  elata, 85, 92, 321
  grandis, 83, 90
Glyphopsyche irrorata, 109, 111
Glyptotendipes, 116, 120
Gnaphalium uliginosum, 94
Golden-crowned kinglet, 173, 178, 179, 183, 194, 197
Goodyera oblongifolia, 95
Great blue heron, 174, 181, 183, 194, 197
Green heron, 175, 184, 194, 197
Green-winged teal, 194
Groundwater
  calculations, 234
  discharge, 232
  exchange, 32
Gymnocarpium dryopteris, 94

H

Habitat(s)
  architecture, 136
  development, 18
  National Wetlands Inventory, 203
  scrub-shrub, 186, 260, 261
  shrub-land as bird, 277
  type
    hydrologic regimes by, 78
    palustrine open water, 145
    seasonal hydrology associated with, 80
Hairy woodpecker, 174, 183, 194, 197
Halesochila taylori, 110, 112
Hammond's flycatcher, 175, 184, 194, 197
Hayesomyia senata, 117, 121
Hedera helix, 92, 322
Hemiptera, 105, 106
Hercaleum lanatum, 94, 321
Hermit thrush, 173, 183, 194, 197
Hibernation habitats, 147
Hieracium pratense, 93
Hippurus vulgaris, 73, 95
Holcus lanatus, 91
Holidiscus discolor, 85, 92
Hooded merganser, 175, 181, 184, 194, 197
House finch, 174, 183, 194, 197
House mouse, 205
House sparrow, 175, 184, 194, 197
HSPF, see Hydrologic Simulation Program — FORTRAN
Hubbard Brook Experimental Forest, 143
Human values, of wetlands, 167
Hutton's vireo, 175, 184, 194, 197
Hydrocotyl ranunculoides, 73, 95
Hydrodyanmics, 304
Hydrologic regimes, by habitat type, 78
Hydrologic Simulation Program — FORTRAN (HSPF), 288
Hydrology, effects of watershed development on, 221–235
  data collection and analysis methods, 223–225
    data collection and analysis for wetland water balance, 224–225
    statistical analysis of development impacts on wetland hydrology, 223–224
  Puget Sound Wetlands and Stormwater Management Research program, 222
  research objectives, 222
  results and discussion, 225–234
    dry period, 230
    hydrologic characteristics of intensively studied wetlands, 230–232
    hydrologic components error analysis, 233–234
    multiple regression analysis, 229–230
    threshold level analysis, 225–229
    urbanization and other factors affecting wetland hydrology, 233
    wetland hydrology by season and wetland, 232–233
    wetland hydrology and water level fluctuation, 225
Hydroperiod, predevelopment, 304
Hydrophyllum tenuipes, 73, 95
Hyla regilla, 149, 162, 163, 164, 165
Hymenoptera, 134
Hypericum
  anagalloides, 82, 93
  formosum, 82, 92
Hypochaeris radicata, 94

I

ICP-MS, see Inductively coupled plasma–mass spectrometry
Ilex aquifolia, 83, 91, 322
Impatiens
  glandulifera, 322
  noli-tangere, 94
Inductively coupled plasma–mass spectrometry (ICP-MS), 49
Insect(s), 97
  abundance, 272
  production, 100
  -substratum relationship, aquatic, 137
  taxa, total and proportional abundance of, 103
Insectivora, 203
Insectivorous urban-avoiders, 276
Invasive weedy plant species, 308, 317
Invertebrates, of wetlands, 98
In-wetland components, 232
Iris pseudacorus, 93
Ischnura cervula, 104, 109, 111, 135
Issues, 4–5

J

Juncus, 101
  acuminatus, 80, 81, 92, 321
  bufonius, 82, 92, 321
  effusus, 72, 75, 80, 81, 91
  ensifolius, 82, 92, 321
  supiniformis, 73, 95

K

Kalmia microphylla, 73, 93

KCSWM, see King County Surface Water Management
Kiefferulus dux, 117, 120
Killdeer, 175, 184, 194, 197
King County Surface Water Management (KCSWM), 288
Kruskal-Wallace (KW) test, 223, 289
KW test, see Kruskal-Wallace test

L

Lamium purpurea, 95
Land
  type, dominant, 231
  use(s)
    categories of wetlands related to, 16
    changes, 287
    total impervious and effective impervious areas associated with, 37
    urbanized, 153
Landscape unit, 301, 303
LDM, see Long-distance migrants
LDR, see Low-density residential
Ledum groenlandicum, 82
Lemna, 101
Lemna minor, 76, 81, 91
Lenarchus
  rho, 110, 112
  vastus, 108, 110, 112
Lentic-breeding amphibian richness, 154
Lepidoptera, 102, 105, 106
Lepidostina cinereum, 110, 112
Libellula forensis, 135
Limnephilus
  cerus, 110, 112
  externus, 110, 112
  fagus, 110, 112
  Fenestratus Gr., 110, 112
  harrimani, 110, 112
  nogus, 110, 112
  occidentalis, 110, 112
  spinatus, 110, 112
  spp., 108
Limnophyes, 115, 119
Linnaea borealis, 85, 93
Literature review and management needs survey, 13–14
LOI, see Loss on ignition
Lolium
  multiflorum, 94
  perenne, 93
Long-distance migrants (LDM), 182
Long-tailed vole, 205, 215
Lonicera
  ciliosa, 92
  involucrata, 73, 76, 78, 83, 90
Loss on ignition (LOI), 59
Lotus corniculatus, 94, 322
Low-density residential (LDR), 225
Ludwigia palustris, 82, 92
Lutra canadensis, 201
Luzula parviflora, 83, 90
Lycopus
  americanus, 92, 321
  uniflorus, 81, 91
Lysichitum americanum, 72, 76, 83, 90
Lythrum salicaria, 72, 95, 322

M

MacGillivray's warbler, 175, 184, 194, 197
Macroinvertebrate distribution, abundance, and habitat use, 97–141
  discussion, 134–137
  methods, 99–102
    field, 99–100
    statistical, 100–102
  results, 102–134
    richness and abundance of aquatic and semiaquatic insects, 102–129
    total insect and chironomid species richness and wetland characteristics, 129–133
    terrestrial arthropod richness and abundance, 133–134
Maianthemum dilatatum, 83, 91
Mallard, 174, 179, 183, 194, 197
Malus fusca, 83, 90
Mann-Whitney (MW) test, 223, 267, 289
Marsh shrew, 205, 215, 217
Marsh wren, 174, 179, 183, 194, 197
Masked shrew, 205, 215
Master drainage planning (MDP), 295
MDP, see Master drainage planning
Melilotus alba, 95
Mentha arvensis, 93, 321
Menyanthes trifoliata, 73, 94
Menziesia ferruginea, 83, 91
Mesosmittia, 115, 119
Metal(s)
  accumulations, 248
  concentrations, for wetland samples, 65
  pollution, from vehicles, 64
  in soils, 63
  trash, dumping of, 66
Metriocnemus, 115, 119
Mice, native, 204, 211
Microsoft EXCEL®, 223
Microtox analysis, 248

Migration, 147
Mimulus guttatus, 73, 93, 321
Model
conceptual, 40, 41
Surface Water Management
Monitoring plan, typical, 19
Montia siberica, 93
Morphology and hydrology, 31–45
hydrology of palustrine wetlands, 33–36
change in wetland storage, 35
evapotranspiration, 35
groundwater, 34–35
precipitation, 34
surface inflows, 34
surface outflow, 35–36
Puget Sounds Wetlands and Stormwater Management Research Program, 32–33
research methods and wetland descriptors, 36–39
watershed characteristics, 36
watershed imperviousness, 37–38
watershed soils, 38–39
wetland morphology, 36
results and conceptual model, 40–41
conceptual model of influences on wetland hydroperiod, 41
water level fluctuation patterns, 40–41
wetland hydrologic functions, 33
wetland hydrology, 39–40
length of summer dry period, 40
seasonal fluctuation in wetland water levels, 39–40
wetland water level measurements and fluctuation, 39
wetlands in urbanizing areas, 33
MRPP, see Multi-Response Permutation Procedures
Multiple regression analysis, 229
Multi-Response Permutation Procedures (MRPP), 102
Murinae, 210
MW test, see Mann-Whitney test
Mycetophilildae, 123, 124, 127, 128
Mycorrhizal fungi, 202
Myosotis
laxa, 73, 92
scorpioides, 93

N

Nancladius, 115, 119
Nasturtium officinale, 93
Natarsia miripes, 118, 121, 125
National Wetland's Inventory (NWI), 170, 203
Nemotaulius hostilis, 110, 112
Nest parasitism, 282
Neuroptera, 102, 105, 106
Nitrification, promotion of, 58
NMDS, see Non-Metric Multidimensional Scaling
NOAA stations, 170
Nonchironomids, 126
Non-Metric Multidimensional Scaling (NMDS), 102, 132
Northern flicker, 174, 183, 194, 197
Northern flying squirrel, 205
Northern pygmy-owl, 176, 184, 194, 197
Norway rat, 205
Nuphar, 101
Nuphar polysepalum, 92
Nutrient
cycling, rapid, 53
toxicity, 4
transfer, rates of among ecosystem components, 8
NWI, see National Wetland's Inventory
Nymphaea odorata, 73, 95, 321

O

Odonata, 102, 104
Odontomesa, 115
Oemleria cerasiformis, 85, 90
Oenanthe, 101
Oenanthe sarmentosa, 81, 91, 321
Olive-sided flycatcher, 175, 181, 184, 194, 197
Oncorhynchus spp., 144
Ondatra zibethicus, 201
Open water (OW) wetland, 57
Oplopanax horridus, 83, 92
Orange-crowned warbler, 173, 183, 194, 198
Orthocladius, 115, 119
Osprey, 181
Ostracerca dimicki, 109, 111
OW wetland, see Open water wetland
Oxidation-reduction potential, 59
Oxyethira, 108
Oxygen
-depleted environment, 54
replenishment of from atmosphere, 60

P

Pacific jumping mouse, 205, 216
Pacific Northwest watersheds, 242

Pacific-slope flycatcher, 173, 178, 179, 183, 194, 198
Paedomorphs, egg masses from breeding, 146
Palustrine open water (POW) habitat type, 145
Palustrine wetlands, see Puget Sound Basin palustrine wetland vegetation, characterization of central
Parachironomus
  cf. *forceps*, 117
  monochromus, 117
Parakiefferiella, 115, 119
Paramerina smithae, 118, 121
Parametriocnemus, 115, 119
Paraphaenocladius, 115, 119
Parasitoid wasps, 134
Paratendipes, 116, 120
Paratendipes albimanus, 117
Particle size distribution (PSD), 59, 253
Permanent residents (PR), 182
Petasites frigidus, 85, 93
Pet predation, on small mammals, 212
Phaenopsectra
  flavipes, 117
  punctimes, 117
Phalaris, 101
Phalaris arundinacea, 9, 72, 73, 74, 75, 76, 77, 83, 86, 90
Phleum pratense, 93
Phoridae, 123, 124, 127, 128, 130, 131
Physocarpus capitatus, 93
Picea sitchensis, 84, 91, 322
Pied-billed grebe, 175, 184, 194, 198
Pileated woodpecker, 175, 181, 184, 194, 198
Pine siskin, 175, 184, 194, 198
Pinus monticola, 94
Plant
  cover, 19
  richness, 259, 290, 291, 292
Plantago
  lanceolata, 93
  major, 93
Plecoptera, 102, 104, 268
Plethodon
  cinereus, 143
  vehiculum, 148, 149, 162, 163, 164, 165
Poa
  palustris, 95
  pratensis, 95
Podmosta delicatula, 104, 109, 111
Podonominae, 114
Polishing, 306
Pollutant(s)
  performance of wetlands in capturing, 13
  trapping, 4
Polycentropus flavus, 110, 112
Polygonum
  amphibium, 73, 94
  hydropiper, 81, 91
Polypedilum
  illinoense, 117
  ophioides, 117
Polypodium glycyrrhiza, 90
Polystichum munitum, 72, 73, 76, 90
Populus
  balsamifera, 76, 85, 90
  tremuloides, 84, 93
Poryophaenocladius, 115, 119
Potamogeton
  diversifolius, 73, 95
  gramineus, 95
  natans, 81, 92
Potentilla, 101
  gramineus, 73
  palustris, 92
POW habitat type, see Palustrine open water habitat type
PR, see Permanent residents
Prairie marshes, 168, 281
Precipitation, 34, 153
Predevelopment hydroperiod, 304
Priority peat wetlands, 302, 312, 313
Procladius bellus, 118, 121
Prodiamesa, 115
Prunus emarginata, 84, 91, 322
PSD, see Particle size distribution
Psectrocladius, 115, 119
Psectrotanypus dyari, 118, 121, 125
Pseudosmittia, 116, 120
Pseudotsuga menziesii, 81, 90
Psocoptera, 102, 104, 106
PSWSMRP, see Puget Sound Wetlands and Stormwater Management Research Program
Psychodidae, 123, 124, 127, 128
Pteridium aquilinum, 73, 84, 90
Ptilostomis ocellifera, 110, 112
Puget Sound Basin palustrine wetland vegetation, characterization of central, 69–95
  discussion, 86–88
  methods, 70–71
  results, 72–86
    abundance and distribution of invasive plant species, 74–75
    community richness, 75–77
    community structure and composition, 72
    habitat character, 72–73
    hydrologic regimes by habitat type, 78–79
    hydrologic regimes of some species, 79–86
    wetland plant associations, 74

Puget Sound Wetlands and Stormwater Management Research Program (PSWSMRP), 3, 222
  design, 13
  study locations, 16
Purple finch, 174, 183, 194, 198
Purple martin, 181
Pysichitum, 101

Q

QA/QC, see Quality assurance/quality control
Quality assurance/quality control (QA/QC), 49

R

Rana
  aurora, 148, 162
  cascadae, 148
  catesbeiana, 144, 149
  pipiens, 144
  pretiosa, 148
Ranunculus
  acris, 73, 95
  repens, 72, 73, 84, 90, 322
Rare species, 307
RC, see Rural controls
Red-breasted nuthatch, 174, 183, 194, 198
Red-breasted sapsucker, 175, 184, 194, 198
Red-eyed vireo, 176, 184, 194, 198
Redox potential, 9
Red-tailed hawk, 175, 184, 194, 198
Red-winged blackbird, 174, 179, 183, 194, 198
Regression model variables, significance of, 230
Reptile communities, 12
Reptilian predators, 202
Research program design, 14
Rhagionidae, 130, 131
Rhamnus purshiana, 73, 84, 90
Rheocricotopus, 116, 120
Rhinanthus crista-galli, 95
Rhododendron groenlandicum, 76, 92
Rhynchospora alba, 73, 94
Ribes
  bracteosum, 95
  divaricatum, 85, 93
  lacustre, 84, 92
  sanguineum, 93
Ring-billed gull, 195
Rodentia, 203
Rorippa
  calycina, 93
  curvisiliqua, 73, 95
Rosa
  gymnocarpa, 85, 91
  nutkana, 95
  pisocarpa, 85, 93
  rugosa, 95
Rubus
  laciniatus, 73, 81, 90
  leucodermis, 92
  parviflorus, 84, 91
  procerus, 74, 84, 90
  spectabilis, 72, 73, 76, 78, 84, 90
  ursinus, 72, 73, 84, 90
Ruby-crowned kinglet, 174, 183, 194, 198
Ruffed grouse, 176, 184, 194, 198
Rufous hummingbird, 173, 183, 194, 198
Rumex
  acetosella, 95
  crispus, 92
  obtusifolius, 73, 94
Runoff, 5
  stormwater, 7
  vegetation changes attributable to increased, 275
Rural controls (RC), 256

S

Sagittaria latifolia, 73, 95
Salicornia sp., 11
Salix, 101
  alba, 85, 91
  hookeriana, 94
  lucida var. *lasiandra*, 78, 84, 90
  pedicellaris, 73, 84, 91
  scouleriana, 72, 73, 78, 84, 90, 321
  sitchensis, 73, 77, 78, 84, 90, 321
Salmonidae, 98
Sambucus racemosa, 76, 84, 90
Sample contamination, 249
Sampling, baseline, 314
Scapanus orarius, 209
Scatopsidae, 123, 124, 127, 128
Sciaridae, 123, 124, 127, 128
Sciomyziidae, 130, 131
Scirpus, 101
  atrocinctus, 86, 91
  microcarpus, 85, 86, 91
Sciurus niger, 210
Scrub-shrub habitat, 186, 187, 260, 261
SCS, see Soil Conservation Service
Scutellaria lateriflora, 84, 91
SDM, see Short-distance migrants
Shannon diversity, 170
Sharp-shinned hawk, 175, 184, 194, 198

Short-distance migrants (SDM), 182
Shrew-mole, 206, 217, 218
Shrews, richness and abundance of, 211
Shrub-land, as bird habitats, 277
Sium suave, 93
Small mammals, see Terrestrial small mammal distribution, abundance, and habitat use
Smilacena
 racemosa, 85, 92
 stellata, 94
Smittia, 116, 120
Snoqualmie, 172
Soil(s)
 characteristics, 60
 drainage classes, 41
 effects of watershed development on, 247
 erosion-prone, 303
 fine-textured, 63
 impacts to wetland, 8–9
 metals in, 63
 moisture, 9
 organic content, median levels of, 247
 permeability, 38
 properties, urbanized wetland, 247
 quality statistics, for wetlands for experiencing significant urbanization change, 61–62
 saturation, 87
 statistics, for treatment wetland, 251–252
 texture, 63
 watershed, 38
Soil Conservation Service (SCS), 38
Soils, water quality and, 47–67
 soils, 59–66
  characteristics of wetlands with nonurbanized watersheds, 66
  collection and methods, 59–60
  research findings, 60–66
 water quality, 48–59
  collection and methods, 48–49
  research findings, 49–51
  seasonal variation, 54–57
  variation with wetland morphology, 57–59
  wetland water quality in context, 51–54
Solanum, 101
 dulcamara, 75, 76, 77, 84, 90, 322
 nigrum, 95
Solidago canadensis, 93
Soluble reactive phosphorus (SRP), 48
Song sparrow, 173, 178, 179, 183, 194, 198
Sora, 176, 184, 194, 198
Sorbus
 americana, 84, 91
 scopulina, 91
Source control best management practices, 301, 306, 313
Southern red-backed vole, 205, 215, 217
Soyedina interrupta, 109, 111
Sparganium, 101
 emersum, 81, 91
 eurycarpum, 73, 94
 spp., 76
Species diversity, 171
Sphaeroceridae, 130, 131
Sphagnum spp., 76, 91, 323
Spirea, 101
Spirea douglasii, 73, 76, 77, 79, 82, 90, 177
Spirodela polyrhiza, 93
Sport fish, stranding of, 317
Spotted sandpiper, 184, 194, 198
Spotted towhee, 173, 178, 179, 183, 194, 198
SRP, see Soluble reactive phosphorus
Stachys cooleyae, 85, 92
Stage excursions, 310
Stellaria
 longifolia, 95
 media, 84, 91
Steller's jay, 174, 183, 194, 198
Stictochironomus, 116, 120
Storm
 drainage system, 241
 events, frequency of, 289
 surges, 8
Stormwater
 impact studies, 14, 15–20
 managers, 4
 ponds, 293
 protection measures, 293
 runoff, 7
  -flood event, 293
  pretreatment of, 301
 treatment, use of wetlands for, 13
Stormwater management guidelines, wetlands and, 299–323
 approach and organization of management guidelines, 300–302
  scope and underlying principles of guidelines, 300–302
  support material, 302
 comprehensive landscape planning for wetlands and stormwater management, 302–309
  comprehensive planning steps, 303–307
  stormwater wetland assessment criteria, 307–309
 definitions, 318–321
 information needed to apply guidelines, 317–318

native and recommended noninvasive plant species for wetlands in central Puget Sound basin, 321-323
wetland protection guidelines, 309–317
general wetland protection guidelines, 309–310
guidelines for protection from adverse impacts of modified runoff quantity discharged to wetlands, 310–313
guidelines for protection from adverse impacts of modified runoff quality discharged to wetlands, 313–315
for protection of specific biological communities, 315–317
Stratiomyiidae, 130, 131
Streptopus
amplexifolius, 93
roseus, 94
Study wetlands, landscape data for, 17
Summer dry period, length of, 40
Surface
inflow, 33, 34, 232
outflow, 35
Surface Water Management Model (SWMM), 7
Surrounding landscape, 18
Suspended solids, 58
Swainson's thrush, 173, 178, 179, 183, 194, 199
Swamps
New Jersey Pine Barrens, 9
Pine Barrens cedar, 11
SWMM, see Surface Water Management Model
Sylvilagus floridanus, 210
Symphoricarpos albus, 92
Syrphidae, 130, 131
SYSTAT®, 223

T

Tanacetum vulgare, 95, 322
Tanypodinae, 114
Tanytarsus, 118
Taraxacum officinale, 94
Taricha granulosa, 162, 163, 164, 165
Taxa richness
EPOT, 269
values, 267
Taxus brevifolia, 93
Terrestrial arthropod richness, 133
Terrestrial small mammal distribution, abundance, and habitat use, 201–218
discussion, 209–213
methods, 202–203
results, 203–209
Thienemanniella, 116, 120
Threatened species, 307
Threshold level analysis, 225
Thuja plicata, 76, 85, 90, 322
Thysanoptera, 102, 105, 106
TIA, see Total impervious area
Tiarella trifoliata, 85, 92
Tipulidae, 123, 124, 127, 128
TKN, see Total Kjeldahl nitrogen
Tolmiea menziesii, 85, 91
Torreyochloa pauciflora, 82, 91
Total impervious area (TIA), 21, 311
Total Kjeldahl nitrogen (TKN), 59, 66
Total petroleum hydrocarbon (TPH), 248
Total phosphorus (TP), 48, 59
Total suspended solids (TSS), 49
Townsend chipmunk, 205, 218
Townsend's warbler, 174, 183, 194, 199
Townsend vole, 202, 205, 215, 216, 217
TP, see Total phosphorus
TPH, see Total petroleum hydrocarbon
Trap installation procedures, 145
Treatment
BMPs, 313
wetlands, 15, 250
Tree swallow, 174, 179, 183, 194, 199
Trichoptera, 268
Trifolium
pratense, 95
repens, 94
Trillium ovatum, 73, 85, 91
Trowbridge shrew, 206, 212, 216, 217, 218
TSS, see Total suspended solids
Tsuga heterophylla, 76, 85, 90
Turbidity, 58
Typha, 101
latifolia, 75, 76, 82, 86, 91, 321
spp., 4

U

UC, see Urban controls
Urban controls (UC), 256
Urbanization, 5–6, 47
Urban plan developments (UPD), 295–296
Urtica dioica, 77, 85, 90
Urticularia
minor, 82, 93
vulgaris, 81, 93
U.S. Environmental Protection Agency (USEPA), 5
USEPA, see U.S. Environmental Protection Agency
U.S. Geological Survey (USGS), 36, 70
USGS, see U.S. Geological Survey

V

Vaccinium
  ovatum, 73, 95
  oxycoccos, 93
  parvifolium, 74, 85, 90
  uliginosum, 73, 94
Vagrant shrew, 206, 215, 216, 217, 218
Vallisneria americana, 94, 321
Varied thrush, 174, 183, 194, 199
Vaux's swift, 176, 181, 184, 194, 199
Vegetation
  community
    changes, 261
    hydrologic changes affecting, 297
    types, 70, 79
  destruction, 309
  effects of water level changes on wetland, 10
  impacts to, 9–11
    hydrologic impacts, 9–10
    water quality impacts, 10–11
  plot, water depth at, 257
  richness, watershed urbanization and, 258
  sample station, 257
  structure
    generalized classifications of, 87
    of wetlands, 282
Vernal wetlands, 312
Veronica, 101
  americana, 73, 85, 90
  scutellata, 85, 91, 321
Versatility ratings, 185
Vicia sativa, 73, 95
Viola glabella, 93
Violet-green swallow, 174, 183, 194, 199
Virginia rail, 175, 184, 194, 199

W

WAC, see Washington Administrative Code
Warbling vireo, 174, 183, 194, 199
Washington Administrative Code (WAC), 314
Washington Department of Ecology, 65
Wastewater additions, 4
Water
  depth, stepwise regression summary for ranges of, 292
  permanence, 154, 157
  storage volumes, 231
  supply facilities, 37
Waterbirds, 282
Water level fluctuation (WLF), 32, 35, 39, 71, 154, 255
  association of with vegetation community types, 79
  expression of, 311
  patterns, 40, 42
  relationship between outlet condition and event, 229
  seasonal, 39, 259
  wetland hydrology and, 225
Water quality, 48
  effects of watershed development on, 238
  hydrologic impacts on, 8
  impacts, 7–8
    direct water quality impacts, 7–8
    hydrologic impacts on water quality, 8
  improvement study, 14
  ranges, found in study wetlands, 316
  statistics
    for treatment wetlands, 245–246
    for wetlands not experiencing significant urbanization change, 50, 55–56
  variables, 52, 53
Water quality and soils, effects of watershed development on, 237–254
  highly urbanized wetlands, 239
  moderately urbanized wetlands, 238–239
  observations for individual treatment wetlands, 243–244
  profile of treatment wetlands, 244–247
  treatment wetlands, 250–253
  treatment of wetlands, 242–243
  urbanized wetland soil profiles, 247–248
Watershed(s)
  amount of forested area in, 227
  changes, categories of wetlands related to, 16
  characteristics, 36, 271
  conditions, 18
  development, 64
  diminished infiltration in wetland, 7
  hydrologic processes, influences on, 43
  imperviousness, 37
  nonurban, 49
  Pacific Northwest, 242
  runoff from urbanizing, 259
  soil characteristics of wetlands with nonurbanized, 66
  soils index (WSI), 38, 41, 228
  urbanization, 152, 258
  water quality assessment, 318
  wetlands in pristine, 177
Water shrew, 206
Weather data, 170
Western tanager, 174, 183, 194, 199
Western wood-pewee, 175, 184, 194, 199
Wetland(s)
  acidic, 54

avian richness among, 187, 188
bird
communities, 188
richness in, 277
species within freshwater, 186
buffer, 304, 309
characteristics, 146
Community Structure, 18
constructed, 305
control, 15
data collected in program, 249
ecological benefits of, 167
ecosystems, 297, 300
environment, structure of, 301
estuarine, 307
facultative, 70
flooding, frequency of, 158
flow-through, 57
forested, 307
highly urbanized, 239
human values of, 167
hydrographs, 227
hydrology, 222, 300
characteristics of intensively studied, 230
functions, 33
urbanization and other factors affecting, 233
identification of terrestrial small mammals at, 202
impacts of urbanization on, 5–6
indicator status, 70
influence of wetland and watershed characteristics on impacts to, 6–7
inland palustrine, 69
invertebrate communities, 98
landscape data for, 17
macroinvertebrate index scores, 271
management areas (WMA), 295
moderately urbanized, 238
morphology, 18, 36, 57, 240
nutrient cycle, 266
obligate bird species, 315
observations for individual treatment, 243
open water, 57
Pacific Northwest, 13
performance of in capturing pollutants, 13
permanently flooded, 136
plant associations, 74
priority peat, 302, 312, 313
profile of treatment, 244
resources, holistic view of managing, 302
response and management, 294
restoration of previously degraded, 307
seasonally flooded, 133
size, relationships between bird species richness and, 180
soils
microtox analysis of, 248
trapping of toxic materials by, 9
sources of impacts to, 6
species richness
in forested, 188
in urban, 280
storage
change in, 35
volume, 231
study, water quality ranges found in, 316
survey, 14
treatment, 15, 242, 250
particle size distributions for, 253
soil statistics for, 251–252
type, management criteria by, 14
vernal, 312
water
balance, data collection and analysis for, 224
quality, 51, 315
watershed, outlet and hydrologic characteristics, 226
-to-watershed ratio, 34, 41, 234
Wetland hydroperiod, management of, 287–298
discussion, 293–297
basin planning, 294–295
master drainage planning and guidelines, 295–297
wetland management areas, 295
methods, 288–289
results, 290–293
Wetland plant communities, relationship of to watershed development, 255–263
discussion, 259–261
methods, 256–258
results, 258–259
water level fluctuation and vegetation richness, 258–259
watershed urbanization and vegetation richness, 258
White-crowned sparrow, 175, 184, 194, 199
Willow flycatcher, 173, 178, 179, 183, 194, 199
Wilson's warbler, 173, 178, 179, 183, 194, 199
Winter wren, 173, 178, 179, 183, 194, 199
WLF, see Water level fluctuation
WMA, see Wetland management areas
Wood duck, 174, 181, 184, 194, 199
WSI, see Watershed soils index

X

Xestochironomus, 116, 120

Y

Yellow-rumped warbler, 175, 184, 194, 199
Yellow warbler, 173, 183, 194, 199

Z

Zapada cinctipes, 109, 111
Zavrelimia
fastuosa, 118, 122
sinuosa, 118, 122, 125, 126
thryptica, 118, 122